T0253556

Analytic Combustion

Anil Waman Date

Analytic Combustion

With Thermodynamics, Chemical Kinetics and Mass Transfer

Second Edition

 Springer

Anil Waman Date
Department of Mechanical Engineering
Indian Institute of Technology Bombay
Mumbai, Maharashtra, India

ISBN 978-981-15-1855-3   ISBN 978-981-15-1853-9   (eBook)
https://doi.org/10.1007/978-981-15-1853-9

This Springer imprint is published by the registered company Springer Nature Singapore Pte Ltd.
The registered company address is: 152 Beach Road, #21-01/04 Gateway East, Singapore 189721, Singapore

Preface to Second Edition

It is almost 9 years since the first edition of *Analytic Combustion* was published by the Cambridge University Press, New York. However, my postretirement reappointment in 2017 as Adjunct Professor by the Director, IIT Bombay (Prof. Devang Khakhar) gave me the opportunity to undertake a fresh review of the book. The review revealed typographic errors in my LaTex files as well as printer's devils in the published copies of the book.

In this revised edition, while the overall structure and the emphasis of the book have been retained, in some chapters new sections are added or the text material has been better cross-referenced. At this point, I must draw special attention of the reader to the two case studies presented in Chap. 11. They pertain to modeling of (i) an experimental wood-burning cookstove from CTARA[1] and (ii) UNDP approved design of the Vertical Shaft Brick Kiln (VSBK) originally from rural China. Both are complex systems that require recall of fundamentals as well as of practical field observations in rural settings to guide modeling considerations. I emphasize this because increasingly, the world over, researchers are seeking to apply formal tools of combustion science to retrofit or to redesign such rural technologies for meeting goals of sustainable development.

In this post-retirement period, I often had to call upon my faculty aswell as non-faculty colleagues and students from Mechanical Engineering Department andfrom CTARA to help me in many indirect ways to prepare the revised manuscript. It is to their credit that they ungrudgingly offered this help as and when required.

---

[1]Centre for Technology Alternatives for Rural areas, IIT Bombay.

Finally, I would like to express my special gratitude to Ms. Swati Meherishi, Senior Editor-Applied Science and Engineering, Springer for her considerable advice and guidance during the preparation of the manuscript for this second edition. I would also like to thank Mr. Lokeshwaran and Mr. Maniarasan Gandhi for engaging with me through several emails to take the book through its production stages.

April–June 2019   Anil Waman Date
                                                                 IIT Bombay, Bombay, India

Preface to First Edition

It is fair to say that a very substantial part (over 90 percent) of the total energy used today in transportation, power production, space heating, and domestic cooking is produced by *combustion* (burning) of solid, liquid, and gaseous fuels. Although the phenomenon of combustion has been known to the earliest man, and although great strides have been made through painstaking experimental and theoretical research to understand this phenomenon and to use this understanding in designs of practical equipment (principally, burners and combustions chambers or furnaces), any claim to a perfect *science of combustion* remains as illusive as ever. Designers of combustion equipment thus rely greatly on experimental data and empirical correlations.

Combustion is a phenomenon that involves the *change in the chemical state of a substance* from a fuel state to a product state via a *chemical reaction* accompanied by the release of *heat energy*. To the extent that a change of state is involved, the laws of Thermodynamics provide the backbone to the study of combustion. Design of practical combustion equipment, however, requires further information in the form of the *rate of change of state*. This information is provided by the empirical sciences of *Heat and Mass Transfer* coupled with *Chemical Kinetics*. In addition, the rate of change is also governed by *Fluid Mechanics*.

The heat released by combustion is principally used to produce mechanical work in engines and power plants, or is used directly in applications such as space-heating or cooking. Combustion can also produce adverse impacts, however, as in a fire or in causing pollution from the products of combustion. Thus, understanding of combustion is necessary for producing useful effects as well as for fire extinction and pollution abatement. In earlier times, pollution was regarded as a very local phenomenon (comprising smoke and particulates, for example). However, recognition of the so-called greenhouse gases (which are essentially products of combustion) and their effect on global climate change has given added impetus to the study of combustion.

The foregoing will inform the reader that the scope for the study of combustion is, indeed, very vast. A book that is primarily written for postgraduate students of mechanical, aeronautical, and chemical engineering must, therefore, inevitably make compromises in coverage and emphasis. Available books on combustion, though reflecting these compromises, cannot be said to have arrived at an agreement on a *standard* set of topics or on the manner of their presentation. Much depends on the background and familiarity of the authors with this vast and intriguing subject. Deciding on coverage for postgraduate teaching is also further complicated by the fact that in a typical undergraduate program, students often have inadequate, or no, exposure to three subjects; namely, mass transfer, chemical kinetics, and thermodynamics of mixtures.

It is with gratitude that I acknowledge that this book draws extensively from the writings of Professor D. B. Spalding (FRS, formerly at Imperial College of Science and Technology, London) on the subjects of combustion, heat, and mass transfer. In particular, I have drawn inspiration from Spalding's *Combustion and Mass Transfer*. The book, in my reckoning, provides an appropriate mix of the essentials of the theory of combustion and their use in understanding the principles guiding the design of practical equipment involving combustion.

The topics in this book have been arrived at iteratively following the experience of teaching an *Advanced Thermodynamics and Combustion* course to dual-degree (BTech + MTech) and MTech students in the Thermal and Fluids Engineering stream in the Mechanical Engineering Department of IIT Bombay. The book is divided into 11 chapters.

Chapter 1 establishes the links between combustion and its four neighbors mentioned above. Chapter 2 deals with the thermodynamics of a pure substance; thus serving to refresh material familiar to an undergraduate student. Chapter 3 deals with the thermodynamics of *inert* (non-reacting) and reacting *gaseous mixtures*. Learning to evaluate properties of mixtures from those of individual species that comprise them is an important preliminary. In Chapter 3, this aspect has been emphasized.

Stoichiometry and Chemical Equilibrium are idealizations that are important for defining the product states resulting from a chemical reaction. The product states are evaluated by invoking implications of the law of conservation of mass and the Second law of thermodynamics. These topics are discussed in Chapter 4.

Product states resulting from real combustion reactions in real equipment do not, in general, conform to those discussed in Chapter 4. This is because the real reactions proceed at *finite rate*. As mentioned previously, this rate is governed by chemical kinetics, diffusion/convection mass transfer, and fluid mechanics (turbulence, in particular). Chapter 5 makes the first foray into the world of real chemical reactions by introducing chemical kinetics. Construction of a global reaction from postulated elemental reactions is discussed in this chapter; besides listing the global rate constants for hydrocarbon fuels.

In practical equipment, the states of a reacting mixture are strongly influenced by the transport processes of heat, mass, and momentum in flowing mixtures. These transport processes, in turn, are governed by fundamental equations of mass, momentum (the Navier–Stokes equations), and energy transfer. Therefore, these three-dimensional equations are derived in Chapter 6 along with simplifications (such as the boundary layer flow model, Stefan tube model, and Reynolds flow model) that are commonly used in a preliminary analysis of combustion phenomena. It is important to recognize that chemical kinetics, heat, and mass transfer, and fluid mechanics are rigorous independent subjects in their own right.

The material in Chapters 4, 5, and 6, in principle, can be applied to the analysis of combustion in practical equipment. However, to yield useful quantitative information, the use of computers with elaborate programs becomes necessary. An introductory analysis can nonetheless be performed by making idealizations and introducing assumptions. In combustion engineering, these idealizations of practical equipment are called *Thermo-Chemical Reactors*. Chapter 7 thus presents the analysis of three types of reactors, namely: Plug-flow reactor, Well-stirred reactor, and Constant-mass reactor. Mathematical models of such reactors can be applied to several combustion devices such as gas-turbine combustion chambers, fluidized-bed combustion, fixed bed combustion, biomass gasifiers, and combustion in internal combustion engines.

When fuels burn, they do so with a visible-to-the-eye zone called a *flame*. A closer examination of combustion in practical equipment, of course, requires the study of properties of flames. Chapter 8 thus deals with *premixed flames* because in many combustion devices, gaseous fuel is premixed with air before entering the combustion space through a *burner*. Phenomena such as flame *ignition, propagation, stabilization, and quenching* can be studied by applying simplified equations governing premixed flames.

Gaseous fuels are also burned *without* premixing with air. In these situations, the air required for combustion is drawn from surroundings by processes of diffusion and entrainment. Flames formed in this manner are called *diffusion flames*. Mathematical treatment of such flames requires the solution of multidimensional equations. Diffusion flames are discussed in Chapter 9.

Burning of solid particles (pulverized coal combustion, for example) or liquid droplets (or, sprays) can be viewed as *inter-phase* mass transfer phenomena because the change in state is accompanied by a change in phase; from solid (or liquid) to gaseous phase. Typically, in solids combustion, this phase change is accompanied by *heterogeneous* combustion reactions at the interface. The products of these reactions (along with volatiles) then burn in the gaseous phase through *homogeneous* reactions. In volatile liquid fuel burning, there is no heterogeneous reaction but phase change takes place (*without* a change in the chemical composition of the fuel) owing to latent heat transfer from the gaseous phase. Treatment of both these types of phenomena can be generalized through simplified models introduced in Chapter 6. Chapter 10 is devoted to the estimation of burning rates of solid and liquid fuels.

Finally, in Chapter 11, examples of prediction of combustion in practical devices are presented. The purpose of this chapter is to introduce the reader to the type of predictions obtained by integrating various modeling techniques.

In each chapter, several solved problems are presented. In addition, there are several exercise problems at the end of each chapter. Although some problems can be solved using a pocket calculator, others will require writing computer codes. The reader is strongly advised to solve these problems to assimilate the text material.

June 2010

Anil Waman Date
IIT Bombay, Bombay, India

Contents

About the Author

Professor Anil Waman Date obtained PhD degree in Heat Transfer from Imperial College of Science and Technology, London. He joined Indian Institute of Technology, Bombay as Asst Prof and became part of the Thermal & Fluids Group in Mechanical Engineering Department in 1973. Following his superannuation, IIT Bombay has sought continued engagement of Prof Date as the First Rahul Bajaj Chair Professor in Mech Engineering (2009-2012), as Emeritus Fellow (2010-2013) followed by further re-appointment as Professor (2013-2015) and as Adjunct Professor (2016-2018). In 2006, Professor Date received Excellence in Teaching award and in 2011, received the Life-Time Achievement Award, both from IIT Bombay. In the year 2000, Professor Date was elected Fellow of the Indian National Academy of Engineering (FNAE). Presently, Prof Date is Honorary Professor in the Mech Engg Dept of IIT Bombay. Prof Date's research interests have been in the fields of heat and mass transfer with and without phase change and chemical reaction, computational fluid dynamics, appropriate technology and technology & development issues.

Symbols and Acronyms

A	Area (m^2) or Pre-exponential Factor
B	Spalding number
c_p	Constant pressure specific heat (J/kg-K)
c_v	Constant volume specific heat (J/kg-K)
D	Mass diffusivity (m^2/s)
Da	Damkolhler number
e	Specific energy (J/kg)
E_a	Activation energy (J/Kmol)
Fr	Froude number
f	Mixture fraction
g	Specific Gibbs function (J/kg)
h	Enthalpy (J/kg)
Δh_R	Heat of reaction (J/kg)
ΔH_c	Heat of combustion (J/kg)
J	Jet momentum (kg-m/s^2)
k	Thermal conductivity (W/m-K) or Turbulent kinetic energy (J/kg)
Le	Lewis number
l	Length scale
M	Molecular weight or Mach number
m	Mass flow or mass transfer rate (kg/s)
n	Number of moles
N	Mass transfer flux (kg/m^2-s)
Nu	Nusselt number
P	Perimeter (m)
Pr	Prandtl number
p	Pressure (N/m^2)
q	Heat flux (W/m^2)
$\dot{Q}$	Heat generation rates (W/m^3)
R	Volumetric reaction rate (kg/m^3-s)
R_u	Universal gas constant (J/kg-K)

R_{st}	Stoichiometric Air-fuel ratio
r	Radius (m)
r_{st}	Stoichiometric Oxygen-fuel ratio
Re	Reynolds number
S	Flame speed (m/s)
s	Specific entropy (J/kg-K)
Sc	Schmidt number
St	Stanton number
T	Temperature (°C or K)
t	Time (s)
u	Specific internal energy (J/Kg)
u, v, w	x-,y-,z-direction velocities (m/s)
u_i	Velocity in x_i, $i = 1, 2, 3$ in direction
V	Volume (m^3)
W	Thermodynamic work (N-m)
x_{vol}	Volatiles mass fraction
Z	Collision frequency or Compressibility factor

Greek Symbols

α	Heat transfer coefficient (W/m^2-K)
δ	Boundary layer thickness or Flame thickness (m)
Δ	Incremental value
ε	Turbulent energy dissipation rate (m^2/s^3) or emissivity or degree of reaction
Ψ	Stream function or weighting factor
Φ	General variable or dimensionless conserved property or equivalence ratio or relative humidity
Γ	General exchange coefficient = μ, ρD or k/cp (kg/m-s)
μ	Dynamic viscosity (N-s/m^2)
ν	Kinematic viscosity (m^2/s)
ω	Species mass fraction
ρ	Density (kg/m^3)
σ	Stefan–Boltzmann constant
τ	Shear stress (N/m^2) or, dimensionless time

Subscripts

ad	Adiabatic
b	Burned state
cr	Critical condition
eff	Effective value

g	Gas or Vapour
l	Laminar or to liquid
m	Mixture or mean value
mix	Mixture
o	Infinite surroundings
ox, fu, pr	Oxidant, Fuel, and Product
p	Particle or Droplet
$react$	Reactants
st	Stoichiometric condition
$stoich$	Stoichiometric condition
$surr$	Surroundings
sys	System
T	Transferred substance state
t	Turbulent
U	Universe
w	Wall or interface state
x_i	$x_i, i = 1, 2, 3$ directions
∞	Infinity state

Acronyms

AFT	Adiabatic flame temperature
BDC	Bottom dead center
BTK	Bull's trench kiln
CFC	Chloro fluro carbon
CFD	Computational fluid dynamics
CMCTR	Constant mass thermo-chemical reactor
EBU	Eddy break-up model
GDP	Gross domestic product
GHG	Greenhouse gas
HHV	Higher heating value
LHS	Left-hand side
LHV	Lower heating value
PFCTR	Plug-flow thermo-chemical reactor
RHS	Right-hand side
RPM	Revolutions per minute
SCR	Simple chemical reaction
STP	Standard temperature and pressure
TDC	Top dead center
UNDP	United Nations Development Program
VSBK	Vertical shaft brick kiln
WSCTR	Well-stirred thermo-chemical reactor

Chapter 1
Introduction

1.1 Importance of Thermodynamics

Ever since humans learned to harness power from sources other than what their own muscles could provide, several important changes have occurred in the way in which societies are able to conduct a civilized life. Modern agriculture and urban life are almost totally dependent on people's ability to control natural forces that are far greater than their muscles could exert. We can almost say that one of the main propellants of modern society is the increased availability of non-muscle energy.

Pumping water to irrigate land and supply water to towns and cities; excavation of ores, oil, and coal, their transport and processing; transport of grains, foods, and building materials as well as passengers on land, water, and in the air are tasks that are exclusively amenable to nonmuscular energy. In fact, the extent of goods and services available to a society (measured in gross domestic product, GDP) is almost directly related to the per capita energy consumption. This is shown in Fig. 1.1 for the years 1997 [37] and 2004 [131]. The figure also shows that over these seven years, the GDP of all countries increased, but the developed countries achieved reduced per capita energy consumption through technological improvements compared to the developing countries.[1]

At the dawn of civilization (about the sixth millennium BCE), heavy tasks were performed by gangs of slaves and animals. Huge structures such as the pyramids (about 2600 BCE) and in the relatively more recent times Taj Mahal (about 1650 CE) were constructed in this manner. The power of the *wind* and the *flowing water* was probably harnessed only by about the sixth century CE through sails and windmills. Early windmills were used mainly for grinding grains and for sawing wood. No fuel was burned in the use of these energy resources.

---

[1] Although the per capita energy consumption of India and China is low (owing to their large populations), the total energy consumption is about 45% of that of the United States which is the largest consumer of absolute as well as per capita energy. In this sense, India with about 16% world's population consumes only 3% of energy whereas the United States with 4.6% population consumes almost 23% of the world's energy.

© Springer Nature Singapore Pte Ltd. 2020
A. W. Date, *Analytic Combustion*,
https://doi.org/10.1007/978-981-15-1853-9_1

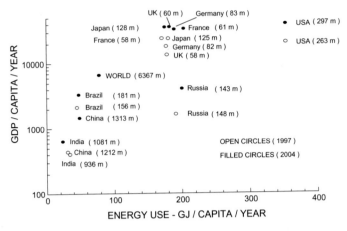

Fig. 1.1 Dependence of GDP on per capita energy consumption—Figures in parentheses indicate population. GDP at constant 1987 dollars (1997 data) and 2004 dollars (2004 data)

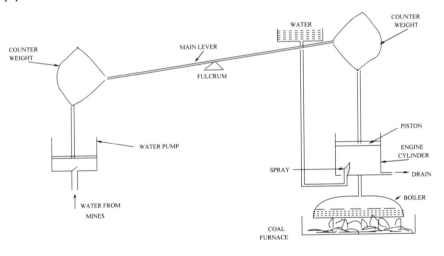

Fig. 1.2 Schematic of Newcomen engine

The most significant development in the use of non-muscle energy came around 1712 when Newcomen in England developed the *steam engine* (see Fig. 1.2) in which the *piston and cylinder* assembly performed the task of *extracting power* and *rejection of heat* through condensation, and the *boiler* provided the *pressurized steam*. The energy required for generating steam was provided by *burning coal* (or any other fuel). The pressurized steam pushed the engine cylinder upward to execute the power stroke. The residual steam in the cylinder was then condensed by means of a water spray, thus enabling the return stroke. This engine developed approximately 15 HP.[2]

---

[2]For further details, see http://technology.niagarac.on.ca/people/mcsele/newcomen.htm.

The engine was used to drain water from mines by means of a reciprocating pump. The steam engine was a great advance because it brought a source of power of unprecedented magnitude dependent neither on muscle power nor on the vagaries of weather and geography.

Because of the simultaneous rise of modern science, improvements in *energy conversion devices* such as the Newcomen engine were brought about. Thus, in 1769, James Watt invented the idea of a *separate condenser*, increasing at least two to threefold the availability of power from the same cylinder and piston assembly for the same amount of fuel consumption. Provision of a separate condenser always kept the engine cylinder *hot*, whereas in the Newcomen engine, the cylinder became cold after every charge of the cold water spray. The number of reciprocations (or the *speed*) of the engine could also now be increased.

Saving of fuel consumption brought the idea of *specific fuel consumption* (kg/s of fuel per kW of power, say, or kg/kJ) and *efficiency* (a dimensionless ratio of the rate of mechanical energy output to the rate of fuel energy input) of the energy conversion devices. This meant systematic measurement of the *energy value* (now called the calorific value or the heat of combustion) of the fuels (kJ/kg) and of the power developed. The *horsepower* unit, for example, was invented by James Watt ($1\,HP \equiv 0.746\,kW$).

Steam engines were used extensively in spinning and weaving machinery, in paper mills, and for pumping water from the mines. In 1807, Robert Foulton in the United States made a commercial *steamboat* that plied on the Hudson River and thus demonstrated the use of steam power for transport. Subsequently, steam was used for transport on land in the form of *steam locomotives*. This gave rise to a network of railways throughout the world.

The steam engine, however, was an *external combustion* engine. By 1768, Street had proposed an *internal combustion* (IC) engine in which *vaporized turpentine* was exploded in the presence of air.[3] Commercial use of the *gas engine* was delayed by almost a century. The gas ($H_2 + CO$) was derived from partial combustion of coal.

Then, in 1870, Otto invented the *gasolene (or Petrol) engine*. In both the gas and the petrol engines, the fuel + air charge was ignited by either an *electric spark* or an *external flame*. Petrol engines were soon followed by heavy-oil engines (invented by Diesel) in which the charge was ignited by a *hot bulb* or by contact, making use of the fact that air, when *compressed* to a high pressure (say about fifteen times the atmospheric pressure), also increased its temperature. These developments occurred in Germany, France, and Britain.

The internal combustion engines were generally more compact and more efficient (Otto's engine had an efficiency of 15%) than the steam engines, and it appeared as though the steam engines would become obsolete. But, in 1884, Parsons[4] revived the use of steam by combining it with the principle of the windmill. Thus was

---

[3]For the early history of gas engines, see http://www.eng.cam.ac.uk/DesignOffice/projects/cecil/engine.html and http://encyclopedia.jrank.org/GaG-GEO/GAS-ENGINE.html.

[4]For a brief history of Steam turbines, see http://www.history.rochester.edu/steam/parsons/, http://www.g.eng.cam.ac.uk/125/1875-1900/parsons.html.

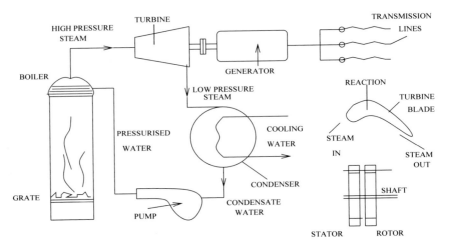

Fig. 1.3 Schematic of a modern steam power plant cycle

born the idea of a *reaction steam turbine*. This machinery reduced the specific fuel consumption nearly by half and a large *steam power plant* became a reality. Such power plants find economic favor even today. The plant cycle is shown in Fig. 1.3. The thermal efficiency of a plant is defined as

$$\eta_{th} = \frac{\text{net work (turbine} - \text{pump)}}{\text{heat input in the boiler}} \tag{1.1}$$

Today's large power plants have efficiency as high as 40%. This means that a 500 MW plant will require 1250 MW heat (or nearly 216 tons of coal/hour) and will reject heat to the tune of 750 MW to the cooling water and, hence, to the environment. That is, allowing for a 7 °C rise in cooling water temperature, nearly 25.6 tons of cooling water per second are required. The environmental consequences will be considered shortly.

In 1910, the *gas turbine* (GT) became a reality. The GT was developed primarily for aircraft propulsion because it offered an alternative to the earlier IC engines in terms of a high power-to-weight ratio. Figure 1.4 shows a GT plant in which propulsive power output is effected either through a shaft coupling air compressor and turbine, or directly through the propulsive force of a jet.

In many modern power stations, the gas and the steam-turbine plants are coupled. This is possible because the exhaust from the gas turbine is typically available at 350 °C to 450 °C. This temperature is in the range of temperatures of superheated steam employed in the steam power plant. As such, the main fuel is burned in the combustion chamber of a GT plant. The hot exhaust from the gas turbine is used to produce steam (instead of burning fuel in a boiler), which is then expanded in a steam turbine. This combined plant typically records efficiencies of the order of 50%.

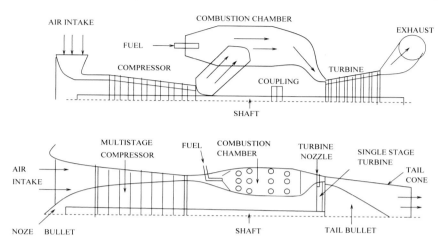

Fig. 1.4 Gas turbine plant: a) industrial GT engine (top figure), b) GT jet engine (bottom figure)

Fossil fuels, however, are not inexhaustible; thus the discovery of *nuclear* energy around 1915 was welcomed. The raw materials for nuclear energy are principally uranium U_{235} and thorium Th_{232}. Easily extractable uranium[5] is available in few countries like USA, Canada, Australia, South Africa, and Nigeria, whereas thorium is more widely available, but its extraction is much more difficult. Uranium is used in power plants that work on the principle of nuclear *fission*, in which heat generated (as a result of the fission reaction) is carried away by means of an inert gas (CO_2) or by light (H_2O) or heavy (D_2O) water. These carriers then pass on the heat to water in a steam generator, and the steam power plant cycle is executed. The next possibility is that of nuclear *fusion* as opposed to fission. Here, the hydrogen atom is split leading to a release of a large amount of energy. Because H_2 is plentiful (from water), there should be no shortage of raw material, in principle. However, it has not been conclusively demonstrated that the energy consumed in extracting hydrogen from water and preparing it for the fusion is less than that which can be released from its reaction with oxygen. The splitting reaction also requires very high temperatures (more than 100 million Kelvin) for sustenance [61]. Extensive use of nuclear energy is also beset with serious environmental concerns.

In view of the difficulties associated with the extensive use of nuclear energy through fission or fusion (there are also intermediate routes, such as the fast breeder reactors), there has emerged an opinion in favor of what are called *renewable* sources of energy such as wind, solar thermal, solar–electric (or, photovoltaic), and biomass. All these are manifestations of the principal source of energy, the *Sun*. The availability of these sources, however, is uncertain (that is, they are seasonal and/or intermittent) and the energy is not highly concentrated in them as in fossil fuels. This means that the plants using these sources must be very large incurring heavy initial (capital) costs,

---

[5]Usable uranium from its raw ore is only 0.25% to 2% and therefore very expensive.

although the running costs may be low. Also, the energy conversion devices working on fossil fuels are not directly suitable for adaptations to these renewable sources. Modifications are needed in several aspects: for example, working fluids, engine cycles, and manufacturing techniques. Intelligent modifications can be devised only on the basis of the science of *thermodynamics*.

At this stage, therefore, we may define thermodynamics as the *science of energy and its conversion*. Many of the developments of the earlier devices were largely intuitive, lacking formalism. This, however, need not be the case anymore.

1.2   Laws of Thermodynamics

At the beginning of the nineteenth century, questions about specific fuel consumption were being thought of in terms of *work* (which the engineer wanted to get out of an engine or a plant) and *heat* which, via the burning of a fuel, the engineer had to employ to produce work. What are the laws governing conversion of heat into work? Watt showed that by providing a separate condenser in the Newcomen engine, the specific fuel consumption could be reduced. *Could the specific fuel consumption be reduced indefinitely and if not, why not?* In providing answers to such questions, the science of thermodynamics was born.

A practical answer to these questions was provided by Sadi Carnot in 1824. Through abstract reasoning, he perceived that the *temperature* provided the key. The difference in temperature was likened to two different levels of water, and the heat was likened to the *flow of water* resulting from the difference in levels. Just as the flow of water provided work in a waterwheel, so did the heat engine provide work because of *flow of heat*. This thought formed the basis of the Second law of thermodynamics.

In one respect, Carnot was wrong, for whereas the same amount of water flows away from the waterwheel, less heat is *rejected* by an engine or a power plant (see the estimates provided in the previous section) than supplied to it. Carnot can hardly be blamed for not noticing the essential difference between the water substance and the heat, because in the engines of his day, the heat *lost* between supply and rejection amounted to less than 5%.

In 1850, Joule provided the answer to the *lost* heat—it had *turned into* work. The flow of heat thus could not be likened to the flow of water. Joule, of course, performed a *reverse* process, in which work was shown to have been completely converted to heat. This discovery formed the basis of the First law of thermodynamics. The explanation of the *lost* heat was thus only by inference that whereas work can be completely converted to heat, the reverse can also be exercised—but only at a premium.

The inference of the *lost* heat, though not completely clear at this stage, becomes clear when Carnot's and Joule's findings are combined, that is, when the First and the Second laws are combined. This recognition was made by Lord Thompson in Scotland and independently by Clausius in Germany. This forms a body of knowledge that is known today as *Classical, or Macroscopic, or Equilibrium thermodynamics*.

These terms essentially imply that one is dealing only with the bulk nature of matter as gas, liquid, or solid. Molecular and atomic structures are ignored. In addition, the main concerns are associated with the *before* and *after* of a process undergone by an identified bulk substance. The process may be heating, cooling, expansion, compression, phase change, or a chemical reaction. The identified substance in thermodynamics is called the *System*. The terms "before" and "after" of a process are identified with the *change in the state* of the system.

The state of a system in thermodynamics is represented in terms of *properties* of the system. The basic properties are: pressure, temperature, volume, and composition. The system is also characterized by several other properties such as energy, entropy, availability, and enthalpy which are derived from the basic properties. The formal definition of thermodynamics can thus be given as:

Thermodynamics is the science of relationships among heat, work and the properties of the system.

Because thermodynamics is concerned with the change in the states of a system, the *in-between* events are left out. Thus, in a chemical reaction, for example, the initial *reactants* and the final *products* represent the change in the state of the material substance (in this case, a mixture of several components or chemical species). In practical equipment, however, the product-state is dependent on the *rate of change*. Thermodynamics *does not deal with the rate of a process*. The sciences that describe the rate of a process are largely empirical. Heat Transfer, Mass Transfer, Fluid Mechanics, and Chemical Kinetics are such empirical sciences that become necessary to design practical equipment. In fact, these rates can be explained by taking a *microscopic* view of matter, in which interactions between molecules and atoms are studied. But then, these interactions are essentially statistical and the ideas developed in *kinetic theory* and *quantum theory* must be invoked. This body of knowledge is called *Statistical Thermodynamics*. The laws of Classical thermodynamics deal only with bulk matter.

1.3 Importance of Combustion

Throughout the world over, since the early 1970s, much concern has been expressed about the impending *energy crisis*. By this term, it is meant that the fuels used to generate most of the energy used by humans are running out. Simultaneously, there has also been another concern namely, the *environmental crisis*. Geographically, economically, and politically speaking, both these crises have global as well as local dimensions [99].

Table 1.1 provides estimates of primary energy use in different countries. Primary energy signifies coal, oil, natural gas, hydropower, and nuclear sources, and electricity is taken as secondary energy (converted to Tons of Oil Equivalent (TOE) based on 38% oil-to-electricity conversion) [118]. These estimates include all sectors of the economy such as domestic, industrial, agriculture, transport, and infrastructure/construction sectors. The table shows that in 2001, for the world as a whole,

Table 1.1 Primary energy consumption (including electricity) in million Tons of Oil Equivalent (MTOE)—2001

Country	Oil	Natural gas	Coal	Nuclear	Hydro	Total
USA	896	578	546	183	48	2251
Japan	248	71	103	73	20	515
France	96	38	12	96	18	258
Germany	132	75	85	39	6	336
UK	77	87	40	20	2	226
China	232	25	519	4	54	834
India	97	25	173	4	16	314
World	3517	2220	2243	601	585	9156

38.4% of the energy came from oil, about 24% each from natural gas and coal, about 6.6% from nuclear, and 6.4% from hydropower sources. For individual countries, however, there is considerable variation. For example, in France, nearly 37% of energy comes from nuclear, whereas for most other countries nuclear energy provides less than 10% and in India and China, less than 2%. In fact, India and China with limited nuclear, hydro, and natural gas energy rely almost exclusively on oil and coal.

These estimates preclude biomass and other nonconventional or renewable energy sources, largely because their use is extremely small in most countries. However, in countries such as India and China, for example, biomass contributes an additional 16% to 20% to energy consumption. Finally, although the share of India and China in total world consumption was 3.4% and 9.1%, respectively, in 2001, it is estimated that by 2030, this share will rise to 8% and 16%, respectively.

Finally, it is important to recognize that with rising populations, the world energy consumption is expected to rise by nearly 35% to 40% over the next thirty years. A large contribution to this increase is likely to come from the expanding economies of India and China.

Accompanying the current and expected increase in energy use is the concern for environment expressed in terms of two effects: (1) Air–Water–Land pollution and (2) Global warming. The first effect can be countered through local pollution abatement measures, but the second effect has global dimensions and is not easily tractable. Global warming (including unavoidable climate change) is caused by accumulation[6] of the so-called *greenhouse gases* (GHGs) CO_2, CH_4, N_2O and chloro fluro carbon (CFCs) and aerosols in the atmosphere.[7] The GHGs, while allowing the solar radiation to warm the earth's surface, essentially trap the reflected outgoing solar radiation from the earth in the lower atmosphere (less than about 6 km to 7 km above the earth); thus raising the earth's temperature still further. It is estimated that

---

[6]CO_2 emitted today has a residence time in the atmosphere of about a century.

[7]Burning 1 kg of coal releases 3.4 kg of CO_2. Thus, a 500 MW power plant will release $216 \times 3.4 \simeq$ 735 tons of CO_2 per hour.

Table 1.2 Carbon emissions—1996 (GDP in constant 1987 U.S. dollars)

Country	Total emissions (mil-tons of C/yr)	Emission per capita (kg of C/capita/yr)	Emission/GDP (kg of C/GDP)
USA	1407	5270	0.26
Russia	496	3340	2.01
Japan	307	2460	0.1
Germany	228	2790	0.15
France	93	1600	0.09
UK	148	2530	0.19
China	871	730	1.62
India	248	270	0.65
Brazil	68	430	0.2
Indonesia	81	410	0.62
World	6250	1090	NA

CO_2 concentrations have increased from about 280 ppm at the start of the industrial era to about 330 ppm today [37]. According to mathematical global warming models, this change implies to $0.5\,°C$ to $1.0\,°C$ rise in the earth's temperature. The actual measured temperature values from 1855 to 1999 have confirmed this temperature rise. Such model predictions have been used to assess the likely earth's temperature up to the year 2100. The most optimistic scenarios predict further temperature rise of about $1.5\,°C$ whereas the most pessimistic scenarios predict about $3.5°$ to $5.0\,°C$ temperature rise.

In climate change literature, CO_2 is considered to be the major carrier of carbon into the atmosphere. The overall effect of GHGs is estimated in terms of carbon emissions. Table 1.2 provides estimates of emissions in different countries. The table shows that the United States and other industrialized countries have high per capita carbon emissions. But, France has relatively lower emissions, however, because of its heavy reliance on nuclear energy. Again, following expected high economic growth, carbon emissions from India and China are expected to rise. Also, high values of emissions per GDP in Russia, China, India, and Indonesia suggest that fossil fuels are not burned efficiently in these countries. Again, carbon emissions from biomass burning are taken to be zero because growth of biomass by photosynthesis is a natural, solar energy driven CO_2 sink.

It is important to correlate CO_2 emissions in terms of *intensity of energy use* indicators. This has been done by Kaya [118]. The correlation reads

$$CO_2 \; \frac{kg}{year} = \text{population} \times \frac{GDP}{\text{population}} \times \frac{\text{energy}}{GDP} \times \frac{CO_2}{\text{energy}}$$
$$- \text{ sequestration.} \tag{1.2}$$

Table 1.3 World reserves of primary energy—2004—1 barrel of oil = 136 kg of oil

Source	Estimated reserves	Time to exhaustion
Oil	1000–1227 billion barrels	100 years
Natural gas	120–180 trillion m^3	250 years
Coal	1000–1100 billion tons	250 years
Uranium	1.3–11.0 million tons	20–170 years

In this correlation, (GDP/population) represents standard of living, (energy/GDP) represents energy intensity of the economy, (CO_2/energy) represents carbon intensity of the primary energy, and sequestration (when applied) represents the amount of CO_2 prevented from entering the atmosphere by capture and sequestration. This implies CO_2 capture at the site of its generation (say, a power plant) by a chemical process or, simply by absorbing CO_2 in deep oil/gas reservoirs, oceans, or aquifers. Using information from Tables 1.1 and 1.2, it is possible to estimate the emission rate (see exercise). The correlation in Eq. 1.2 suggests that to reduce carbon emissions, population, energy intensity as well as standard of living may have to be reduced unless sequestration becomes economically attractive. Unfortunately, estimates at present predict an increase between 20% to 45% in the cost of net energy production if sequestration is applied.

Fossil fuels are finite. Therefore, it is of interest to estimate current reserves of the main primary energy sources. Table 1.3 shows *proven* estimates of primary energy reserves available in 2004 and the expected time of exhaustion if the resource was used at the current level of consumption.[8] The lifetime (or the time of exhaustion) of a reserve is estimated from

$$\text{Life Time (yrs)} = R^{-1} \left[\ln \left\{ R \times \left(\frac{Q_T}{C_0} \right) + 1 \right\} \right] \tag{1.3}$$

where R is the percentage rate of growth of consumption per year, Q_T is the total reserve (J, tons, m^3, barrels), and C_0 is the present rate of consumption per year (J, tons, m^3, barrels/yr).

At a time when the fossil fuels are running out, it is important that the maximum energy extractable from a fuel in a given device is indeed extracted. In other words, engineers seek to obtain nearly 100% combustion efficiency. In practical devices, however, it is not just sufficient to burn fuels with high efficiency but it is also necessary to ensure that the products of combustion have a particular temperature and chemical composition. Temperature control is necessary to achieve downstream process control, as in many metallurgical, chemical, and pharmaceuti-

---

[8]There is considerable difficulty in estimating energy reserves. *Proven reserves* implies economic viability of extraction and processing. This does not include known possibilities of availability of a resource. For example, it is known that uranium is available from the sea bed and deeper earth's crust, but its economic viability is not known.

cal industries. Composition control is also necessary for meeting the strict environmental pollution norms. According to the U.S. 2000 National Ambient Air Quality
Standards (NAAQS) [37], the permissible levels of pollutants are: CO (9–35 ppm),
NO_2 (0.053 ppm), O_3 (0.12 ppm), SO_2 (0.03–0.14 ppm), and particulate matter PM
(50–150 $\mu g/m^3$). Excessive amounts of CO can have neurological effects, and can
cause visual acuity, headache, dizziness, or nausea and can lead to acid rain afflict-
ing flora and fauna. NO_2 and SO_2 can cause pulmonary congestion or nasal and eye
irritation or coughs and can result in weathering and corrosion due to formation of
nitric and sulfuric acids. Excessive PM can lead to asthma, and throat and lung can-
cers, and can have other brain and neurological effects. PM can also cause visibility
impairment owing to light scattering of small particles. Ozone O_3 is a photo-oxidant
having a strong oxidizing capacity in the presence of sunlight. It can cause vegetation
damage and can affect crop yields. All the above pollutants are released from fossil
fuel combustion and, therefore, their concentrations in products of combustion must
be restricted.

Combustion is also important in fire prevention. Forest fires and fires at the oil
wells and mines are some examples. Such fires not only destroy valuable energy
source, but also contribute to carbon emissions. Combustible material (wood, paints,
other synthetic materials) used in buildings can catch fire from electric sparks or
any other cause. Plastics and other organic materials are used extensively in modern
society. Even clothing made from cotton or synthetic yarns can catch fire. Prevention
and extinction of fires is an important task performed by engineers.[9]

Exercises

1. Calculate the values of specific fuel consumption (kg/kJ) and efficiency in each
   of the following:

   (a) A man working on a hand loom can produce 6.5 N-m of work per second
       for 8 h in a day. His daily food intake is 1.5 kg (energy value of food =
       3700 kcal/kg)
   (b) The oil engine of a seagoing vessel develops 1500 HP. Its rate of Diesel
       consumption is 4.2 kg/min (calorific value = 45 MJ/kg)
   (c) A dynamo delivers 120 A DC at 110 V. The dynamo is driven by a prime
       mover that is fed with wood at the rate of 350 kg in 8 h (calorific value =
       14.5 MJ/kg)

2. From Fig. 1.1, estimate the *average* percentage change in total GDP per year at
   1987 prices and in total energy use per year of India, China, and the United States
   between 1997 and 2004. (To determine 2004 GDP at 1987 prices, divide 2004
   GDP by 1.29.)
3. From the data given in Table 1.2, estimate the Kaya index of CO_2 emissions for
   India, China, and the United States for the year 1996. Neglect sequestration.

---

[9]It is estimated that there are more than 220,000 forest fires per year in different parts of the world,
destroying more than 6.5 million hectares of forest. In China, for example, anywhere between 20
and 100 million tons of coal are lost due to coal fires at the mines (Anupama Prakash, http://www.
gi.alaska.edu/~prakash/coalfires/introduction.html).

4. In the year 2004, reserves of oil in India were about 760 million tons. The consumption rate then was 120 million tons per year of which 72% was imported. Estimate the lifetime of India's reserves if the percentage growth rate of consumption was $R = 1\%, 5\%$, and 20% per year.
5. Consider a 500 MW coal thermal power plant having overall 35% efficiency. If the calorific value of coal is 25 MJ/kg and allowable temperature rise of water in the condenser is 7 °C, estimate the coal consumption rate (tons/hour), rate of cooling water requirement (tons/hour), and CO_2 production rate (tons/year). Also, calculate specific fuel consumption (kg/kJ).

Chapter 2
Thermodynamics of a Pure Substance

2.1 Introduction

The main purpose of this chapter is to briefly recapitulate what is learned in a first course in classical thermodynamics.[1] As defined in Chap. 1, *thermodynamics is the science of relationships among heat, work, and the properties of the system.* Our first task, therefore, will be to define the keywords in this definition: *system, heat, work, and properties.* The *relationships* among these quantities are embodied in the First and the Second laws of thermodynamics. The laws enable one to evaluate the change in the *states* of the system, as identified by the changes in its properties. In thermodynamics, this change is called a *process*, although, in common everyday language, the processes maybe identified with terms such as cooling, heating, expansion, compression, phase change (melting, solidification, evaporation, condensation), or chemical reaction (such as combustion, catalysis, etc.). As such, it is important to define two additional terms: *state* and *process.*

The last point is of particular importance, because the statements of thermodynamic problems are often couched in everyday language. However, with the help of the definitions given in the next section, it will become possible not only to deal with all processes in a generalized way but also to recognize whether a given process can be dealt with by thermodynamics or not. Thus, adherence to definitions is very important in thermodynamics. Establishing a connection between the problem statement in everyday language and its mathematical representation is best helped by drawing a sketch by hand.[2] This will be learned through several solved problems.

---

[1] Only those portions of a typical undergraduate curriculum that are relevant to combustion calculations are given emphasis.

[2] Sometimes it is also possible to experience the process for oneself. After all, thermodynamics is a science that is very much built on the *observations* and *experiences* of everyday life.

© Springer Nature Singapore Pte Ltd. 2020
A. W. Date, *Analytic Combustion*,
https://doi.org/10.1007/978-981-15-1853-9_2

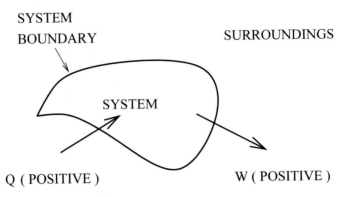

Fig. 2.1 Definition of system, system boundary, and surroundings

2.2   Important Definitions

2.2.1   System, Surroundings, and Boundary

With reference to Fig. 2.1, a *system* is any identifiable and prescribed *collection of matter* (solid, gas, or liquid). The system is separated from its surroundings by a *system boundary*. Everything outside the boundary is called the system's *surroundings*. As we shall see shortly, a system boundary may be a real *physical* boundary or it may even be an *imaginary* boundary. The latter type of boundary is often invoked to facilitate the application of the laws of thermodynamics to the identified matter.

2.2.2   Work and Heat Interactions

In thermodynamics, *work* and *heat* are defined only at the system boundary as interactions between the system and its surroundings or *vice versa*. As such, both are viewed as *energies in transit*.[3]

Because these interactions are directional, they are assigned a sign. By convention, when heat (Q) flows *into the system*, it is assigned a positive value. In contrast, when work (W) is *delivered from the system* to the surroundings, it is assigned a positive value. In Fig. 2.1, both interactions are thus positive. When they occur in opposite directions, they take their negative values.

---

[3] In this sense, the everyday statement *heat contained in a body* has no meaning in thermodynamics.

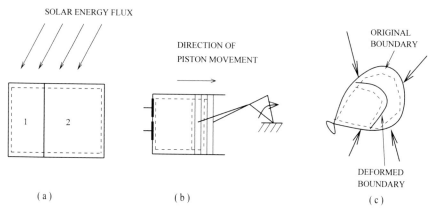

Fig. 2.2 Examples of a closed system: a) all boundaries rigid, b) one rigid shifting boundary, c) flexible boundary

Sometimes, the surroundings are also considered to be a system.[4] As such, a work interaction that is positive with respect to the system will be negative with respect to the surroundings. The magnitude of W, however, is the same with respect to both the system and the surroundings.

2.2.3 Closed (Constant-Mass) System

A *closed system* is defined as a system in which a *fixed* amount of matter is confined within its boundaries and across which heat and work interactions take place. The system boundaries may be rigid, flexible, or rigid but shifting. To eliminate any ambiguity about the identified mass, a system boundary is often shown by a *dotted* line.

Figure 2.2 shows three examples of the closed system. In Fig. 2.2(a), two rooms separated by a wall form one closed system. The system is receiving solar energy; hence, there is a *positive* heat interaction. In Fig. 2.2(b), execution of an expansion stroke of an internal combustion (IC) engine is shown. As such, the valves are closed and the piston is moving to the right. The piston, though rigid, represents a shifting boundary. In this case, the system is delivering work (hence, *positive*) to the surroundings and some heat may be lost (hence, *negative*) to the cooling water in the surroundings. In Fig. 2.2(c), a balloon is compressed. As a result, the system boundary deforms. This is an example of a system with flexible boundaries in which the work interaction is *negative*. In all cases, the identified mass within the boundaries remains fixed, although the *volume* occupied by this mass may change.

---

[4]This type of reference to surroundings will be made later in defining *universe* (see Sect. 2.7.2).

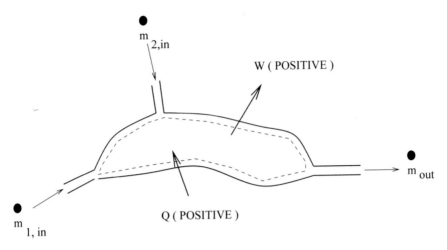

Fig. 2.3 Schematic of an open system

2.2.4 Open (Constant-Volume) System

An *open* system is defined as a system of *fixed volume*, across the boundaries of which mass and energy (heat and work) may flow in and/or out.

Figure 2.3 shows the schematic of an open system. The dotted line in this figure identifies the fixed volume.[5] The mass may flow in or out through multiple inlets and outlets, respectively. These flows (usually represented as flow rates, $\dot{m}$) may be of the same substance or different substances, and they may be in different phases. Although the volume is fixed, a further implication is that the *mass of the system* may change with time. Practical equipment such as a heat exchanger (multiple inlets and outlets), a steam turbine, or a compressor are examples of open systems.

2.2.5 In-Between Systems

In practical situations, one often encounters systems that at first sight, appear to be neither constant-mass nor constant-volume systems. Figure 2.4 shows such a system in which a balloon is blown by a man. When the balloon is fully inflated, its mass, as well as its volume within the dotted lines, has changed, as shown in Fig. 2.4(a). Thermodynamics cannot deal with such a system, because the identified system is neither closed nor open. One, therefore, must carefully observe that to blow up a balloon, the man first inhales air and stores it in the cavity in his mouth. He then simply pushes the air into the balloon (using the muscle power of his cheek) and his mouth cavity shrinks.

---

[5]This is also called a *control volume*.

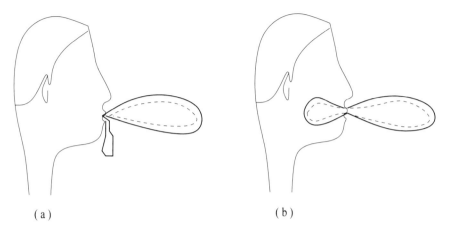

(a)   (b)

Fig. 2.4 Blowing up a balloon. a) In-between system, b) constant-mass system

This observation tells us that the appropriate system to consider is (balloon + mouth cavity) and not just the balloon. For the appropriately defined new system, the mass of the system remains constant at all times; hence, it is a closed system. This is shown in Fig. 2.4(b).

2.2.6 Thermodynamic Equilibrium

In thermodynamics, properties of a system are defined only when the system is in equilibrium. *Thermodynamic equilibrium* implies that there are *thermal* (constant and uniform temperature in the system), *mechanical* (constant and uniform pressure), and *chemical* (constant and uniform composition) equilibria *simultaneously*.[6]

The idea of a thermodynamic equilibrium can be quite confusing, partly because of erroneous use of the language. To clarify matters, let us examine how a system will reach a state of equilibrium. Consider a closed system interacting with its surroundings through work (W) and heat (Q). At t = 0, let us suppose that the interactions are shut off; that is, $Q = W = 0$ for $t > 0$. Further, suppose that this state of affairs continues for a long time. After a very long time (in fact, infinite time), the system will achieve perfect thermodynamic equilibrium.

A system in which $Q = W = 0$, is called an *isolated system*. We may thus say that an isolated system reaches perfect equilibrium after a long time, but the exact time is not known. Does this mean that the properties of a system can never be defined? Engineers, as always, take a pragmatic view in such matters.

To illustrate, consider a closed vessel filled completely with water at ambient temperature in which the water has attained *stillness*. At first thought, it might appear that the system is in equilibrium because the temperature and composition (H_2O) are uniform everywhere. However, is pressure everywhere the same? Clearly not,

---

[6]Here, *constant* implies constancy with time, whereas *uniform* implies spatial uniformity. That is, p, T, and composition of the system are same in every part of the system.

because hydrostatic variations with height must prevail. As such, the system is not in mechanical equilibrium and, therefore, not in perfect equilibrium. Engineers, however, take the system to be in perfect equilibrium by assuming that in a vessel of small height, the pressure variation is indeed small.

Similarly, consider a 10 L bucket filled completely with cold water (density: 1 kg/L) at 30 °C. A 2 kW heater is dipped in the bucket. The bucket is perfectly insulated from all sides. The heater is now kept switched on for only 10 min. The engineer evaluates the final temperature T_f of water from the application of the First law of thermodynamics as $10 \times 4.18 \times (T_f - 30) = 2 \times 10 \times 60$, to deduce that $T_f = 58.7 °C$. The moot question is: Is this temperature an equilibrium temperature, or some kind of an average temperature? Clearly, immediately after the heater is switched off, the temperature will not be uniform everywhere and therefore will not be a thermodynamic equilibrium temperature. However, if the water was left as is, temperature gradients in the water will cause density gradients, which in turn will set up natural convection currents. Hence, the water will naturally mix and, after a long time, the temperature everywhere will be 58.7 °C. In that event, the temperature will be an equilibrium temperature. At short times after switch off, however, it will be only an average temperature. Such an average temperature has no meaning in thermodynamics, although its magnitude equals thermodynamic temperature.

2.2.7 Properties of a System

Thermodynamics uses several properties that are either measured directly (mainly pressure, volume, temperature, and composition) or simply inferred (for example, internal energy, enthalpy, entropy, availability) from the measured ones. Further, the properties are categorized into two types: *intensive* and *extensive*.

Intensive properties: These properties have the same value whether one considers the whole system or only a part of it. Thus, pressure p (N/m^2), temperature T (°C or K), specific volume v (m^3/kg), specific internal energy u (kJ/kg), specific enthalpy h (kJ/kg), specific entropy s (kJ/kg-K), and the like are intensive properties. Except for temperature, all intensive properties are referred to with lowercase letters.

Extensive properties: The value of an extensive property for the entire system is the sum of values of that property for all smaller elements of the system. Thus, mass M (kg), volume V (m^3), internal energy U (kJ), enthalpy H (kJ), entropy S (kJ/K), and so forth are extensive properties. They are designated by uppercase letters. Unlike other intensive properties, pressure and temperature have no extensive-property counterparts.

2.2.8 State of a System

Specifying values of *intensive properties* (p, T, v, u, h, s, etc.) of a system is termed as specifying the *state* of the system. Because intensive properties are defined only when the system is in equilibrium, it follows that the state also cannot be defined unless the system is in equilibrium.

To define the state of a system completely, one must specify all measured, as well as inferred, intensive properties. However, the intensive properties are simply too numerous. The relevant question then is: What is the lowest number of properties that would suffice to define the state of the system completely? The answer to this question can be given using the past experience of carrying out thermodynamic calculations. As such, without any elaboration, we state a *rule*:

If the system experiences N types of reversible work modes, then the minimum number of *independent* intensive properties to be specified is N + 1.

Thus, if a system simultaneously experiences two types of work interactions (say, $W_{expansion}$ and $W_{capillary}$), then one will need to specify only three intensive properties to specify the state. These are: p, v, or T (from expansion work), and surface tension σ (from capillary work).

Most often, we are interested in a system that experiences only the expansion/compression work. Such a system will require only two independent properties to be specified. In fact, a substance that can be *idealized* as having only one associated work mode is called a *simple* substance. When this work mode is of the expansion/compression type, the substance is called a *simple compressible* substance. For such a substance, the choice of the two properties, however, must be made with care.

For example, if water at 1 bar and 100 C is considered then specifying intensive properties p and T will not suffice because the substance may be pure steam, pure water, or a mixture of water and steam. As will be shown shortly, pressure and temperature do not take *independent* values in different states comprising different proportions of steam and water. In fact, p and T will be the same in all states of the system. In such a case, the appropriate choice will be p and v, or T and v.

Finally, if there are three properties, x_1, x_2, and x_3, then mathematically speaking, the two-property rule can be stated as

$$x_3 = f(x_1, x_2) \quad \text{or} \quad F(x_1, x_2, x_3) = 0. \tag{2.1}$$

This equation is also called the *equation of state* for a simple compressible substance because it helps in defining the state of a system. The state can be represented on a graph (called the state diagram) or, in the form of a table (as in a steam table), or by an algebraic equation (as in $p\,v = R_g\,T$, for example).

2.3 Behavior of a Pure Substance

2.3.1 Pure Substance

In thermodynamics, the term *pure substance* carries a special meaning. A pure substance is a system that is (a) homogeneous in composition, (b) homogeneous in chemical aggregation, and (c) invariable in chemical aggregation.

This definition thus clarifies that two systems comprising pure steam and pure steam + water are both pure substances having chemical formula H_2O that will not change with time. Likewise, another system comprising ice + water will also be a pure substance.

However, now consider air which consists of N_2 and O_2 in the ratio 79/21 by volume when in *gaseous* state at normal temperatures and pressures. At very low temperatures, however, one component will condense out, leaving the other in a gaseous state. Thus, the system will not satisfy condition (b). Similarly, at very high temperatures, O_2 and N_2 may *dissociate* to form O, N, NO, and so on, and hence, condition (c) is not satisfied. Thus, air is not a pure substance in general. However, because in most practical applications extreme temperatures are not encountered and air essentially preserves its chemical composition, it is treated as a pure substance.

2.3.2 Typical Behavior

The behavior of a pure substance is *measured* in terms of three independent properties: pressure, temperature, and volume. For our discussion, we consider the *simple compressible* substance that experiences only the *compression or expansion* (that is, volume change) work mode. For measurement, a *constant-pressure* experiment is conducted, as shown schematically in Fig. 2.5.

The experiment is conducted in an insulated piston–cylinder assembly with a mass placed above the piston, so the system will experience constant-pressure. The cylinder is heated from below. During the heating process, the temperature T and specific volume v of the system will change. After a short period of heating, the cylinder bottom is quickly insulated and the system is allowed to settle down to a new equilibrium state. In this new state, T and v are again recorded. The cylinder-bottom insulation is now removed, and the heating process is repeated. The values of T and v thus recorded are plotted Fig. 2.6 as state points. Because a simple compressible substance is considered, the states of the system are completely defined by only two properties, T and v.

Let the applied mass create a pressure, p_1 (say). Initially, let the system be in a solid state, as shown by point **a** in Fig. 2.6. As the system is heated, its temperature will rise till point **b**. At **b**, the substance has reached its *melting point* and, therefore, any further addition of heat will be absorbed as *latent heat of melting or fusion* at constant temperature. Between points **b** and **c**, the system will comprise solid

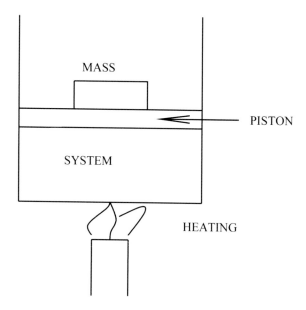

Fig. 2.5 Constant-pressure experiment

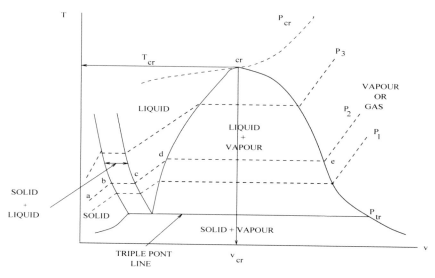

Fig. 2.6 Behavior of a pure substance—the state diagram

and liquid in different proportions but still in equilibrium. Both the melting point $(T_b = T_c)$ and latent heat (λ) are properties of the system. At **b**, the substance is still in solid state and, therefore, its enthalpy is denoted by the symbol h_s. This is also called the *solidus* enthalpy. At **c**, the substance is complete liquid and hence its enthalpy is denoted by h_l. Thus,

$$\lambda = h_l - h_s = h_c - h_b. \tag{2.2}$$

When all the latent heat is absorbed (point **c**), the substance has turned into liquid and again its temperature will rise through further heating until point **d**. Again, the system will undergo volume change without temperature increase up to point **e**. Between **d** and **e**, the system will comprise liquid and vapor in different proportions but still in equilibrium. The heat supplied between **d** and **e** is called the *latent heat of vaporization* and is commonly denoted by the symbol h_{fg}. The enthalpy of the substance at **d**, being in liquid state, is denoted by h_f and that at **e**, being in complete vapor or gaseous state, is denoted by h_g. Thus,

$$h_{fg} = h_g - h_f = h_e - h_d. \tag{2.3}$$

With further heating, the system will continue to remain in gas/vapor phase indefinitely with the rise in its temperature and specific volume.[7]

The same experiment can be performed at various pressures $p_2, p_3, p_4, \ldots$ to arrive at points **a, b, c, d,** and **e** at each pressure. The locus of points **b** is called the *solidus line*, and the locus of points **c** is called the *liquidus line*. Similarly, the locus of points **d** is called the *saturated liquid line*, and that of point **e** is called the *saturated vapor line*. The region between saturated vapor and saturated liquid lines is called the *two-phase* region. The region to the right of the saturated vapor line is called the *superheated vapor region*. In this region, the temperature is greater than the *saturation temperature* $T_e = T_d$ for the pressure under consideration. Similarly, in the pure liquid region between points **c** and **d**, if $T < T_d$ at the pressure corresponding to the saturation temperature T_d, the state is called the *subcooled liquid* state.

The *state diagram* thus represents all states of matter in their practical range of utility. The diagram is unique for each pure substance. In most power-producing devices, we are typically interested in the region to the right of the saturated liquid line. For completeness, we note, however, that the variations of T and v at two pressures: namely, the *critical pressure* p_{cr} and the *triple-point pressure* p_{tr} have special significance.

At the critical point, $h_{fg} = 0$. That is, a liquid at this pressure is instantly converted to all vapor without any latent heat of vaporization. In this state, the meniscus separating liquid and vapor that is observed at subcritical pressures is no longer seen. Similarly, at the triple point formed at the junction of solidus and saturated liquid lines, all three phases (solid, liquid, and vapor) exist simultaneously. For, $p < p_{tr}$, a solid is directly converted to vapor without formation of liquid. This is called the *sublimation* process. Triple points are ideal *fixed points* for calibration of thermometers. Table 2.1 gives approximate values of critical points and Table 2.2 gives approximate values of the triple point for a few substances [89].

---

[7]At very high temperatures, the system (gas) will turn into *plasma*, when the gas will shed electrons and become ionized. The gas will then be a mixture of neutral atoms, ionized atoms, and electrons. Such ionized state is not of interest in this book.

Table 2.1 Critical point of pure substances

Substance	T_{cr}	p_{cr}	v_{cr}
Units	C	bar	m^3/kg
Freon-12	+112	91.28	0.00174
Freon-134	+101.5	40.6	
NH_3	+132.5	43.9	0.00426
N_2	−146.3	33.5	0.00322
O_2	−118.3	50.35	0.00133
CO	−141	35.9	0.00332
CO_2	+31.2	73.8	0.00217
Cl_2	+144	77.13	0.00175
He	−268	2.29	0.0145
H_2	−240	12.97	0.03245
CH_4	−82	45.8	0.0062
C_3H_8	+97	42	0.00454
SO_2	+57	77.8	0.0019
Air	−140.6	37.74	0.00286
H_2O	+374.14	221	0.003155

Table 2.2 Triple point of pure substances

Substance	T_{tr}	p_{tr}	v_{tr}
Units	C	bar	m^3/kg
N_2	−207	0.123	0.001157 (l) 0.001487 (g)
Hg	−38.3	0.006	
H_2O water	+0.01	0.0027	
Carbon/graphite	+3430	3.5	
CO_2 dry ice	−56.6	5.1	

In practice, to facilitate thermodynamic calculations, the state diagram is often plotted in terms of any two convenient inferred variables rather than in terms of the measured variables p, v, and T. For example, in steam work, the diagram is plotted in terms of enthalpy h and entropy s. It is called Mollier's diagram. Similarly, in air-conditioning and refrigeration work, it is convenient to use the variables h and p. How variables h and s are constructed from the measured variables p, v, and T will be dealt with in a later section.

2.4 Law of Corresponding States

In several machines, such as IC engines, gas turbines, compressors, and blowers, the substance used is in the *gaseous* state. It is, therefore, of great interest to represent the behavior of a substance in its gaseous state by algebraic equations to facilitate thermodynamic calculations. In this state, the intermolecular distances (and, therefore, the intermolecular forces) are greatly influenced by temperature and pressure. This implies that the volume occupied by molecules in a given volume of gas will vary greatly with pressure and temperature. Many equations of the gaseous state covering range of pressures and temperatures above and below the critical point have been proposed. Van der Waals (1879) is credited for pioneering such an equation. It reads

$$\left(p + \frac{a}{\bar{v}^2} \right) (\bar{v} - b) = R_u T, \tag{2.4}$$

where a and b are gas-specific constants (see Table 2.3) and $\bar{v}$ is specific molar volume (m³/kmol). The values of constants a and b can be determined by recognizing that at the critical point we have two conditions: $\partial p / \partial v \mid_{cr} = \partial^2 p / \partial v^2 \mid_{cr} = 0$ (see Fig. 2.6). Thus, from Eq. 2.4, it can be shown that

$$a = 3 \, p_{cr} \, \bar{v}_{cr}^2 = \frac{27}{64} \frac{R_u^2 \, T_{cr}^2}{p_{cr}} \qquad b = \frac{\bar{v}_{cr}}{3} = \frac{R_u \, T_{cr}}{8 \, p_{cr}}. \tag{2.5}$$

To make further progress, we define three *reduced* quantities with suffix r:

$$p_r \equiv \frac{p}{p_{cr}}; \qquad v_r \equiv \frac{v}{v_{cr}}; \qquad T_r \equiv \frac{T}{T_{cr}}. \tag{2.6}$$

In addition, the *compressibility factor*, denoted by Z, is defined as

$$Z \equiv \frac{p \, v}{R_g \, T}. \tag{2.7}$$

Then, combining Eqs. 2.4 to 2.7, it can be shown that

$$Z^3 - \left(\frac{p_r}{8 \, T_r} + 1 \right) Z^2 + \left(\frac{27 \, p_r}{64 \, T_r^2} \right) Z - \frac{27 \, p_r^2}{512 \, T_r^3} = 0. \tag{2.8}$$

Table 2.3 Van der Waals gas constants

Substance	a	b
	N-m⁴/kmol²	m³/kmol
Air	1.35×10^5	0.0365
O_2	1.38×10^5	0.0318
N_2	1.367×10^5	0.0386
H_2O	5.51×10^5	0.0304
CO	1.479×10^5	0.0393
CO_2	3.656×10^5	0.0428
CH_4	2.286×10^5	0.0427
H_2	0.248×10^5	0.0266

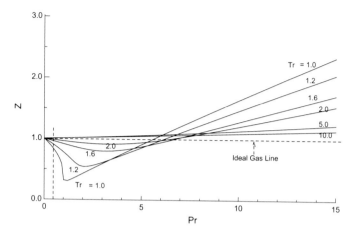

Fig. 2.7 The law of corresponding states

This equation shows that $Z = F(p_r, T_r)$ only and Z is not a function of v_r. It does not contain a or b and is thus applicable to any gas. This is known as the *law of corresponding states*. The law represents a remarkable unification of reduced property data for several gases.[8]

Equation 2.8 is plotted in Fig. 2.7 for $1 < T_r < 10$ and $0 < p_r < 15$. When $Z = 1$, we have what is called an *ideal gas*.[9] The figure shows that $Z \simeq 1$ when $p_r < 0.5$ and $T_r > 1.5$. In fact, in this range, Z varies between 0.97 and 1.03 and, therefore, can be taken to approximate an ideal gas. Such an approximate ideal gas is referred as the *perfect gas*. Thus, the perfect gas behavior is taken as

---

[8] Van der Waals was awarded the Nobel Prize in 1910 for this discovery.

[9] In this sense, the van der Waals Eq. 2.4 is simply a modification of the ideal gas law $p\,\bar{v} = R_u\,T$ via gas-specific constants a and b. Constant b represents the volume occupied by molecules per mole of gas, which is subtracted from the container molar volume $\bar{v}$. The estimate of this *unavailable* volume is, therefore, related to the hard-sphere molecular diameter d. The constant a, on the other hand, is related to the idea that intermolecular forces are appreciable only between a molecule and its nearest neighbors. Molecules are attracted equally in all directions, but the molecules in the outer layers will experience only a net *inward force*. A molecule approaching the container wall is thus slowed down by the inward force and the pressure exerted on the wall is somewhat smaller than what it would have been in the absence of attractive forces. This reduction in pressure is estimated to be proportional to n^2 where n is the number of molecules ($N = n \times N_{Avo}$) per unit volume. Thus constants a and b are given by

$$b = \frac{2}{3} N_{Avo}\, d^3,$$ (2.9)

$$\text{and } a = \alpha n^2 \bar{v}^2,$$ (2.10)

where α is constant of proportionality and N_{Avo} is Avogadro's number (6.02×10^{26} molecules/kmol).

$$pv \simeq R_g T \quad \text{for} \quad p_r < 0.5 \quad \text{and} \quad T_r > 1.5. \quad\quad (2.11)$$

Clearly, in the region of perfect gas behavior, the gas pressure is considerably lower than the critical pressure and the temperature is higher than the critical temperature. The main attribute of an ideal (or perfect) gas is that the intermolecular distances are much greater than the molecular diameter. As such, the intermolecular force fields exert little influence on the *collision properties* of the molecules.[10]

From Table 2.1, we can now deduce that air, for example, will behave as a perfect gas when $p < 0.5 \times 37.74$ (18.9 bar) and $T > 1.5 \times 132.4$ (198.6 K). Most practical equipment operating with air operate in this range of pressure and temperature and, hence, air is commonly treated as a perfect gas in practical work. Similarly, steam will behave as a perfect gas when $p < 0.5 \times 221$ (110 bar) and $T > 1.5 \times 647$ (975 K). Most practical equipment operating on steam has temperatures that are much smaller than 975 K, although the specified pressure range is routinely encountered. It is for this reason that steam is not treated as a perfect gas in practical work.

2.5   Process and Its Path

As mentioned earlier, in engineering we encounter a variety of processes, such as heating or cooling, humidifying or dehumidifying, melting or solidification, evaporation or condensation, mixing or separation, and so forth. In most of these processes, there is only one relevant work mode and, therefore, the two-property rule applies to define the states of the system. In general, we say that a process has occurred when a change of state takes place. Usually, the state will change because of heat and/or work interactions. All processes occur over a *finite* time. Our interest is to define any process in a generalized way.

2.5.1   Real and Quasi-static Processes

Let, $z_1 = F(x_1, x_2)$ be the initial *equilibrium* state of a system. Heat and/or work interactions take place over a finite time after which they are stopped, so the system becomes an isolated system. Such a system will again reach equilibrium and it will assume a new state. Let this new state be designated $z_2 = F(x_1, x_2)$. The relevant question now is: What is the state of the system between z_1 and z_2? Obviously, the intermediate states cannot be defined unless the system was in equilibrium at every instant during the change from z_1 to z_2. Real processes rarely pass through equilibrium states. Such an instantaneous equilibrium may be assumed, however, if the *rate of change* was slow. As will be shown, many real processes are indeed

---

[10]Collision properties determine the rates of chemical reactions, as discussed in Chap. 5.

very slow. Alternatively, the real process may be somewhat *modified* and made to progress slowly, but still produce state z_2. In thermodynamics, such slow processes are called *quasi-static* processes. Before giving a formal definition of such a process, two examples are considered.

In IC engines, for example, the linear speed of the piston is typically of the order of 20 m/s. Consider the expansion or the compression strokes. At the prevailing temperatures during these strokes, the speed of sound ($\sqrt{\gamma R_g T}$) will be greater than the piston speed by several orders of magnitude. As such, the pressure wave (which travels at the speed of sound) will travel very fast and, therefore, equalize the pressure in every part of the system as soon as the change in pressure takes place. If the temperature (and composition) of the system is also taken to be nearly uniform, then clearly the system will be in equilibrium at every instant and the states can be defined throughout the process.

However, consider the situation shown in Fig. 2.8(a). The system consists of a *gas* in an insulated and rigid rectangular box partitioned by a diaphragm. The system is shown by dotted lines. On the other side of the diaphragm, vacuum prevails. Initially, let the gas be in state z_1. The diaphragm is then *burst*. The gas will, therefore, expand and fill the entire box (state z_2).

The bursting process, of course, is very fast—almost instantaneous. As such, although states z_1 and z_2 can be represented, the in-between states cannot be, because they are not in equilibrium. This ignorance of the in-between states is acknowledged in Fig. 2.8(c) by a zigzag-dotted line. However, we can take a clue from the earlier

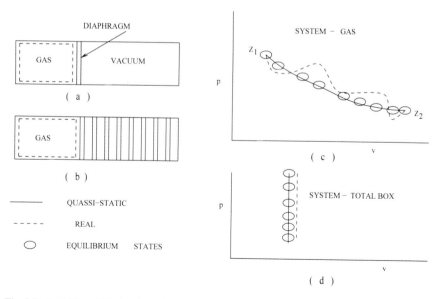

Fig. 2.8 Real (dotted line) and quasi-static (solid line) processes

IC engine example and ask: Can the bursting process be slowed down so that it can be treated as a quasi-static process?

A thought experiment can be readily devised. It will consist of fitting several diaphragms, one behind the other, in the space occupied by vacuum (see Fig. 2.8(b)). Each diaphragm is burst in succession after a finite time. In this case, the final state z_2 will be the same as before, but before each diaphragm is burst, the system will have time to attain equilibrium and, therefore, in-between states can also be represented. This is shown by the series of discrete circles in Fig. 2.8(c). Of course, if an infinitely large number of diaphragms were used, then the discrete points will merge into a line. In a quasi-static process, it is common practice to assume that the real process has been split into a large number of smaller ones and, hence, the process is drawn by a *solid line*. The solid line is called the *path* of the process. The path of a quasi-static process can be traced so much so that the path may even be described by an algebraic relation.

In general, because a system undergoes a state change as a result of work and/or heat interaction, the path traced by a process must be a function of direction and magnitude of Q and W. Alternatively, we may say that Q and W depend on the path traced. Therefore, in thermodynamics, Q and W are called *path functions*.

In the preceding discussion, the identified system was a gas that underwent abrupt or *unresisted* expansion. For such an expansion, the work done is zero. This can be convincingly argued by taking the *entire box* as the system. For this newly defined system, W $= 0$ (rigid walls and, hence, no volume change) and the system will undergo a *constant-volume* process as shown in Fig. 2.8(d). But, the real process is shown by a dotted line to acknowledge that the state of the system cannot be represented in the absence of equilibrium.

Finally, in the diaphragm example, the quasi-static process has been constructed by modifying the *real* process. No such modification was necessary in the IC engine example. In conclusion, therefore, we have the following definition:

Any slow process (with or without modification of the real process) that can be represented on a state diagram is called a quasi-static process.

2.5.2 Reversible and Irreversible Processes

The previous examples tell us that no real processes can be represented on a state diagram. However, quasi-static processes can be, through assumption(s) that may or may not involve physical modification of the real process. In thermodynamics, the quasi-static processes (in which reference is made to the system only) are considered *equivalent* to *reversible* processes. The limitations of this equivalence, however, can be understood from the formal definition of a reversible process. It reads [115]:

A process is reversible if means can be found to restore the system and all elements of its surroundings to their respective initial states.

This definition informs us that unlike the quasi-static process, the reversible process makes reference to both the system and its surroundings. However, it may be argued that a quasi-static process, being very slow, has the greatest *chance* of meeting the stipulations of a reversible process. For example, it might appear that the heat and work interactions occurring during the compression stroke of an IC engine could be reversed identically to restore the system and the surroundings to their original states. In reality, however, this is most unlikely, even though the process is slow. This is because, although the work supplied can be recovered through expansion and heat lost through the walls can also be supplied from *external means*, the process of friction between the piston and the cylinder walls cannot be, because friction will be present in the process in both directions. Thus, to restore the system and its surroundings to their original state before expansion/compression, more heat or work may have to be added/subtracted by external means. As such, even a slow process can never be strictly reversible. If *friction is assumed negligible*, however, then the slow quasi-static process can be considered equivalent to a reversible process. The idea of a reversible process is simply an *ideal* construct used in thermodynamics. It can never be truly realized. This is even more true for the very rapid bursting process of unresisted expansion considered earlier.

Apart from friction and unresisted expansion, there are several other processes, such as mixing of two unlike fluids, heat transfer across a finite temperature difference, flow of electric current through a resistor, and plastic deformation of a material, in which the process cannot be executed in a truly reversible manner. These processes are, therefore, always *irreversible*. Thus, by extension of the definition of a reversible process, we can say:

Any process that is not reversible is called an irreversible process.

In this sense, all real processes are irreversible because the causes (mainly *dissipative* in character) that impart irreversibility are almost always present in them.

Finally, in thermodynamics, the work interaction can be evaluated only for a quasi-static reversible process.[11] Thus, the work of expansion/compression in which volume change is involved is

$$W = \int_{z_1}^{z_2} p \, dV. \tag{2.12}$$

Similarly, the heat interaction Q is calculable only a quasi-static reversible process. How this is done will become apparent in Sect. 2.7.1.

2.5.3 Cyclic Process

A *cyclic* process is one in which the initial state z_1 and the final state z_2 are identical. That is $z_2 = z_1$. Figure 2.9 shows such a process on a two-property diagram. The

---

[11]In mechanics, formal definition of work is given by work (J) = force (N) × displacement (m) in the direction of the force. The thermodynamic definition of work done by a system is consistent with this definition if and when it can be demonstrated that *the sole effect external to the system could be reduced to raising (or lowering) of a weight*. In carrying out this demonstration, one is free to *change the surroundings*.

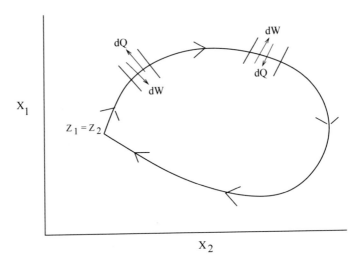

Fig. 2.9 Cyclic process comprising infinitesimally small reversible or irreversible subprocesses; x_1 and x_2 are any intensive properties

total process, however, may comprise several subprocesses, in which the directions and magnitudes of the heat and work interactions are not necessarily same. Some or all processes may be reversible (quasi-static) or irreversible.

The concept of a cyclic process is useful because the formal mathematical statements of the First and the Second law of thermodynamics are made for a cyclic process.

2.6 First Law of Thermodynamics

As stated in Chap. 1, the First law of thermodynamics originated following Joule's celebrated stirrer experiment. Joule had, in fact, performed a *cyclic* process in a closed insulated system to deduce that work had completely turned into heat (see Sect. 2.6.2). The mathematical statement of the First law is, therefore, made for a closed system undergoing a *cyclic* process. The statement reads (see Fig. 2.9):

If a closed system undergoes a cyclic process, then

$$\oint_{path} d\,Q = \oint_{path} d\,W \,, \tag{2.13}$$

where dQ and dW are heat and work interactions during infinitesimally small subprocesses of the cyclic process.

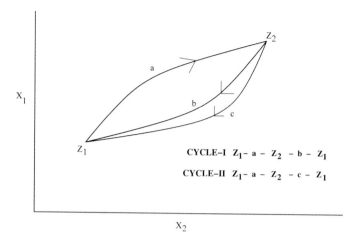

Fig. 2.10 Consequence of First law for a finite process

Because the statement pertains to a law, no proof needs to be given. The law simply states that since the initial (z_1) and the final (z_2) states are identical, the integral sum of heat and work interactions over all subprocesses must also be the same.[12] The statement has been verified through a large number of observations. We now explore the consequences of the law for a closed system undergoing a *finite* process, and for an *open system*.

2.6.1   First Law for a Finite Process—Closed System

With reference to Fig. 2.10, let a cyclic process comprise only two processes: sub-process (1) $z_1 - a - z_2$ and subprocess (2) $z_2 - b - z_1$. Let this be called cycle-I. Similarly, let cycle-II comprise subprocess (1) and subprocess (3) $z_2 - c - z_1$. Let the heat and work interactions for each subprocess be designated Q and W, respectively. The magnitudes and directions of Q and W will, of course, be different in each subprocess because Q and W are path functions. Then, applying the First law (Eq. 2.13) in which the cyclic integral is replaced by an algebraic sum, we have

For cycle-I: $W_{z_1-a-z_2} + W_{z_2-b-z_1} = Q_{z_1-a-z_2} + Q_{z_2-b-z_1} \; ;$

---

[12] Alternatively, the law can also be stated as

$$\oint_{path} (d\,Q - d\,W) = 0.$$

For cycle-II: $W_{z_1-a-z_2} + W_{z_2-c-z_1} = Q_{z_1-a-z_2} + Q_{z_2-c-z_1}.$

Manipulation of the preceding expressions shows that

$$(Q - W)_{z_2-b-z_1} = (Q - W)_{z_2-c-z_1} = -(Q - W)_{z_1-a-z_2}. \qquad (2.14)$$

Three important comments pertaining to this equation are in order.

1. The first equality shows that $(Q - W)$ is the same for both the return paths along b and c. Thus, $(Q - W)$ is *independent of the path*. This is also confirmed by the second equality as $-(Q - W)_{z_1-a-z_2} = (Q - W)_{z_2-a-z_1}.$
2. From this deduction, we conclude that although Q and W are independent *path functions*, $(Q - W)$ is *not a path function*. Therefore, it must be only a *state function*. That is, $(Q - W)$ must be a function of some property belonging to *initial* state z_2 and *final* state z_1 of the finite processes along paths a, b, and/or c.
3. This *unknown* property of the state is *defined* as *energy* and is denoted by the symbol E.

Thus, Eq. 2.14 is written as

$$(Q - W)_{z_2-a,b,c-z_1} = E_1 - E_2 \quad \text{or} \quad (Q - W)_{z_1-a,b,c-z_2} = E_2 - E_1, \qquad (2.15)$$

or, in general,

$$(Q - W)_{path} = E_{final} - E_{initial} = \Delta E. \qquad (2.16)$$

This is the formal statement of the First law of thermodynamics for a finite process underwent by a closed system.

Equation 2.16 indicates that energy is a man-made concept. It represents the difference between two energies (Q and W) that are defined at the system boundary and, therefore, are energies in transit. In this sense, the First law not only creates the need for defining a property called energy, but also enables evaluation of *change* in energy. Thus, a quantitative meaning can be assigned only to ΔE but not to E. The energy E can have any value. Because E is a state function, it belongs to the system and, therefore, it is a property of the system similar to pressure, temperature, or volume. E has the same units as Q and W; that is, Joules.[13]

There are various types of energy. Some energies that are contained in a system can be *externally* observed. Examples are kinetic energy ($KE = 0.5 \, mV^2$ where m and V are system mass and velocity, respectively) and potential energy ($PE = mgZ$ where g is acceleration due to gravity and Z is the vertical height). Later we shall encounter chemical energy (ChE).[14] There are others that cannot be observed. They

---

[13]The reader will now appreciate why a loose statement such as *heat contained in body (meaning system)* is unacceptable in thermodynamics. A system does not contain heat. It contains energy.

[14]Other possibilities include magnetic, electrical, or capillary energies, although we are not concerned with these in this book.

are called *hidden* or *internal* energy, and are denoted by symbol U. In fact, E is related to U by

$$E = U + KE + PE + ChE + \cdots . \tag{2.17}$$

In thermodynamics, it is assumed that U, though hidden, is manifested through the temperature of the system. In many applications, KE and PE are very small and ChE not of consequence, in a pure substance.[15] Hence, the most familiar statement of the First law is given by

$$Q - W = \Delta U, \tag{2.18}$$

or, for an infinitesimally small process,[16]

$$dQ - dW = dU . \tag{2.19}$$

For an infinitesimally small process, $dW = m \times p \times dv$, where m is the mass of the system. Evaluating $dQ = m \times dq$ and $dU = m \times du$, Eq. 2.19 can be stated in terms of intensive properties as

$$dq - pdv = du . \tag{2.20}$$

2.6.2 Joule's Experiment

Joule performed a cyclic process in two subprocesses, as shown in Fig. 2.11. In the first, a rigid insulated vessel contained water at ambient temperature. Work to this system was supplied by means of a stirrer by lowering a weight ($W = mgZ$) that was connected by a string to the pulley mounted on the stirrer shaft. In this process, 1 say, $Q_1 = 0$ and W_1 is negative. Hence,

$$- (-W_1) = \Delta U . \tag{2.21}$$

Thus, the internal energy of water (manifest in raised temperature) had increased. The vessel (now, without insulation) was then instantly placed in an *insulated* cold bath. The hot contents of the vessel, therefore, cooled (decreasing its internal energy)

---

[15]Because E can have any value, to facilitate calculations, E is assigned a *datum value* at some reference state. Thus, PE may be assigned a zero value at the mean-sea-level (say) or U may be assigned zero value at zero degrees C or K. These deductions apply only to a pure substance or to the components of an *inert mixture*. Later, in Chap. 4, however, we will find that for a *reacting mixture*, ChE cannot be assigned a datum value in this way and special treatment becomes necessary.

[16]In some books, Eq. 2.19 is written as $d'Q - d'W = dU$ to remind us that dQ and dW, being path functions, are *inexact differentials*, whereas dU (or dE) is an exact differential. In this book, it is assumed that the reader is aware of this difference.

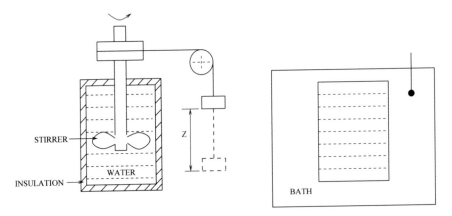

Fig. 2.11 Joule's stirrer experiment

and the bath temperature increased. This temperature increase, ΔT, enabled evaluation of $\Delta U = (m \times c_p \times \Delta T)_{bath}$. In this process, $W_2 = 0$ and Q_2 is negative. Therefore,

$$(-Q_2) = \Delta U . \qquad (2.22)$$

However, ΔU in the second process, though equal in magnitude to that of the first process, bore opposite (negative) sign. Algebraic addition of the above equations, therefore, showed that $W_1 = Q_2$, or that the stirrer work supplied had completely turned into heat.[17] The equations also explain our convention that work, when supplied to the system, and heat, when rejected by the system, are treated as negative.

2.6.3 Specific Heats and Enthalpy

Unlike other system properties that make no reference to any process, *specific heats* are system properties that are *defined* for infinitesimally small processes. Thus, through heat interaction dq (kJ/kg), if the system temperature changes by dT, then the specific heat is defined as

$$c \equiv \frac{dq}{dT} \quad \frac{kJ}{kg - K} .$$

As dq is a *path function*, the values of c for different processes will be different. In general, replacing dq from Eq. 2.20,

---

[17] Joule evaluated work in lbf-ft and heat in Btu. Thus, $W_1/Q_2 = J$ (a constant) had units of lbf-ft/Btu and $J = 778$ has come to be known as Joule's *mechanical equivalent of heat*. In SI units, however, work is evaluated in N-m and heat in Joules. As 1 N-m = 1 Joule, the mechanical equivalent of heat in SI units is 1 N-m/Joule.

$$c \equiv \frac{pdv + du}{dT} .$$

(2.23)

To facilitate engineering calculations, however, the values of c are documented for only two processes: a constant-volume process and a constant-pressure process.

Specific Heat at Constant Volume c_v

For a *constant-volume process*, dv = 0. Therefore, from Eq. 2.23, we have

$$c_v \equiv \frac{du}{dT} |_v .$$

(2.24)

Enthalpy

In thermodynamics, to facilitate a compact definition of specific heat at constant pressure c_p, a new property, called enthalpy, is *defined* and is denoted by h (kJ/kg). Thus,

$$h \equiv u + pv .$$

(2.25)

Specific Heat at Constant Pressure c_p

From the above definition, for an infinitesimally small process, dh = du + pdv + vdp. For a constant-pressure process, however, dp = 0. As such, from Eq. 2.23,

$$c_p \equiv \frac{dh}{dT} |_p .$$

(2.26)

2.6.4 Ideal Gas Relations

We have already introduced the notion of a perfect gas, for which the compressibility factor (see Eq. 2.7) $Z \sim 1$ and, hence, $pv = R_g T$. Therefore, from Eq. 2.25, the enthalpy of a perfect gas can also be evaluated as

$$h = u + R_g T .$$

(2.27)

Hence, dh = du + R_gdT, or

$$\frac{dh}{dT} = \frac{du}{dT} + R_g .$$

(2.28)

In thermodynamics, an *ideal gas* is one whose enthalpy h and internal energy u are functions of *temperatures* only and depend on no other property, such as pressure or volume.[18] As such, $u = u(T)$, $h = h(T)$ and $\partial(u, h)/\partial v \mid_T = \partial(u, h)/\partial p \mid_T = 0$. Making use of definitions of specific heats (Eqs. 2.26 and 2.24), we have

$$c_p = c_v + R_g .$$

(2.29)

The values of c_p (and, hence, of c_v) for different gases are tabulated in Appendices A and D as a function of temperature. In thermodynamic analysis, a perfect gas is taken to be an ideal gas; both obeying the equation of state Eq. 2.11. For a perfect gas, however, c_p and c_v are taken as *constant*. This is perfectly legitimate if the temperature range in the application under consideration is small.

To facilitate calculations, it is convenient to write Eq. 2.29 in terms of γ, the ratio of specific heats. Thus

$$\gamma \equiv \frac{c_p}{c_v} .$$

(2.30)

Then, Eq. 2.29 can also be expressed as

$$\frac{R_g}{c_p} = \frac{\gamma - 1}{\gamma} \quad \text{or} \quad \frac{R_g}{c_v} = \gamma - 1 .$$

(2.31)

2.6.5   First Law for an Open System

We derived the First law for a constant-mass closed system in Eq. 2.16. Here, we make use of this equation to derive the First law for an open system which is a constant-volume system with inflows and outflows of mass (see Fig. 2.3). Heat and work interactions take place with the control volume (CV). Also, because the sum of the rates of inflows may not always equal the rate of outflow(s), there will be accumulation (or depletion) of mass within the CV resulting in *unsteadiness*. Thus, the CV mass (and consequently, CV energy) will be a function of time. As such, Eq. 2.16, which is derived for a constant-mass system, cannot be applied to the CV.

To circumvent this difficulty, we define two constant-mass systems (see Fig. 2.12). At the initial instant t, let the system be defined by A-B-C-D-E-F and at instant $t + \Delta t$, let the system be defined by a-b-c-d-e-f. At both times, the system masses are the same, but the mass of CV (m_{CV}) is different. In other words, at time t, the system will comprise volume (CV + AFfa + BCcb), whereas at time $t + \Delta t$, the system volume is (CV + DdeE). Thus, the masses and energies associated with these volumes are

---

[18]This state of affairs will prevail when the gas molecules are sufficiently far apart so that the molecular force field plays no part. As such, a perfect gas at very low pressures and high temperatures will behave as an ideal gas.

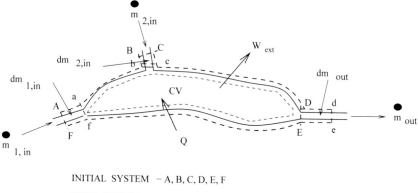

INITIAL SYSTEM − A, B, C, D, E, F

FINAL SYSTEM − a, b, c, d, e, f

Fig. 2.12 Derivation of First law for an open system

$$m_t = m_{CV} + dm_{1,in} + dm_{2,in}$$

$$= m_{CV} + \left(\sum_{k=1}^{k=2} \dot{m}_{k,in}\right) \Delta t \; , \tag{2.32}$$

$$m_{t+\Delta t} = (m + \Delta m)_{CV} + dm_{out}$$

$$= m_{CV} + \Delta m_{CV} + \dot{m}_{out}\, \Delta t, \tag{2.33}$$

$$E_t = E_{CV} + \left(\sum_{k=1}^{k=2} \dot{m}_{k,in}\, e_{k,in}\right) \Delta t \; , \text{and} \tag{2.34}$$

$$E_{t+\Delta t} = E_{CV} + \Delta E_{CV} + \dot{m}_{out}\, e_{out}\, \Delta t \; , \tag{2.35}$$

where symbol e represents specific energy (kJ/kg) and Δm and ΔE represent accumulated mass and energy, respectively, in time Δt.

Mass-Conservation Equation

From Eqs. 2.32 and 2.33, as $m_t = m_{t+\Delta t}$,

$$m_{CV} + \left(\sum_{k=1}^{k=2} \dot{m}_{k,in}\right) \Delta t = m_{CV} + \Delta m_{CV} + \dot{m}_{out}\, \Delta t \; ,$$

or, letting $\Delta t \to 0$, we have

$$\frac{d\, m_{CV}}{dt} = \sum_{k=1}^{k=2} \dot{m}_{k,in} - \dot{m}_{out} \; . \tag{2.36}$$

This is called the mass-conservation equation.[19]

Energy Equation

We now apply Eq. 2.16 to the two constant-mass systems and evaluate each term separately. Then, from Eqs. 2.34 and 2.35, the change in energy will be given by

$$E_{t+\Delta t} - E_t = \Delta E_{CV} + (\dot{m}\, e)_{out} - \left(\sum_{k=1}^{k=2} \dot{m}_{k,in}\, e_{k,in} \right) \Delta t . \qquad (2.37)$$

If $\dot{Q}$ is the *rate of heat exchange* with the surroundings then

$$Q = \dot{Q}\, \Delta t . \qquad (2.38)$$

In an open system, the evaluation of the work interaction W requires careful thought. This is because, in addition to the usual *external* work interaction, denoted by W_{ext} in Fig. 2.12, there is also the *flow-work* W_{flow}. The flow-work is the work required to displace the system boundary over time Δt. It is evaluated as follows.

Let $v_{1,in}$ and $v_{2,in}$ be the specific volumes at the inflow sections. The total volumes swept out in time Δt will, therefore, be $(v\, dm)_{1,in}$ and $(v\, dm)_{2,in}$, respectively. Now, let the pressures at the inflow sections be $p_{1,in}$ and $p_{2,in}$. Then, the flow-work done in the inflow sections will be $\sum -(p\, v\, dm)_{k,in}$ for $k = 1, 2$. The negative sign indicates that the work is *supplied* to the system. Similarly, outflow-work will be $(p\, v\, dm)_{out}$, and it will be positive. Thus, the net work interaction in time Δt is

$$W = W_{ext} + (p\, v\, dm)_{out} - \sum_{k=1}^{k=2} (p\, v\, dm)_{k,in}$$

$$= \dot{W}_{ext}\, \Delta t + (p\, v\, \dot{m})_{out}\, \Delta t - \sum_{k=1}^{k=2} (p\, v\, \dot{m})_{k,in}\, \Delta t . \qquad (2.39)$$

Now, from Eq. 2.16, $Q - W = \Delta E = E_{t+\Delta t} - E_t$. Therefore, from Eqs. 2.37–2.39, and letting $\Delta t \to 0$, we have

$$\dot{Q} - \dot{W}_{ext} = \frac{dE_{CV}}{dt} + \dot{m}_{out}\, (e + p\, v)_{out} - \sum_{k=1}^{k=2} \dot{m}_{k,in}\, (e + p\, v)_{k,in} . \qquad (2.40)$$

From Eq. 2.17, the specific energy $e = u + 0.5\, V^2 + gz$. Therefore, Eq. 2.40 can also be written as

---

[19]If there were more than one (say, three) outlet, the mass-conservation equation will read as

$$\frac{d\, m_{CV}}{dt} = \sum_{k=1}^{k=2} \dot{m}_{k,in} - \sum_{l=1}^{l=3} \dot{m}_{l,out} .$$

$$\dot{Q} - \dot{W}_{ext} = \frac{dE_{CV}}{dt}$$
$$+ \dot{m}_{out} \left(u + p\,v + \frac{V^2}{2} + gz \right)_{out}$$
$$- \sum_{k=1}^{k=2} \dot{m}_{k,in} \left(u + p\,v + \frac{V^2}{2} + gz \right)_{k,in} \quad (2.41)$$

or, from the definition of enthalpy in Eq. 2.25,

$$\dot{Q} - \dot{W}_{ext} = \frac{dE_{CV}}{dt}$$
$$+ \dot{m}_{out} \left(h + \frac{V^2}{2} + gz \right)_{out}$$
$$- \sum_{k=1}^{k=2} \dot{m}_{k,in} \left(h + \frac{V^2}{2} + gz \right)_{k,in} . \quad (2.42)$$

Equations 2.40–2.42 are different ways of writing the First law of thermodynamics for an open system.

The following comments are important with respect to the energy and the mass-conservation equations of an open system.

1. For a *steady-flow* process, $dm_{CV}/dt = 0$.
2. For a *steady-state* process, $dE_{CV}/dt = 0$.
3. For a *steady-flow*, steady-state process, $dm_{CV}/dt = dE_{CV}/dt = 0$.

2.7 Second Law of Thermodynamics

For a finite process, the First law (see Eq. 2.16) cannot tell us if the system has undergone a reversible or an irreversible process because $(Q - W)$ is independent of the path followed by the process. Therefore, the law can also be written[20] as

$$(Q - W)_{rev} = (Q - W)_{irrev} = \Delta E \quad (2.43)$$

The Second law of thermodynamics enables a distinction to be made between a reversible and an irreversible process. The formal statement of the law is again made for a cyclic process. It reads:

---

[20]This manner of writing the First law is an acknowledgment that as long as our interest lies in determining state change (ΔE or ΔU), the real irreversible process can always be replaced by a reversible one. We have already noted that to replace an irreversible process by a quasi-static reversible process, the former process may have to be modified.

If a closed system undergoes a cyclic process, then

$$\oint \frac{dQ}{T} < 0 \qquad \textbf{for an irreversible process and} \qquad (2.44)$$

$$= 0 \qquad \textbf{for a reversible process .} \qquad (2.45)$$

Again, as this is a law, no proof needs to be given. Unlike the First law (Eq. 2.13), the Second law makes no reference to the work interaction. Only heat interaction is considered. In addition, reference is also made to *temperature* of the system. The First law statement made no reference to any system property.

We now consider the consequences of the law for a closed system undergoing finite reversible and irreversible processes, as well as for the open system.

2.7.1 Consequence for a Finite Process—Closed System

Reversible Process

To deduce the consequence for a finite reversible process, consider Fig. 2.10 again. Now, let the three subprocesses along a, b, and c be *reversible*. Then, applying Eq. 2.45, we have

$$\text{for cycle-I :} \quad \int_{z_1-a-z_2} \frac{dQ}{T} \Big|_{rev} + \int_{z_2-b-z_1} \frac{dQ}{T} \Big|_{rev} = 0, \text{ and}$$

$$\text{for cycle-II :} \quad \int_{z_1-a-z_2} \frac{dQ}{T} \Big|_{rev} + \int_{z_2-c-z_1} \frac{dQ}{T} \Big|_{rev} = 0,$$

where dQ are infinitesimal heat interactions during the subprocesses. Manipulation of the above expressions shows that

$$\int_{z_1-a-z_2} \frac{dQ}{T} \Big|_{rev} = -\int_{z_2-b-z_1} \frac{dQ}{T} \Big|_{rev} = -\int_{z_2-c-z_1} \frac{dQ}{T} \Big|_{rev}$$

or

$$\int_{z_1-a-z_2} \frac{dQ}{T} \Big|_{rev} = \int_{z_1-b-z_2} \frac{dQ}{T} \Big|_{rev} = \int_{z_1-c-z_2} \frac{dQ}{T} \Big|_{rev}. \qquad (2.46)$$

Thus, we can conclude that although dQ is a path function, $\int (dQ/T) \big|_{rev}$ is *not a path function*. Therefore, it must be a *state function*. The integral simply depends on a *change* in some system property corresponding to the initial and final states. This property is called *entropy* and is denoted by the symbol S (kJ/K). Thus, the Second law for a finite reversible process is given by

$$\int_{z_1}^{z_2} \frac{dQ}{T} \Big|_{rev} = S_2 - S_1 = S_{final} - S_{initial} = \Delta S, \qquad (2.47)$$

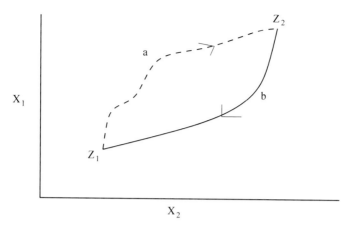

Fig. 2.13 Second law—irreversible process

or, in terms of specific entropy s (kJ/kg-K),

$$\int_{z_1}^{z_2} \frac{dq}{T} \Big|_{rev} = s_2 - s_1 = \Delta s \ . \tag{2.48}$$

Thus, we can say that just as the First law defines energy, the Second law defines entropy. Both E and S are man-made properties. Again, like E, the absolute value of S has no meaning. Further, if the process were infinitesimally small, then

$$\frac{dq}{T} \Big|_{rev} = ds \ . \tag{2.49}$$

Irreversible Process

Consider Fig. 2.13 in which a cyclic process comprising two subprocesses is shown. The first process ($z_1 - a - z_2$) is an *irreversible* process (hence, shown by a dotted line), whereas the second ($z_2 - b - z_1$) is *reversible*. For this case, Eq. 2.44 will apply. Therefore,

$$\int_{z_1-a-z_2} \frac{dQ}{T} \Big|_{irrev} + \int_{z_2-b-z_1} \frac{dQ}{T} \Big|_{rev} < 0$$

or

$$\int_{z_1-a-z_2} \frac{dQ}{T} \Big|_{irrev} < -\int_{z_2-b-z_1} \frac{dQ}{T} \Big|_{rev}$$

$$< \int_{z_1-b-z_2} \frac{dQ}{T} \Big|_{rev} \tag{2.50}$$

or

$$\int_{z_1}^{z_2} \frac{dQ}{T}\Big|_{irrev} < (S_2 - S_1) = \Delta S \;, \tag{2.51}$$

or, in terms of specific entropy,

$$\int_{z_1}^{z_2} \frac{dq}{T}\Big|_{irrev} < (s_2 - s_1) = \Delta s \;. \tag{2.52}$$

The following comments pertaining to Eqs. 2.48 and 2.52 are important:

1. For the *same* entropy change,

$$\int_{z_1}^{z_2} \frac{dq}{T}\Big|_{rev} > \int_{z_1}^{z_2} \frac{dq}{T}\Big|_{irrev} . \tag{2.53}$$

2. Analogous to Eqs. 2.45 and 2.44 for a cyclic process, the equations for a finite process are

$$s_2 - s_1 > \int_{z_1}^{z_2} \frac{dq}{T} \qquad \text{for an irreversible process} \tag{2.54}$$

$$= \int_{z_1}^{z_2} \frac{dq}{T} \qquad \text{for a reversible process .} \tag{2.55}$$

3. Similarly, for an infinitesimal process,

$$ds > \frac{dq}{T} \qquad \text{for an irreversible process} \tag{2.56}$$

$$= \frac{dq}{T} \qquad \text{for a reversible process .} \tag{2.57}$$

4. The above equations can also be written as

$$s_2 - s_1 = \int_{z_1}^{z_2} \frac{dq}{T} + \Delta s_p \tag{2.58}$$

$$ds = \frac{dq}{T} + ds_p \;, \tag{2.59}$$

where Δs_p and ds_p are *entropy production* during finite or infinitesimal change, respectively. The heat transfer may be reversible or irreversible. Entropy production is zero for a reversible process. For an *isentropic process* ($s_2 - s_1 = ds = 0$), entropy production may be positive or negative, depending on the direction of heat transfer. Similarly, for an *adiabatic process* ($dq = 0$), entropy production may be positive or negative, depending on the direction of the work interaction.

2.7.2 *Isolated System and Universe*

We have already introduced an isolated system as one for which $Q = W = 0$. Thus, consider a system exchanging Q and W with the surroundings for $t < 0$. Now, at $t = 0$, let the system be isolated with system entropy S_0. Although the interactions have been stopped, the system will still change its state in *time*, ultimately reaching equilibrium. The question is, in what manner will the system approach equilibrium? For the isolated system, $dQ = 0$ for all times. Therefore, from integration of Eq. 2.51 we have

$$S(t) - S_0 \geq 0 \qquad\qquad (2.60)$$

where S(t) is the instantaneous entropy. Its variation with time will appear as shown in Fig. 2.14(a). Equation 2.60 tells us that the difference ($S - S_0$) can only *increase* (irreversible change) with time or be zero (a reversible change). However, will this be an indefinite increase? Fig. 2.14(a) tells us that at short times, the entropy change will be positive (as indicated by the inequality sign) but, after a long time, the rate of change will be zero. In other words, in this state, $dS/dt = 0$, and a state of *equilibrium* is said to have been reached. In addition, S(t) will have reached its maximum value *above* S_0. Further, Eq. 2.60 also tells us that the approach to equilibrium can *never be* as shown in Fig. 2.14(b). That is, for an isolated system, a decrease of entropy below S_0 is impossible.

This characteristic approach to equilibrium leads us to define the concept of *universe*; a concept that is widely used to determine if any general system has undergone a reversible or an irreversible process.

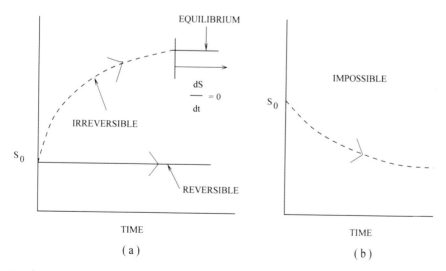

Fig. 2.14 Isolated system and equilibrium

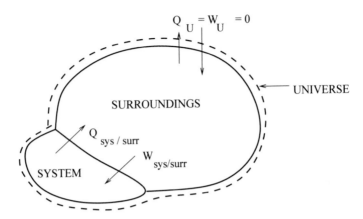

Fig. 2.15 Definition of a universe

It was mentioned in Sect. 2.2.1 that *surroundings* are everything outside the system. It was also commented in a footnote in Sect. 2.2.2 that surroundings can also be treated as a system. *In thermodynamics, the new dual system comprising system + surroundings is called the universe.* It is shown in Fig. 2.15. The universe, by definition, is a dual system for which $Q = W = 0$ for all times and, therefore, it is also an isolated system.

For the universe, therefore, Eq. 2.60 will again apply. Hence, the entropy change denoted by ΔS_U is given by

$$\Delta S_U = \Delta S_{sys} + \Delta S_{surr} \geq 0 . \tag{2.61}$$

In other words, the entropy change of the universe will continue to increase with time as long as the system and its surroundings continue to exchange Q and W. When these *internal* exchanges have stopped, $dS_U/dt = 0$.

From this discussion, we also conclude that any *isolated* system can be broken up into subsystems that exchange Q and W at their boundaries. Further discussion of equilibrium is given in Sect. 2.7.4.

2.7.3 First Law in Terms of Entropy and Gibbs Function

Consider the First law Eq. 2.43 again, but in terms of *per unit mass* and for a process undergoing infinitesimally small change. Then,

$$(dq - dw)_{rev} = (dq - dw)_{irrev} = de . \tag{2.62}$$

Most often, we encounter systems in which potential and kinetic energy changes are either zero or negligible compared with changes in internal energy. Then, the First law can also be written as

$$(dq - dw)_{rev} = (dq - dw)_{irrev} = du . \tag{2.63}$$

This equation states that $(dq - dw)$ of any irreversible (real) process can be replaced by $(dq - dw)$ of a reversible process giving the same change of state du.[21] Recall that $dw_{rev} = p\,dv$ and, from Eq. 2.57, $dq_{rev} = T\,ds$. Thus, the First law can also be written in terms of entropy change and will read as

$$T\,ds - p\,dv = du . \tag{2.64}$$

This statement of the First law enables us to evaluate entropy change of a real process by *replacing* it by a reversible process. Incidentally, because $du = dh - (p\,dv + v\,dp)$, Eq. 2.64 can also be written as

$$T\,ds + v\,dp = dh . \tag{2.65}$$

Gibbs Function

Like enthalpy (see Eq. 2.25), a Gibbs function g is defined as

$$g \equiv h - T\,s . \tag{2.66}$$

Therefore, using Eq. 2.65

$$\begin{aligned} dg &= dh - T\,ds - s\,dT \\ &= v\,dp - s\,dT . \end{aligned} \tag{2.67}$$

This equation helps us to define different types of *equilibrium*.

2.7.4 Thermal Equilibrium

Consider a *constant-volume* system exchanging heat Q with *very large* surroundings at T_0 (see Fig. 2.16a). In thermodynamics, a very large system is called a *reservoir* which has the attribute that any exchange of heat with it does not alter its temperature. As such, the surroundings will remain at T_0 throughout the process. Because a constant-volume system is considered, no work interaction with the surroundings takes place. Also, the state of the system will be characterized by its temperature T_{sys}.

---

[21] However, this does not imply that $dq_{rev} = dq_{irrev}$ or $dw_{rev} = dw_{irrev}$.

Initially, at $t = 0$, let the system temperature $T_{sys} < T_0$. Then, heat transfer will be from the surroundings to the system and will continue till $T_{sys} = T_0$. Let the heat transfer be dQ for an infinitesimally small process. Then, $dS_{sys} = dQ/T_{sys}$ and $dS_{surr} = -dQ/T_0$. Hence, from Eq. 2.61, we deduce that

$$dS_U = dS_{sys} + dS_{surr}$$

$$= |dQ| \left[\frac{1}{T_{sys}} - \frac{1}{T_0} \right]$$

$$= |dQ| \left[\frac{T_0 - T_{sys}}{T_0 T_{sys}} \right] \geq 0$$

$$\frac{dS_U}{dt} = \left| \frac{dQ}{dt} \right| \left[\frac{T_0 - T_{sys}}{T_0 T_{sys}} \right] \geq 0 . \tag{2.68}$$

This equation states that the entropy of the universe (that is, system + surrounding) will continue to increase with time because $T_0 \geq T_{sys}$ for all times. Several important observations need to be made concerning Eq. 2.68:

1. dQ will be finite but its magnitude will decrease as $T_{sys} \rightarrow T_0$, ultimately reaching zero value when $T_{sys} = T_0$. As such, the quantity in square brackets will always be positive or zero. The heat transfer occurs spontaneously in response to the temperature difference. Because dS_U is proportional to dQ, dS_U/dt, like

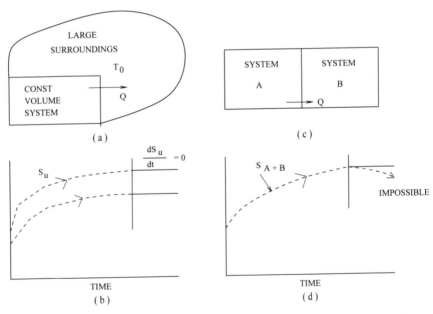

Fig. 2.16 Equilibrium of a finite system: a) constant-volume system, b) behavior with time, c) constant-volume-constant-m c_p systems; d) stable equilibrium

dQ/dt, will decrease with time. In thermodynamics, we say that S_U increases for a *spontaneous* heat transfer process. This is shown by the rising curve in Fig. 2.16b.

2. When $T_{sys} = T_0$, dQ = 0 and $dS_U/dt = 0$. The universe now attains *thermal equilibrium*. No further change in S_U or the state of the system or the surroundings can now occur. We, therefore, say that S_U of an isolated system or the universe in equilibrium is maximum and remains constant.

3. A relevant question is: Can S_U ever decrease *after the system has reached equilibrium*? To answer this question, let us consider two *finite* systems A and B in contact with each other such that system (A + B) is an *isolated system* (see Fig. 2.16c). Initially, let (A + B) be in equilibrium so that $T_A = T_B = T_i$. We further assume that $(mc_p)_A = (mc_p)_B$. Also, A and B are *constant-volume* systems, so work interaction between them is absent.

Now, suppose by *sheer chance*, heat transfer dQ takes place from B to A. As a result, the system temperatures will be $T_B = T_i - dT$ and $T_A = T_i + dT$. Thus,

$$dS_{A+B} = dS_A + dS_B$$

$$= dQ \left[\frac{1}{T_i + dT} - \frac{1}{T_i - dT} \right]$$

$$= dQ \left[\frac{-2\,dT}{T_i^2 - dT^2} \right]$$

$$\simeq -2\,dQ \frac{dT}{T_i^2} < 0 \qquad dT << T_i \; . \tag{2.69}$$

In light of Eq. 2.61, this result is clearly *absurd*. As such, no *spontaneous* heat transfer can occur between systems in thermal equilibrium and, therefore, the entropy of the universe cannot decrease after equilibrium is attained.

4. Another way of stating the preceding fact is that even if this scenario was made possible by chance, the system will quickly *revert* back to $dS_U = 0$. Thus, after equilibrium has been attained, we say that systems A and B are in *stable* equilibrium (see Fig. 2.16d).

5. The foregoing results obtained for a constant-volume system can be generalized as follows.

$$dS_U > 0 \qquad \text{spontaneous change.} \tag{2.70}$$

$$dS_U = 0 \qquad \text{equilibrium.} \tag{2.71}$$

$$dS_U < 0 \qquad \text{stable equilibrium.} \tag{2.72}$$

2.7.5 Equilibrium of a General Closed System

In the previous examples, we deliberately dealt with constant-volume system so that the work interaction was absent. In general, however, both heat and work interactions will take place. Thus, let the system and its surroundings be two arbitrary *finite* systems. However, the system + surroundings form an isolated system with $Q = W = 0$ and hence, $\Delta U_{sys+surr} = 0$. Also, because $W = 0$, $\Delta V_{sys+surr} = 0$. Therefore,

$$\Delta V_{sys} = -\Delta V_{surr} \qquad \Delta U_{sys} = -\Delta U_{surr} \ . \qquad (2.73)$$

From Eq. 2.61,

$$\Delta S_{sys} \geq - \Delta S_{surr}$$
$$\geq - \left(\frac{|\Delta Q|}{T_0} \right), \qquad (2.74)$$

where ΔQ is the heat transferred from the surroundings to the system and, therefore, negative with respect to the surroundings. Replacing ΔQ from the First law and using Eq. 2.73,

$$T_0 \, \Delta S_{sys} \geq - (p_0 \, \Delta V_{surr} + \Delta U_{surr})$$
$$\geq p_0 \, \Delta V_{sys} + \Delta U_{sys} \ . \qquad (2.75)$$

To study equilibrium, we now *postulate* that during the process, $p = p_0$ and $T = T_0$—that is, the pressure and temperature of the system and the surroundings remain the same during the process. Then, from the First law Eqs. 2.64 and 2.67, we have

$$T_0 \, \Delta S_{sys} \,|_{p=p_0, T=T_0} \geq [p \, \Delta V + \Delta U]_{sys}$$
$$\geq [\Delta G - V \, \Delta p + T \, \Delta S + S \, \Delta T]_{sys}$$
$$\geq [\Delta G - V \, \Delta p + T_0 \, \Delta S + S \, \Delta T]_{sys} \qquad (2.76)$$

or, canceling $T_0 \, \Delta S_{sys}$,

$$[\Delta G - V \, \Delta p + S \, \Delta T]_{sys} \leq 0 \ . \qquad (2.77)$$

This equation establishes the conditions for a spontaneous change (indicated by <) and equilibrium (indicated by =) of a general system. The system may have undergone *any* finite process such that $p_{sys} = p_0$ and $T_{sys} = T_0$.

2.7.6 Phase-Change Processes

For a pure substance, constant (p, T) conditions prevail during vapourization of saturated liquid into saturated vapor because of latent heat transfer h_{fg}. In combustion of liquid fuels, the fuel first vaporizes and then burns in the gas phase. During these evaporation processes, Eq. 2.77 will reduce to

$$\Delta G_{sys}|_{p,T} = 0 \qquad \text{or} \qquad g_f = g_g \qquad (2.78)$$

and, from definition of the Gibbs function in Eq. 2.66, $h_f - T\, s_f = h_g - T\, s_g$. Using Eq. 2.3, therefore,

$$h_{fg} = h_g - h_f = T\, (s_g - s_f) = T\, s_{fg} \,. \qquad (2.79)$$

This equation is routinely used in steam work. Its basis lies in the equilibrium condition for a constant p and T process.

Clausius–Clapeyron Equation

A further deduction can be made by considering phase-change at (p + dp, T + dT) in the neighborhood of (p, T). Then

$$g_f + dg_f = g_g + dg_g. \qquad (2.80)$$

Subtracting Eq. 2.78 from this equation, we have $dg_f = dg_g$, or, from Eq. 2.67, $v_f\, dp - s_f\, dT = v_g\, dp - s_g\, dT$. Therefore, using Eq. 2.79,

$$\frac{dp}{dT} = \frac{s_g - s_f}{v_g - v_f} = \frac{s_{fg}}{v_{fg}} = \frac{1}{T}\frac{h_{fg}}{v_{fg}} \,. \qquad (2.81)$$

This is known as the *Clausius–Clapeyron equation*. Further, because $v_g \gg v_f$, $dp/dT \simeq (1/T)\,(h_{fg}/v_g)$, or, replacing v_g by the ideal gas relation,[22]

$$\frac{dp}{p} = \frac{h_{fg}}{R_g}\frac{dT}{T^2} \,. \qquad (2.82)$$

For a small change from state 1 to state 2, assuming that h_{fg} remains constant, this equation can be integrated to give

$$\frac{p_2}{p_1} = \exp\left\{ -\frac{h_{fg}}{R_g}\left(\frac{1}{T_2} - \frac{1}{T_1}\right) \right\} \,. \qquad (2.83)$$

This equation will be used in Chap. 10 to determine the vapor pressure of fuel vapor at T_2 knowing p_1 and T_1.

---

[22]Recall that steam does not the obey ideal gas relation. However, a fuel vapor mixed with air is assumed to behave as an ideal gas.

Finally, in Chap. 4, we shall find that Eqs. 2.77 and 2.78 establish the condition for a *chemical equilibrium*.

2.7.7   Second Law for an Open System

To derive Second law equation for an open system, let the CV of Fig. 2.12 interact with infinite surroundings at temperature T_0. Then, writing Eq. 2.61 in *rate form*,

$$\Delta \dot{S}_U = \Delta \dot{S}_{CV} + \Delta \dot{S}_{surr} \geq 0 . \tag{2.84}$$

However,

$$\Delta \dot{S}_{CV} = \frac{dS_{CV}}{dt} + (\dot{m}\, s)_{out} - \sum_{k=1}^{k=2} (\dot{m}\, s)_{k,in} \tag{2.85}$$

and

$$\Delta \dot{S}_{surr} = -\frac{\dot{Q}}{T_0} . \tag{2.86}$$

Hence, using Eq. 2.42,

$$
\begin{aligned}
T_0 \, \Delta \dot{S}_{surr} &= -\dot{Q} \\
&= -\dot{W}_{ext} + \frac{dE_{CV}}{dt} + \dot{m}_{out}\left(h + \frac{V^2}{2} + gz\right)_{out} \\
&\quad - \sum_{k=1}^{k=2} \dot{m}_{k,in}\left(h + \frac{V^2}{2} + gz\right)_{k,in} .
\end{aligned} \tag{2.87}
$$

Substituting the last two equations in Eq. 2.84, we have

$$
\begin{aligned}
\dot{W}_{ext} &\leq \frac{d}{dt}\left[T_0\, S_{CV} - E_{CV}\right] \\
&\quad + \sum_{k=1}^{k=2} \dot{m}_{k,in}\left(h - T_0\, s + \frac{V^2}{2} + gz\right)_{k,in} \\
&\quad - \dot{m}_{out}\left(h - T_0\, s + \frac{V^2}{2} + gz\right)_{out} .
\end{aligned} \tag{2.88}
$$

The equality sign in this equation thus represents the *maximum work* ($\dot{W}_{ext,max}$) extractable from an open system. If $\dot{W}_{ext,actual}$ represents the *actual work*, then the *rate of irreversibility* $\dot{I}$ is defined as

$$\dot{I} \equiv \dot{W}_{ext,max} - \dot{W}_{ext,actual}. \quad . \tag{2.89}$$

Using this equation, engineers can readily assess the extent of unwanted irreversibilities present in any open system equipment. Finally, Eq. 2.88 is sometimes expressed in terms of *availability function* b, which is defined as

$$b \equiv h - T_0 \, s \tag{2.90}$$

Exercises

1. In the examples given below, the underlined words define the system, and the process undergone by the system is then described. In each case, identify (i) whether the system is closed or open, (ii) the sign of work (W) and heat (Q) interactions, and (iii) whether the work interaction can be demonstrated as being equivalent to raising or lowering of a weight.

   (a) The air in a bicycle tire and the connected reciprocating plunger pump. The plunger is pushed down, forcing the air into the tire. Assume adiabatic boundaries.
   (b) The water and water vapor in a rigid container. The container is placed above a flame and the pressure and temperature of its contents rise.
   (c) The system in (b) bursts and its contents explode into the cold atmosphere.
   (d) The liquid in a Adiabatic rigid vessel. The liquid comes to rest from an initial state of turbulent motion.
   (e) The gas in a rigid insulated container is stirred by a stirrer.
   (f) In the previous case, what will be Q and W if only the gas is considered as the system?

2. Determine the perfect gas range of pressure and temperature of R-134, ammonia (NH_3), and helium (He).
3. Show that the perfect gas law can also be expressed as

$$u = u_0 + \frac{p \times v}{\gamma - 1} \, ,$$

   where u_0 is an arbitrary datum value at absolute zero. If $u_0 = 1935$ and $\gamma = 1.29$, evaluate Q and W when a gas changes state from 7 bar and 3.1 m^3/kg to 10.1 bar in a *constant internal energy* process.
4. The heat transferred during a quasi-static process of an ideal gas with constant specific heats in which only expansion work is done can be represented as dq =

$C \times dT$, where C is a constant. Show that the equation for this process can be represented as $p \times v^k = $ constant. Determine k in terms of cp, cv, and C.

5. From Tables 2.1 and 2.3, estimate the hard-sphere diameters[23] of CO, CO_2, H_2, and N_2.

6. Plot a T-v state diagram for CO_2 at $p_r = 0.5, 1.0, 5.0, 10.0$ in the range $1.0 < T_r < 10$ using van der Waals Eq. 2.4

7. Many generalized equations for behavior of pure substances have been proposed in the literature. One such equation is from Dieterici. It reads:

$$p = \left(\frac{R_g T}{v - b} \right) \exp \left\{ - \left(\frac{a}{R_g T v} \right) \right\} .$$

Determine

(a) Values of a and b in terms of critical pressure and temperature.
(b) The law of corresponding states.
(c) The perfect gas region.

8. Evaluate $(p_{cr} v_{cr})/(R_g T_{cr}) = $ constant from the Van der Waals and Dietirici equations. Compare the value of the constant for a few gases using Table 2.1.

9. A system consists of a gas initially at 1 bar and 25 °C. The gas is contained in a cylinder (70 cm diameter) fitted with a leak-proof frictionless piston. 40 kJ of heat are added to the gas causing slow expansion at constant pressure such that the final temperature is 85 °C. If the internal energy is given by U (J) = 100 T (K), find the distance traveled by the piston during the process.

10. A closed system changes from state 1 to state 2 while receiving heat Q from condensing steam at T_s and losing heat to the surroundings at T_0 and p_0. The system also delivers work W while displacing the surroundings. By applying the First and the Second laws, show that any irreversibility in the process gives rise to wastage of steam energy by an amount

$$Q - \frac{T_s}{T_s - T_0} (\Delta E + p_0 \, \Delta V - T_0 \, \Delta S) .$$

11. 1 kg of saturated liquid water at 10 bar is mixed with 0.5 kg of steam at 10 bar and 300 °C in a closed system such that the pressure remains constant. The mixing is caused by a stirrer supplying 1000 N-m at 20 rpm for 30 s. The heat loss from the system is estimated at 250 kJ ($T_{surr} = 27$ °C). Calculate the change in entropy of the universe.

12. A 1-liter rigid insulated container contains N_2 at 50 °C and 10 bar. It is desired to increase the pressure to 15 bar in 10 min by means of an electrical heater placed in the container. Calculate the power consumption. Assume $cp_{N_2} = 1.038$ kJ/kg-K. Also, evaluate the change in entropy of the universe.

13. 1 kg of air is expanded adiabatically in a piston–cylinder arrangement from 15 bar and 800 K to 500 K. If the process was isentropic, then the final temperature

---

[23] In Chap. 5, it will be inferred that hard-sphere diameters are functions of temperature.

would have been 475 K at the same final pressure. Determine the final pressure and evaluate the change in entropy and work done in the actual adiabatic process. Take $\gamma_{air} = 1.4$.

14. Compressed air at pressure p_0 and temperature T_0 fills a large steel container. The air is used to inflate a rubber boat that is initially flat. Show that after inflation, the temperature T_f and pressure pf of air in the boat are related by the following relation:

$$T_f = \gamma \, T_0 \left[1 + \frac{p_a}{p_f} (\gamma - 1) \right]^{-1},$$

where p_a is atmospheric pressure and $\gamma = cp/cv$. Assume that the filling process is adiabatic.

15. If $s = s\,(T, p)$, show that

$$T \, ds = cp \, dT - T \left(\frac{\partial v}{\partial T} \right)_p dp \ .$$

Derive a similar expression for Tds if $s = s\,(T, v)$.

16. Using the preceding results, show that for a van der Waals gas,

$$s - s_0 = c_v \ln \frac{T}{T_0} + R_g \ln \left(\frac{v - b}{v_0 - b} \right)$$

$$\text{and} \quad c_p - c_v = R_g \left[1 - \frac{2 \, a \, (v - b)^2}{R_g \, T \, v^3} \right]^{-1}.$$

17. CO_2 at $p_r = 1.35$ and $T_r = 1.0$ undergoes frictionless adiabatic expansion such that the final volume is 8 times the original volume. Estimate the change in temperature and specific entropy assuming (a) an ideal gas, (b) a van der Waals gas.

18. A compressor receives steam at 3.4 bar and x = 0.95. The steam is delivered dry and saturated at 6 bar. The steam flow rate is 4 kg/s. The inlet and outlet pipe diameters of the compressor are 20 cm each. Assuming adiabatic compression, calculate the power required to drive the compressor (mechanical efficiency = 0.92). Also, evaluate the rate of irreversibility.

19. Oxygen at 12 bar and 15 °C flows with a velocity of 70 m/s through a pipe that is connected to a 60-liter insulated tank containing oxygen at 1 bar and 5 °C. A valve is opened, allowing oxygen to flow into the tank until it contains 5 times the original mass. Calculate the final temperature and pressure of oxygen in the tank. Assume $\overline{cp}_{O_2} = 27$ kJ/kmol-K.

Chapter 3
Thermodynamics of Gaseous Mixtures

3.1 Introduction

In air-conditioning and combustion applications, we encounter essentially *gaseous* mixtures. In air-conditioning, dehumidification, or humidification, the substance comprises a binary mixture of air + water vapor. The substance is called an *inert* mixture because no chemical reaction takes place between air and water vapor. In contrast, combustion mixtures are *multicomponent* mixtures because several chemical species (for example, $C_m H_n, O_2, CO, CO_2, H_2O, NO_x, SO_x$) comprise them. They are not inert because under the *right conditions of temperature and pressure*, they can react chemically. Hence, they are called *reacting mixtures*. If the right conditions are not obtained, however, then they are inert mixtures.

Our interest in this chapter is to define mass, volume, and energy properties of gaseous mixtures. In the previous chapter, we noted that the state of a *pure compressible* substance requires specification of only two independent intensive properties. In contrast, the state of a *mixture* requires additional information that describes the *composition* of the mixture. Here, composition refers to the *proportions of the components* constituting the mixture.

In most applications of interest in this book, pressures are substantially low and temperatures are substantially high so the perfect gas range is obtained. As such, we shall assume that the mixture substance, as well as its components, behave as a *perfect gas*. The intensive properties p, v, and T of a perfect gas are related by Eq. 2.11. Here, this relation is rewritten in different ways to introduce some of the quantities that are routinely used in mixture calculations:

$$p_{mix}\, v_{mix} = R_{mix}\, T \ , \tag{3.1}$$

$$p_{mix}\, V_{mix} = m_{mix}\, R_{mix}\, T \ , \tag{3.2}$$

$$p_{mix}\, v_{mix} = \frac{R_u}{M_{mix}}\, T \ , \tag{3.3}$$

$$p_{mix}\, V_{mix} = n_{mix}\, R_u\, T \ , \text{and} \tag{3.4}$$

© Springer Nature Singapore Pte Ltd. 2020
A. W. Date, *Analytic Combustion*,
https://doi.org/10.1007/978-981-15-1853-9_3

$$p_{mix} \, \bar{v}_{mix} = R_u \, T \; , \tag{3.5}$$

where the subscript *mix* refers to mixture; p (N/m^2) refers to mixture pressure, V (m^3) to mixture volume, R_u (J/kmol-K) and R_{mix} (J/kg-K) are universal and mixture gas constants, respectively, M_{mix} (kg/kmol) is mixture molecular weight, $n_{mix} = m_{mix}/M_{mix}$ represents number of kmols, and $\bar{v}_{mix} = V_{mix}/n_{mix}$ (m^3/kmol) represents specific volume per kmol. In thermodynamics of mixtures, all intensive properties with an over bar are called *molar* properties and are defined on a *per kmol* basis. Equation 3.1 can also be written in terms of mixture density as $p_{mix} = \rho_{mix} \, R_{mix} \, T$, as $v_{mix} = 1/\rho_{mix}$.

3.2 Mixture Composition

In different engineering applications, mixture composition is expressed in terms of fractions by mass or volume (or, number of moles), to reflect proportions of individual species of the mixture.

3.2.1 Mass Fraction

The total mass of the mixture is clearly the sum of the masses of the individual species. Thus

$$m_{mix} = m_1 + m_2 + m_3 + \cdots + m_j + \cdots$$
$$= \sum_j m_j \; . \tag{3.6}$$

The *mass fraction* ω_j of species j is defined as[1]

$$\omega_j \equiv \frac{m_j}{m_{mix}} \; . \tag{3.7}$$

From Eq. 3.6, therefore,

---

[1]Recall the definition of *specific humidity* (W) used in psychometry:

$$W = \frac{\text{Mass of water vapor}}{\text{Mass of dry air}} .$$

Clearly, for this binary mixture of air and water vapor,

$$\omega_{vap} = \frac{W}{1 + W} \quad \text{and} \quad \omega_{air} = 1 - \omega_{vap}.$$

$$\sum_j \omega_j = \sum_j \frac{m_j}{m_{mix}} = 1 \ . \tag{3.8}$$

Dalton's Law

Equation 3.7 can also be written as

$$\begin{aligned}
\omega_j &= \frac{m_j}{V_{mix}} \times \frac{V_{mix}}{m_{mix}} \\
&= \left[\frac{m_j}{V_{mix}} \right] \times \frac{1}{\rho_{mix}} \\
&= \frac{\rho_j}{\rho_{mix}} \ . \tag{3.9}
\end{aligned}$$

It might appear strange that the quantity in the square brackets, which is the ratio of the *species mass* m_j to the *total mixture volume* V_{mix}, can be replaced by *species density* ρ_j. However, this replacement is made on the basis of *Dalton's law*, which states:

As long as perfect gas behavior is assumed and perfect equilibrium prevails, each species j of the mixture behaves as if it existed at the temperature of the mixture and occupied the total volume V_{mix} of the mixture.[2]

Thus, following Eq. 3.8,

$$\rho_{mix} = \sum_j \rho_j \ . \tag{3.10}$$

3.2.2 Mole Fraction and Partial Pressure

Another way of writing Eq. 3.6 is

$$\begin{aligned}
m_{mix} &= m_1 + m_2 + m_3 + \cdots + m_j + \cdots \\
&= n_1 M_1 + n_2 M_2 + n_3 M_3 + \cdots + n_j M_j + \cdots \\
&= \sum_j n_j M_j \ . \tag{3.11}
\end{aligned}$$

However, $m_{mix} = n_{mix} M_{mix}$. Therefore,

$$M_{mix} = \sum_j \left[\frac{n_j}{n_{mix}} \right] M_j \ , \tag{3.12}$$

where the quantity in the square brackets is called the *mole fraction* and denoted by x_j. Hence,

---

[2]The law is compactly stated as: **Any gas is a vacuum to any other gas mixed with it** [115].

$$x_j = \frac{n_j}{n_{mix}} \tag{3.13}$$

$$\text{and } \sum_j x_j = 1 \ . \tag{3.14}$$

In chemical reaction engineering calculations, mixture composition is expressed in terms of mole fractions rather than mass fractions.

In air-conditioning and refrigeration work, the mixture composition is again expressed in terms of mole fraction, but via the notion of a *partial pressure*. Thus, using Dalton's law, the perfect gas relation for species j is written as

$$p_j \, V_{mix} = m_j \, R_j \, T = n_j \, R_u \, T \ , \tag{3.15}$$

where p_j is the partial pressure of j and R_j is the gas constant of j. Dividing this equation by Eq. 3.4, we have

$$\frac{p_j}{p} = \frac{n_j}{n_{mix}} = x_j \tag{3.16}$$

or, from Eq. 3.14

$$p = \sum_j p_j \ . \tag{3.17}$$

The notion of a partial pressure thus can be understood from the *Gibbs–Dalton law* which states [115]:

The pressure of a gaseous mixture is the sum of the pressures that each component would exert if it alone occupied the volume of the mixture at the temperature of the mixture.

3.2.3 Molar Concentration

In combustion engineering, mixture compositions are often expressed in terms of *molar concentration* which is defined by symbol [j] as

$$[j] \equiv \frac{\rho_j}{M_j} = \frac{p_j}{R_u \, T} = \frac{1}{\overline{v}_j} \quad \left(\frac{kmol}{m^3} \right) \ . \tag{3.18}$$

3.2.4 Specifying Composition

Specifying values of ω_j, x_j, [j], or the ratio p_j/p for each species j of a mixture is termed *specifying the composition of the mixture*. Once the composition is known in any one system of specification, it can be specified in any other system through appropriate conversions. Thus, if ω_j is known then, $x_j = \omega_j M_{mix}/M_j$ and so on. Hence, following Eq. 3.14,

$$M_{mix} = \left[\sum_j \frac{\omega_j}{M_j} \right]^{-1} . \tag{3.19}$$

3.3 Energy and Entropy Properties of Mixture

In another version [115], the Gibbs–Dalton law is expressed as

The internal energy, enthalpy, and entropy of a gaseous mixture are, respectively, equal to the sums of the internal energies, enthalpies, and entropies that each component of the mixture would have if each alone occupied the volume of the mixture at the temperature of the mixture.

Thus, the extensive *internal energy* U_{mix} of the mixture is written as

$$U_{mix} = U_1 + U_2 + U_3 + \cdots + U_j + \cdots$$
$$= m_1 u_1 + m_2 u_2 + m_3 u_3 + \cdots + m_j u_j + \cdots$$
$$= \sum_j m_j u_j \tag{3.20}$$
$$= \sum_j n_j M_j u_j = \sum_j n_j \bar{u}_j , \tag{3.21}$$

where molar specific internal energy $\bar{u}_j = M_j u_j$ (J/kmol). But, $U_{mix} = m_{mix} u_{mix} = n_{mix} \bar{u}_{mix}$. Therefore, it follows that the intensive internal energy of the mixture can be written as

$$u_{mix} = \frac{U_{mix}}{m_{mix}} = \sum_j \omega_j u_j \qquad \bar{u}_{mix} = \frac{U_{mix}}{n_{mix}} = \sum_j x_j \bar{u}_j . \tag{3.22}$$

Similar expressions can be developed for enthalpy and entropy.

$$h_{mix} = \sum_j \omega_j h_j \qquad \bar{h}_{mix} = \sum_j x_j \bar{h}_j . \tag{3.23}$$

$$s_{mix} = \sum_j \omega_j s_j \qquad \bar{s}_{mix} = \sum_j x_j \bar{s}_j . \tag{3.24}$$

Likewise, using definitions of specific heat (Eqs. 2.24 and 2.26), it is possible to write

$$(m\, c_v)_{mix} = (m\, c_v)_1 + (m\, c_v)_2 + \cdots + (m\, c_v)_j + \cdots . \tag{3.25}$$
$$(m\, c_p)_{mix} = (m\, c_p)_1 + (m\, c_p)_2 + \cdots + (m\, c_p)_j + \cdots . \tag{3.26}$$

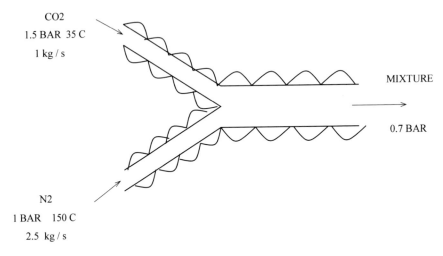

CO2
1.5 BAR 35 C
1 kg / s

MIXTURE

0.7 BAR

N2
1 BAR   150 C
2.5 kg / s

Fig. 3.1 Steady flow apparatus (Problem 3.1)

Dividing by $m_{mix} = n_{mix} M_{mix}$, it follows that

$$c_{v,mix} = \sum_j \omega_j c_{v,j} \qquad \bar{c}_{v,mix} = \sum_j x_j \bar{c}_{v,j} . \qquad (3.27)$$

$$c_{p,mix} = \sum_j \omega_j c_{p,j} \qquad \bar{c}_{p,mix} = \sum_j x_j \bar{c}_{p,j} . \qquad (3.28)$$

The following problem will illustrate the use of the preceding formulas:

Problem 3.1 CO_2 at 1.5 bar and 35 °C is mixed with N_2 at 1 bar and 150 °C in a rigid and insulated steady flow apparatus. The flow rates at entry are $\dot{m}_{CO_2} = 1$ kg/s and $\dot{m}_{N_2} = 2.5$ kg/s. The mixture thus formed exits the apparatus at 0.7 bar. The mixing occurs adiabatically. Determine (a) the mixture exit temperature and (b) the change in entropy. Assume steady state and take $c_{p,N_2} = 0.8485$ kJ/kg-K and $c_{p,CO_2} = 1.0366$ kJ/kg-K.

The problem situation can be visualized as shown in Fig. 3.1. Clearly, the situation represents an open system; hence, Eqs. 2.36 and 2.42 readily apply. Because the apparatus is rigid and insulated, $\dot{W}_{ext} = \dot{Q} = 0$. We further assume that the pipe diameters are such that velocities are small and thus the kinetic energy contributions are negligible. Also, the elevations of inflow and exit planes are the same so the potential energy contributions are also negligible. Then, under steady state, the mass and energy conservation equations will read as

$$\dot{m}_{mix} = \dot{m}_{CO_2} + \dot{m}_{N_2} = 1 + 2.5 = 3.5 \quad \text{kg/s}$$

and

$$\dot{m}_{mix} \, h_{mix} = \dot{m}_{CO_2} \, h_{CO_2} + \dot{m}_{N_2} \, h_{N_2} \ .$$

Dividing this equation by $\dot{m}_{mix}$, we have

$$h_{mix} = \omega_{CO_2} \, h_{CO_2} + \omega_{N_2} \, h_{N_2} \ ,$$

where $\omega_{CO_2} = 1/3.5 = 0.2857$ and $\omega_{N_2} = 2.5/3.5 = 0.7143$. Using definition of enthalpy $h = c_p \, (T - T_{ref})$ where T_{ref} may take any value,

$$c_{p,mix} \, (T - T_{ref}) = \omega_{CO_2} \, c_{p,CO_2} \, (T - T_{ref}) + \omega_{N_2} \, c_{p,N_2} \, (T - T_{ref}) \ .$$

However, using the definition of mixture specific heat (see Eq. 3.28), it follows that T_{ref} becomes inconsequential and hence,

$$c_{p,mix} \, T_{mix} = \omega_{CO_2} \, c_{p,CO_2} \, T_{CO_2} + \omega_{N_2} \, c_{p,N_2} \, T_{N_2} \ ,$$

where $c_{p,mix} = \sum c_{p,j} \, \omega_j = 0.2857 \times 1.0366 + 0.7143 \times 0.8485 = 0.9022$ kJ/kg-K. Therefore, T_{mix} evaluates to

$$T_{mix} = \frac{0.2857 \times 1.0366 \times 35 + 0.7143 \times 0.8485 \times 150}{0.9022}$$

$$= 112.25 \, ^\circ C$$

This is answer (a). To evaluate the entropy change, we use Eq. 2.65. Then, $T \, ds = dh - v \, dp$ and, assuming perfect gas behavior

$$ds = \frac{dh}{T} - \frac{v}{T} \, dp = c_p \, \frac{dT}{T} - R_g \, \frac{dp}{p}$$

or, upon integration from state 1 to state 2,

$$s_2 - s_1 = c_p \ln \left(\frac{T_2}{T_1} \right) - R_g \ln \left(\frac{p_2}{p_1} \right) \ .$$

We apply this relation to the species CO_2 and N_2. To do this, we need to evaluate the *partial pressures* of the species in state 2, which is the exit mixture state. The pressures in the initial state 1 are of course known for each species.

To evaluate p_2, we first evaluate the mole flow rate of CO_2 and N_2 as

$$\dot{n}_{CO_2} = \frac{\dot{m}_{CO_2}}{M_{CO_2}} = \frac{1}{44} = 0.02273 \quad \text{kmol/s} \ .$$

$$\dot{n}_{N_2} = \frac{\dot{m}_{N_2}}{M_{N_2}} = \frac{2.5}{28} = 0.08929 \quad \text{kmol/s} \ .$$

Therefore, the total moleflow rate is $0.02273 + 0.08929 = 0.112$ kmol/s. From these evaluations, we deduce that $x_{CO_2} = 0.02273/0.112 = 0.2029$ and $x_{N_2} = 1 - 0.2029 = 0.797$. As $p_j = x_j\, p$, in the exit mixture state

$$p_{CO_2,2} = 0.2029 \times 0.7 = 0.142 \quad \text{bar.}$$
$$p_{N_2,2} = 0.797 \times 0.7 = 0.558 \quad \text{bar.}$$

Therefore,

$$(s_2 - s_1)_{CO_2} = 1.0366 \ln \left(\frac{273 + 112.25}{273 + 35} \right) - \frac{8.314}{44} \ln \left(\frac{0.142}{1.5} \right)$$
$$= 0.232 - (-0.445) = 0.677 \quad \text{kJ/kg-K}$$

and

$$(s_2 - s_1)_{N_2} = 0.8485 \ln \left(\frac{273 + 112.25}{273 + 150} \right) - \frac{8.314}{28} \ln \left(\frac{0.558}{1} \right)$$
$$= -0.0794 - (-0.173) = 0.09387 \quad \text{kJ/kg-K.}$$

Hence,

$$\Delta \dot{S} = \dot{m}_{CO_2} \times (s_2 - s_1)_{CO_2} + \dot{m}_{N_2} \times (s_2 - s_1)_{N_2}$$
$$= 1 \times (0.667) + 2.5 \times (0.09387)$$
$$= 0.9117 \quad \text{kJ/K-s.}$$

This is answer (b).

3.4 Properties of Reacting Mixtures

3.4.1 Stoichiometric Reaction

In the previous problem, we assumed that at the prevailing temperature and pressure, no chemical reaction between CO_2 and N_2 takes place. That is, the mixture was inert. We now assume that T and p are such that the mixture components may *react chemically* with each other to form, entirely or partially, another set of components. The reacting species are called *reactants*, whereas the species resulting from a reaction are called *products*. As mentioned in Chap. 1, the central problem of combustion science is to *discover* the exact composition of the *products* from the known composition of the reactants. This requires postulation of *model* reactions. The simplest reaction model is that of a *stoichiometric reaction*.

Thus, consider the stoichiometric reaction of methane (CH_4). It is written as

$$CH_4 + 2\,O_2 = CO_2 + 2\,H_2O \ . \tag{3.29}$$

In this equation, the species on the left-hand side are reactants, and the ones on the right-hand side are products. Also, because the compositions of the reactants and the products are different, *state change* has occurred. In both the states, equilibrium prevails. In other words, therefore, Eq. 3.29 states that 1 mole of CH_4 ($n_{CH_4} = 1$) combines with 2 moles of O_2 ($n_{O_2} = 2$) to form 1 mole of CO_2 ($n_{CO_2} = 1$) and 2 moles of steam H_2O ($n_{H_2O} = 2$). Thus, in this reaction, the number of moles on the left-hand and right-hand sides are the same. Now, consider the stoichiometric reaction of propane (C_3H_8). It reads

$$C_3H_8 + 5\,O_2 = 3\,CO_2 + 4\,H_2O \ . \tag{3.30}$$

In this reaction, the number of reactant moles is 6, whereas the number of product moles is 7. Notice that in both reactions, though, the number of atoms of elements C, H, and O are exactly the same on both sides. In writing such reactions, therefore, the symbols of species are written down first and the multiplying number of moles are then specified in a way that conforms to the *principle of conservation of matter*. Thus, while the number of moles may or may not change, the number of atoms of elements must not change.

The presence of the $=$ sign between reactants and products is an acknowledgement that the reaction, while proceeding from left to right has gone to *completion*. The products are then called *complete products*. Of course, in a real reaction, the composition of the products will depend on several factors. In Eqs. 3.29 and 3.30, they are *postulated* to prevent any ambiguity.[3]

Reactions (3.29) and (3.30) can also be expressed in mass units, rather than in molar units. Each species j will have mass $m_j = n_j\,M_j$. Thus,

$$16 \text{ kg } CH_4 + 64 \text{ kg } O_2 = 44 \text{ kg } CO_2 + 36 \text{ kg } H_2O. \tag{3.31}$$
$$44 \text{ kg } C_3H_8 + 160 \text{ kg } O_2 = 132 \text{ kg } CO_2 + 72 \text{ kg } H_2O \ . \tag{3.32}$$

In both reactions, the total masses of reactants and the products have remained the same. This again is a confirmation of the principle of conservation of mass. Thus, in a chemical reaction, the number of moles of reactants and products may or may not remain constant, but their masses must.

In the preceding reactions, combustion of fuel species takes place with oxygen. Most often, however, the reaction is with oxygen present in *air* and the nitrogen,

---

[3]In fact, in a real reaction, whether the reaction will proceed from left to right or in the reverse direction is also *a propri* not known. This directional sense can be established only on the basis of the Second law of thermodynamics, as will be discovered later in this chapter.

being inert, takes no part in the reaction.[4] In this case, the methane and propane reactions, for example, are written as

$$CH_4 + 2\,(O_2 + 3.76\,N_2) = CO_2 + 2\,H_2O + 7.52\,N_2,$$
$$C_3H_8 + 5\,(O_2 + 3.76\,N_2) = 3\,CO_2 + 4\,H_2O + 18.8\,N_2. \qquad (3.33)$$

The methane reaction states that to burn 1 mole of methane, 2×4.76 moles of air are just sufficient to form complete products. In other words, $16\,kg$ of methane require $2\,(32 + 3.76 \times 28) = 274.56\,kg$ of air to form 44 kg of CO_2, and $36\,kg$ of H_2O, and $210.56\,kg$ of N_2 remain unreacted. The propane reaction can be expressed likewise.

3.4.2 Fuel–Air Ratio

In IC engines, gas turbine combustion chambers, or in power plant boilers, typically hydrocarbon fuels (C_mH_n) are burned in the presence of air. However, the amount of air supplied may not be in exact *stoichiometric* proportion. The stoichiometric proportion is defined as

$$\frac{F}{A}\Big|_{stoic} = \frac{\text{mass of fuel}}{\text{mass of air required for complete products}}. \qquad (3.34)$$

Thus, for the methane reaction in Eq. 3.33,

$$\left(\frac{F}{A}\right)_{stoic} = \frac{1 \times 16}{2 \times (1 \times 32 + 3.76 \times 28)} = \frac{16}{274.56} = 0.058275 \ .$$

Similarly, for the propane reaction,

$$\left(\frac{F}{A}\right)_{stoic} = \frac{1 \times 44}{5 \times (1 \times 32 + 3.76 \times 28)} = \frac{44}{686.4} = 0.0641 \ .$$

Sometimes, instead of the (F/A) ratio, the air–fuel (A/F) ratio is specified. For methane, $(A/F)_{stoic} = 1/0.058275 = 17.16$ and for propane, $(A/F)_{stoic} = 1/0.0641 = 15.6$.

In general, the stoichiometric reaction for a hydrocarbon fuel is written as

$$C_mH_n + a_{stoic}\,(O_2 + 3.76\,N_2) = m\,CO_2 + \frac{n}{2}\,H_2O + 3.76\,a_{stoic}\,N_2 \qquad (3.35)$$

---

[4]100 moles of air contain 21 moles of O_2 and 79 moles of N_2. Therefore, 4.76 moles of air will contain 1 mole of O_2 and $79/21 = 3.76$ moles of N_2.

where a_{stoic} is called the *stoichiometric coefficient*. Its value is determined from the C-H-O-N *element balance*. Thus,

$$\text{For C}: \quad m_{react} = m_{prod}.$$

$$\text{For H}: \quad n_{react} = \frac{n}{2} \times 2 = n.$$

$$\text{For O}: \quad 2\,a_{stoic} = 2\,m + \frac{n}{2}.$$

$$\text{For N}: \quad 2 \times 3.76\,a_{stoic} = 2 \times 3.76\,a_{stoic}. \tag{3.36}$$

These equations show that the C, H, and N element balances give no new information. However, the O-balance shows that

$$a_{stoic} = m + \frac{n}{4} \tag{3.37}$$

Thus, for methane, m = 1 and n = 4; hence, $a_{stoic} = 1 + 4/4 = 2$. Similarly, for propane, m = 3 and n = 4 and $a_{stoic} = 3 + (8/4) = 5$. These values can be verified in Eq. 3.33. In fact, the fuel–air ratio is also related to the stoichiometric coefficient, as

$$\frac{F}{A}\Big|_{stoic} = \frac{m_{fuel}}{m_{air,stoic}} = \frac{12\,m + n}{a_{stoic} \times (32 + 3.76 \times 28)}. \tag{3.38}$$

3.4.3 Equivalence Ratio Φ

To decipher if the reactant fuel–air mixture is *rich* (meaning more fuel than stoichiometric proportion) or *lean*, it is customary to define the *equivalence ratio* Φ as

$$\Phi = \frac{(F/A)_{actual}}{(F/A)_{stoic}} \tag{3.39}$$

Thus, if Φ > 1, the mixture is rich; if Φ < 1, the mixture is lean; and if Φ = 1, the mixture is stoichiometric. The value of $(F/A)_{actual}$ is determined from Eq. 3.38 by replacing a_{stoic} by a_{actual}. From Eqs. 3.37 and 3.39, a_{actual} can be evaluated as

$$a_{actual} = a(\text{say}) = \frac{a_{stoic}}{\Phi} = \frac{m + n/4}{\Phi}. \tag{3.40}$$

3.4.4 Effect of Φ on Product Composition

So far we have postulated stoichiometric reactions (that is Φ = 1) that yield complete products. In reality, in combustion devices, the air–fuel mixture is rarely stoichiometric. In fact, in most practical combustion devices, the mixture is *lean* ($\Phi < 1$). For completeness, however, we must also study the case of a *rich* ($\Phi > 1$) mixture because of its practical use in fire extinction.

For $\Phi \neq 1$, the *real* products will depend on several factors, such as p, T, the *rate of reaction* and the fluid dynamic phenomenon of *mixing*. This is true even for a stoichiometric reaction. However, here the products will again be *postulated*. Experience shows that to ensure that this postulation is close to the *real* products, the cases of $\Phi < 1$ and $\Phi > 1$ must be dealt with differently.

Lean Mixture: $\Phi \leq 1$

For this case, the postulated reaction is

$$
C_m H_n + a\,(O_2 + 3.76\,N_2) = n_{CO_2}\,CO_2 + n_{H_2O}\,H_2O
$$
$$
+ n_{O_2}\,O_2 + 3.76\,a\,N_2 \ . \tag{3.41}
$$

Compared with the stoichiometric reaction in Eq. 3.35, product O_2 now appears because air supplied is in excess (that is, $a > a_{stoic}$) of that required for a stoichiometric reaction. To determine the number of moles of each product species, we again carry our element balance. Thus,

$$
\text{For C}:\ \ m = n_{CO_2}\ .
$$
$$
\text{For H}:\ \ n = 2\,n_{H_2O}\ .
$$
$$
\text{For O}:\ \ 2\,a = 2\,n_{O_2} + 2\,n_{CO_2} + n_{H_2O}\ .
$$
$$
\text{For N}:\ \ 2 \times 3.76\,a = 2 \times 3.76\,a\ . \tag{3.42}
$$

These equations specify $n_{CO_2} = m$ and $n_{H_2O} = n/2$. Using O-balance equation and Eq. 3.40, n_{O_2} is given by

$$
n_{O_2} = a - n_{CO_2} - \frac{n_{H_2O}}{2}
$$
$$
= a - \left(m + \frac{n}{4}\right) = a - a_{stoic}
$$
$$
= a_{stoic}\left(\frac{1}{\Phi} - 1\right) = a_{stoic}\left(\frac{1 - \Phi}{\Phi}\right)\ . \tag{3.43}
$$

Thus, the total product moles are

$$
n_{tot,pr} = n_{CO_2} + n_{H_2O} + n_{O_2} + 3.76\,a
$$
$$
= m + \frac{n}{2} + a_{stoic}\left(\frac{1 - \Phi}{\Phi}\right) + 3.76\left(\frac{a_{stoic}}{\Phi}\right)
$$
$$
= m + \frac{n}{2} + a_{stoic}\left(\frac{4.76 - \Phi}{\Phi}\right)\ . \tag{3.44}
$$

Therefore, in terms of mole fractions, the product compositions are

$$x_{CO_2} = \frac{n_{CO_2}}{n_{tot,pr}} = \frac{m}{n_{tot,pr}} ,$$

$$x_{H_2O} = \frac{n_{H_2O}}{n_{tot,pr}} = \frac{n/2}{n_{tot,pr}} ,$$

$$x_{O_2} = \frac{n_{O_2}}{n_{tot,pr}} = \frac{a_{stoic}}{n_{tot,pr}} \left(\frac{1 - \Phi}{\Phi} \right) ,$$

$$x_{N_2} = 1 - x_{CO_2} - x_{H_2O} - x_{O_2} . \tag{3.45}$$

Rich Mixture: $\Phi > 1$

For this case, to achieve closeness to real products, a *two-step* reaction mechanism is postulated.

$$C_m H_n + a (O_2 + 3.76 N_2) = n_{CO_2} CO_2 + n_{H_2O} H_2O$$
$$+ n_{CO} CO + n_{H_2} H_2$$
$$+ 3.76 a N_2 , \tag{3.46}$$

where CO and H_2 appear because of the presence of another reaction called the *water–gas* reaction. It reads

$$CO + H_2O \rightleftharpoons CO_2 + H_2 . \tag{3.47}$$

The water–gas reaction with a $\rightleftharpoons$ symbol is a *two-way* reaction, in which all four species coexist simultaneously. Such reactions are also called *equilibrium reactions*, whereas the reaction shown in Eq. 3.46 is called the *main* reaction, a part of whose products are formed as a result of the equilibrium reaction. To a first approximation, however, the exact proportions of the product species depend on the assumption of *chemical equilibrium* and, in turn, on the prevailing pressure and temperature. Because we will be discussing the idea of chemical equilibrium in Chap. 4, complete analysis of this $\Phi > 1$ case will be taken up there. Here, we present a partial analysis to indicate why the *water–gas* reaction is needed to determine the product composition in reaction 3.46.

It is interesting to note that O_2 does not appear in the product state in Reaction (3.46). This is because less air than required for a stoichiometric reaction has been supplied. As such, O_2 in air is completely consumed. Thus, carrying out the element balance on the reaction, we have

$$\text{For C}: \quad m = n_{CO_2} + n_{CO} .$$
$$\text{For H}: \quad n = 2 n_{H_2O} + 2 n_{H_2} .$$
$$\text{For O}: \quad 2 a = 2 n_{CO_2} + n_{CO} + n_{H_2O} .$$
$$\text{For N}: \quad 2 \times 3.76 a = 2 \times 3.76 a . \tag{3.48}$$

Table 3.1 Mole and mass fractions (Problem 3.2)

Species	x_j	ω_j
CO_2	0.08533	0.13095
N_2	0.752	0.7343
H_2O	0.096	0.06027
O_2	0.0667	0.0744

The total number of products in reaction Eq. 3.46 is 5 but we have only 4 equations. Thus, we need one more equation to determine 5 product moles. This additional equation is provided by the water–gas reaction (see Chap. 4).

Problem 3.2 Octane (C_8H_{18}) is burned with 50% *excess air*. Calculate (A/F) and product composition in terms of x_j and ω_j

For octane m = 8 and n = 18. Therefore, $a_{stoic} = m + n/4 = 8 + (18/4) = 12.5$. But, we have 50% excess air. Therefore, $a_{actual} = 1.5 \times a_{stoic} = 1.5 \times 12.5 = 18.75$. In other words, we have a *lean mixture* with $\Phi = a_{stoic}/a_{actual} = 12.5/18.75 = 0.667$. Thus, Eq. 3.41 will apply and we have

$$C_8H_{18} + 18.75 \, (O_2 + 3.76 \, N_2) = 8 \, CO_2 + 9 \, H_2O + 70.5 \, N_2 + 6.25 \, O_2 \ .$$

Therefore,

$$\frac{A}{F} = \frac{\sum n_j \, M_j}{n_{fu} \, M_{fu}}$$
$$= \frac{18.75 \times (32 + 3.76 \times 28)}{1 \times (12 \times 8 + 18)}$$
$$= 22.58 \ .$$

The total number of product moles is

$$n_{pr,tot} = 8 + 9 + 70.5 + 6.25 = 93.75$$

Thus, the product mole fractions will be $x_j = n_j/n_{pr,tot}$. The values of x_j are given in Table 3.1. The mass fractions will be given by $\omega_j = x_j \, M_j/M_{mix}$ where $M_{mix} = \sum x_j \, M_j$. Thus,

$$M_{mix} = 0.08533 \times 44 + 0.752 \times 28 + 0.096 \times 18 + 0.0667 \times 32 = 28.673 \ .$$

The evaluated mass fractions are also shown in Table 3.1. Incidentally, the mass fractions can also be evaluated in another way. Thus, the product mass will be given by

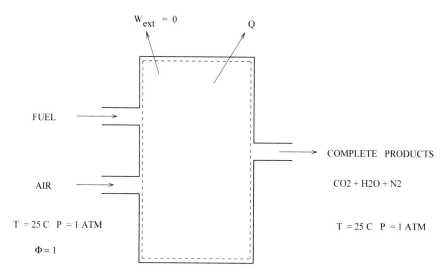

Fig. 3.2 Apparatus for measurement of heat of reaction

$$m_{prod} = \sum n_j \, M_j = 8 \times 44 + 9 \times 18 + 70.5 \times 28 + 6.25 \times 32 = 2688 \, \text{kg}.$$

Then, $\omega_{CO_2} = (8 \times 44)/2688 = 0.13095$. Other mass fractions can be evaluated likewise.

3.4.5 Heat of Combustion or Heat of Reaction

When a chemical reaction between fuel and air takes place, a change of state takes place and *heat is released*. Both the actual products of combustion and the amount of heat released depend on several factors, such as the equivalence ratio Φ, the prevailing pressure p and temperature T and also the manner in which *mixing* of fuel and air is brought about.[5] To *characterize* a fuel in terms of the heat released, therefore, we must eliminate factors that depend on the operating conditions of a combustion device. These factors are Φ, p, T, and mixing. As such, the documented values of heat released pertain to specific conditions.

The heat released is measured under specified conditions in an *open system* apparatus, as shown in Fig. 3.2. In this system, the reactants (fuel and air) are supplied at *standard* conditions, namely $p_{ref} = 1$ atm (1.01325 bars) and $T_{ref} = 25 \,^\circ\text{C} = 298$ K. The fuel and air are in *stoichiometric* proportions, so $\Phi = 1$. The constant-volume apparatus has rigid walls ($W_{ext} = 0$). It is further assumed that the products of com-

---

[5]The mixing process is, of course, determined by fluid mechanics prevailing in the particular combustion equipment.

bustion exit the apparatus *without* any pressure loss. Thus, the system experiences a *constant-pressure* process. It is also assumed that *complete* products (CO_2 and H_2O) are formed and the heat released from the chemical reaction is withdrawn through the walls as heat transfer Q. As a result, the exiting products are also at standard temperature and pressure (STP) conditions. Finally, steady flow and steady state are assumed to prevail.

For such a process, the equations of mass conservation (2.36) and energy (2.42) will read as[6]

$$\dot{m}_{pr} = \dot{m}_{fu} + \dot{m}_a \, , \tag{3.49}$$

$$\dot{Q} = \dot{m}_{pr} \, h_{pr} \, (T_{ref}) - \left[\dot{m}_{fu} \, h_{fu} \, (T_{ref}) + \dot{m}_a \, h_a \, (T_{ref}) \right] \, , \tag{3.50}$$

or, dividing this equation by $\dot{m}_{pr}$,

$$q_{STP} = \frac{\dot{Q}}{\dot{m}_{pr}} = h_{pr} \, (T_{ref}) - \left[\omega_{fu} \, h_{fu} \, (T_{ref}) + \omega_a \, h_a \, (T_{ref}) \right] \, . \tag{3.51}$$

In general,

$$q_{STP} = \sum_j (\omega_j \, h_j)_{pr} \, (T_{ref}) - \sum_j (\omega_j \, h_j)_{react} \, (T_{ref}) \tag{3.52}$$

$$= \left[h_{pr} - h_{react} \right]_{T_{ref}, P_{ref}} = \Delta h_R \, . \tag{3.53}$$

$\Delta h_R \, (\equiv q_{STP} \text{ kJ} / \text{kg of product})$ is called the *heat of reaction*. It is *always negative* because it is measured as withdrawn heat.[7] Δh_R is the change in specific enthalpy owing to *composition change* only and, therefore, called the heat of reaction.

Because the heat of reaction is a useful practical quantity, the negative sign is unhelpful. To circumvent this, engineers define the *heat of combustion* ΔH_C (kJ/kg of fuel) or simply *calorific value*. It is given by

$$\Delta H_C = - \Delta h_R \times \frac{\dot{m}_{pr}}{\dot{m}_{fu}} = - \Delta h_R \times \left[1 + \frac{A}{F} \right] \quad \text{kJ/kg of fuel.} \tag{3.54}$$

The calorific value is thus positive and is a characteristic of the fuel under consideration. Its magnitude is typically documented for STP and $\Phi = 1$ conditions only. However, it can also be evaluated at any other specified temperature T_{spec} and Φ by simply evaluating all enthalpies in Eq. 3.53 at T_{spec} instead of at T_{ref}. Such values, however, are not documented. It is also possible to show that $\Delta H_C \, (T_{spec})$ is related to $\Delta H_C \, (T_{ref})$.

---

[6]Henceforth, subscript pr will refer to product, fu to fuel, and a to air.

[7]Δh_R is negative because the hydrocarbon reaction is *exothermic*. If the reactant species require *addition* of heat to bring about a reaction, then Δh_R will be positive and the reaction is called *endothermic*.

It is customary to define two calorific values, *higher* and *lower*. If the water in the product state is in *liquid* form, the evaluated ΔH_C is called *higher calorific value* or, higher heating value (HHV). If it is in *vapor* form, it is called a *lower calorific value* or lower heating value (LHV). Because in most applications, the water will be in vapor form, the lower calorific value is of engineering interest. Evaluation of ΔH_C can be carried out only when the notion of *enthalpy of formation* is invoked.

3.4.6 Enthalpy of Formation

Unlike the pure substance, whose enthalpy and internal energy are specified with respect to a *datum* temperature, specification of h and u of the species participating in a chemical reaction must be carried out in a special way. The need for this can be understood as follows.

Suppose we choose to treat all species in the manner of a pure substance. Let the datum temperature for all reactant and product species be taken as $T_{ref} = 25\,°C$ and the datum enthalpy be assigned zero value (or any other constant value). Then, surprisingly, in Eq. 3.53, $q_{STP} = \Delta h_R = h_{pr} - h_{react} = 0$ and there will be no heat of reaction. Obviously, this is a wrong conclusion. It can be avoided by defining the *enthalpy of formation*.

Thus, in a reacting system, enthalpy $\overline{h}_j\,(T)$ of a species j at any temperature T is defined[8] as

$$\overline{h}_j\,(T) = \overline{h^0_{f\,j}}\,(T_{ref}) + \Delta\overline{h}_{s,j}\,(T), \tag{3.55}$$

where $\overline{h^0_{f\,j}}\,(T_{ref})$ is called the enthalpy of formation which is defined at STP conditions $T_{ref} = 25\,°C$ and $p_{ref} = 1$ atm. The last term is called the *sensible enthalpy change* that would be required to raise the temperature of species j from T_{ref} to T and is evaluated as

$$\Delta\overline{h}_{s,j}\,(T) = \int_{T_{ref}}^{T} \overline{c}_{p,j}\,dT\ , \tag{3.56}$$

where $\overline{c}_{p,j}$ is a function of temperature. The enthalpy of formation is assigned values following a *convention*.

1. At $T = T_{ref}$, from Eq. 3.55, $\overline{h}_j\,(T = T_{ref}) = \overline{h^0_{f\,j}}\,(T_{ref})$, because from Eq. 3.56, $\Delta\overline{h}_{s,j}\,(T_{ref}) = 0$.
2. $\overline{h^0_{f\,j}}\,(T_{ref}) = 0$ for all *naturally occurring* substances. Species such as C, N_2, H_2, O_2, and the like are, by convention, taken as naturally occurring.
3. For a species such as CO_2, however, $\overline{h^0_{f\,CO_2}}\,(T_{ref}) \neq 0$, because CO_2 is not a naturally occurring substance but is taken to have been formed from the reaction between C and O_2. Thus, in the reaction

---

[8]The enthalpy values of reacting species are always tabulated in molar units.

$$C + O_2 = CO_2 \, ,$$

the heat of reaction is evaluated as

$$\Delta \overline{h}_R = \overline{h}_{CO_2} (T_{ref}) - \left[\overline{h}_C (T_{ref}) + \overline{h}_{O_2} (T_{ref}) \right]$$
$$= \overline{h^0_{f CO_2}} (T_{ref}) - \left[\overline{h^0_{f C}} (T_{ref}) + \overline{h^0_{f O_2}} (T_{ref}) \right]$$
$$= \overline{h^0_{f CO_2}} (T_{ref}) - [0 + 0]$$
$$= \overline{h^0_{f CO_2}} (T_{ref}) \, . \tag{3.57}$$

This derivation shows that the enthalpy of formation of CO_2 equals the heat of reaction of the product (CO_2). Because the latter is negative for this exothermic reaction, $\overline{h^0_{f CO_2}} (T_{ref})$ is also negative.

Values of heats of formation for several species are tabulated in Appendix A. The use of these property tables is demonstrated through examples in Sect. 3.5.

3.4.7 Entropy of Formation

Consider a closed *constant-volume adiabatic* system comprising two reactants (say, CO and O_2) initially at STP. Because no work or heat interactions are possible, the system is isolated. Under these conditions, no chemical reaction will take place. Assume that the system mixture is ignited and product species are formed (CO_2, in the present case). In fact, immediately after ignition, all three species will be present. Thus, the system pressure and temperature will rise. From Eq. 2.65,

$$ds = \frac{(dh - vdp)}{T} \, .$$

But, assuming perfect gas relation,

$$ds = c_p \frac{dT}{T} - R_g \frac{dp}{p} \, .$$

This equation will hold for each species j of the mixture. Integrating the equation from T_{ref} to T and using molar units,

$$\overline{s}_{j,T} = \overline{s^0_j} (T) - R_u \ln \left(\frac{p_j}{p_{ref}} \right) , \tag{3.58}$$

$$\overline{s^0_j} (T) = \overline{s^0_{f j, T_{ref}, P_{ref}}} + \int_{T_{ref}}^{T} \overline{c}_{p,j} \frac{dT}{T} , \tag{3.59}$$

where $\overline{s}^0_{f\,j,T_{ref},p_{ref}}$ is called the *entropy of formation* of species j at T_{ref} and p_{ref}. It is useful to tabulate the value of $\overline{s}^0_j$ (T) at $p_j = p_{ref}$, as shown in Appendix A. These values are denoted by $\overline{s}^0$.

3.4.8 Adiabatic Flame Temperature

With reference to Fig. 3.2, let the reaction vessel be *insulated*. Thus, $\dot{Q} = 0$. Also, let the fuel and air be in stoichiometric proportions, so $\Phi = 1$ and assume STP conditions. As a result of the heat released during the chemical reaction, the products will now be hot. Because the process is adiabatic, the product temperature is called the *adiabatic flame temperature* (AFT) T_{ad}. Under steady flow and steady-state conditions, the First law for an open system will, therefore, read as

$$\dot{m}_{pr}\,h_{pr}\,(T_{ad}) - \left[\dot{m}_{fu}\,h_{fu}\,(T_{ref}) + \dot{m}_a\,h_a\,(T_{ref})\right] = 0 \,, \tag{3.60}$$

where $\dot{m}_{pr}$ is given by Eq. 3.49. The value of h_{pr} (T_{ad}) and, therefore, of T_{ad}, will depend on the postulated products. The latter, in turn, will depend on Φ (see solved Problem 3.7 in Sect. 3.5). Typically, the values of T_{ad} are documented for complete products for $\Phi = 1$ without dissociation. For most hydrocarbon fuels, T_{ad} ranges between 2100 K and 2400 K. Appendix B gives curve-fit coefficients for evaluation of T_{ad} as a function of Φ for different hydrocarbons in the range $0.4 \leq \Phi \leq 1.4$.

3.4.9 Constant-Volume Heat of Reaction

Many combustion processes take place at constant volume. The well-known bomb calorimeter is an example of such a process. Thus, consider a rigid *closed system* containing reactants (fuel + air) having mass m and pressure and temperature p_i and T_i, respectively. Let the reactants be ignited.[9] After combustion, let the products be cooled to initial temperature T_i. The heat extracted is then given by

$$q_v = \frac{Q_v}{m} = (u_{pr} - u_{react})_{T_i} \,, \tag{3.61}$$

where q_v is called the heat of reaction at constant volume. However, because $u = h - pv = h - R_g\,T$, we have

---

[9]In Chap. 8, it will be shown that the ignition energy is very small indeed and, therefore, not accounted here.

$$q_v = h_{pr} - h_{react} - (R_{mix}|_{pr} - R_{mix}|_{react}) T_i,$$
$$= \Delta h_R - (R_{mix}|_{pr} - R_{mix}|_{react}) T_i . \tag{3.62}$$

Thus, the heat of reaction at constant volume is related to the heat of reaction at constant pressure Δh_R (see Eq. 3.53).

If the mixture was ignited in an *adiabatic* closed system, then the final temperature reached is called the AFT at constant volume. It is determined by setting $u_{pr} (T_{ad}) = u_{react} (T_i)$. Note that in this case, the final pressure p_f will also rise. Because the mass and volume remain constant, p_f is given by

$$\frac{p_f}{p_i} = \frac{(R_{mix} \times T_{ad})_{pr}}{(R_{mix} \times T_i)_{react}} . \tag{3.63}$$

Usually, T_{ad} at constant volume is greater than T_{ad} at constant pressure. This is because the increased pressure does no work in a constant-volume process.

3.5 Use of Property Tables

The thermochemistry data required for problem-solving are given in Appendix A.

Problem 3.3 Calculate molar and mass specific enthalpy of CO_2 at $227\,^\circ C$ (500 K).

Referring to Table A.4 for CO_2, we note that $\overline{h}^o{}_f (T_{ref}) = -393520.0\,kJ/kmol$ and $\Delta \overline{h}_s (500K) = 8303.2\,kJ/kmol$. Therefore, using Eq. 3.55,

$$\overline{h} (500\ K) = -393520.0 + 8303.2 = -385216.8 \quad kJ/kmol\ of\ CO_2 .$$

Dividing this by the molecular weight (44 kg/kmol), we have

$$h = - \frac{385216.8}{44} = - 8754.93 \quad kJ/kg\ of\ CO_2 .$$

Problem 3.4 A gas stream at 1200 K comprises CO, CO_2, and N_2 with mole fractions $x_{CO} = 0.1$, $x_{CO_2} = 0.2$, and $x_{N_2} = 0.7$. Determine the mixture enthalpy on molar and mass bases.

Referring to Tables A.4, A.5, and A.13 and using Eqs. 3.23 and 3.55, it follows that

$$\overline{h}_{mix} = x_{CO}\, (\overline{h^0}_f + \Delta \overline{h}_s)_{CO} + x_{CO_2}\, (\overline{h^0}_f + \Delta \overline{h}_s)_{CO_2} + x_{N_2}\, (\overline{h^0}_f + \Delta \overline{h}_s)_{N_2}$$
$$= 0.1\,(-110530 + 28432.9) + 0.2\,(-393520.0 + 44468.8) + 0.7\,(0 + 28146)$$
$$= - 58317.75 \quad kJ/kmol$$

Also, $M_{mix} = 0.1 \times (28) + 0.2 \times (44) + 0.7 \times (28) = 31.2$. Therefore, $h_{mix} = \bar{h}_{mix}/M_{mix} = -58317.75/31.2 = -1869.16\,kJ/kg$.

Problem 3.5 Determine HHV and LHV of methane at STP and $\Phi = 1$.

The stoichiometric Methane reaction is

$$CH_4 + 2\,(O_2 + 3.76\,N_2) = CO_2 + 2\,H_2O + 7.52\,N_2\,.$$

For HHV, the water in the product state will be in liquid form, whereas for LHV, it will be in vapor form. To evaluate LHV first, we refer to Table C.1 and note that $h^0{}_{f,CH_4} = -74600$ and at STP, $\Delta\bar{h}_s = 0$. Hence, referring to Tables A.11 and A.13,

$$H_{react} = 1 \times (-74600) + 2\,(0 + 3.76 \times 0) = -74600\,kJ/kmol\ of\ fuel.$$

Similarly, referring to Tables A.4, A.7 and A.13,

$$H_{pr} = 1 \times (-393520.0) + 2 \times (-241830) + 7.52 \times 0 = -877180.0\,kJ/kmol\ of\ fuel.$$

Therefore,

$$\Delta\bar{h}_C = -\,\Delta\bar{h}_R = -\,(-877180.0 - (-74600)\,) = 802580.0\,kJ/kmol\ of\ fuel,$$

or $\Delta\bar{h}_C\,(LHV) = 802580.0/16 = 50161.25\,kJ/kg$.

For HHV, $\bar{h}_{H_2O,liquid} = \bar{h}_{H_2O,vapor} - \bar{h}_{fg}\,(25\,°C)$. From steam tables, $\bar{h}_{fg}\,(25\,°C) = 18 \times 2242.8 = 43961.4\,kJ/kmol$. Therefore,

$$H_{pr} = 1 \times (-393520.0) + 2 \times (-241830 - 43961.4) + 7.52 \times 0$$
$$= -965102.8\,kJ/kmol\ of\ fuel,$$

and,

$$\Delta\bar{h}_C = -\,\Delta\bar{h}_R = -\,\{-965102.8 - (-74600)\} = 890502.8\,kJ/kmol\ of\ fuel,$$

or $\Delta\bar{h}_C\,(HHV) = 890502.8/16 = 55656.43\,kJ/kg$.

These values are close to the experimental values quoted in Table C.1. Therefore, our postulated product composition can be taken to be nearly accurate for engineering purposes.

Problem 3.6 Determine the LHV of *liquid* n-decane ($C_{10}H_{22}$) at STP and $\Phi = 1$.

The stoichiometric coefficient for n-decane is $a_{stoic} = 10 + (22/4) = 15.5$. Therefore,

$$C_{10}H_{22} + 15.5\,(O_2 + 3.76\,N_2) = 10\,CO_2 + 11\,H_2O + 58.28\,N_2\,.$$

Thus, from Table C.1,

$$H_{react} = 1 \times (\overline{h^0}_{f,vap} - \overline{h}_{fg})_{C_{10}H_{22}}$$
$$= 1 \times (-300900 - 272.68 \times 142) = -339620.5 \text{ kJ/kmol of Fuel.}$$

The product enthalpy is

$$H_{pr} = 10 \times (-393520.0) + 11 \times (-241830) = -6595330 \text{ kJ/kmol of Fuel,}$$

or, Δh_C $(LHV) = (6595330 - 339620.5)/142 = 44054.3$ kJ/kg. The experimental value in Table C.1 is 44241.99 kJ/kg.

Problem 3.7 Determine the LHV of methane at 500 K, $p = 1$ atm and $\Phi = 1$.

To determine LHV, the values of H_{pr} and H_{react} must be determined at T = 500 K. Thus,

$$\Delta \overline{h}_C (T) = H_{react} (T) - H_{pr} (T)$$
$$= \sum (n_j \overline{h}_j)_{react} (T) - \sum (n_j \overline{h}_j)_{pr} (T)$$
$$= \sum \left\{ n_j (\overline{h^0}_{f,j} (T_{ref}) + \Delta \overline{h}_{s,j} (T)_{react}) \right\}$$
$$- \sum \left\{ n_j (\overline{h^0}_{f,j} (T_{ref}) + \Delta \overline{h}_{s,j} (T)_{pr}) \right\}$$
$$= \left[\sum n_j \overline{h^0}_{f,j} |_{react} (T_{ref}) - \sum n_j \overline{h^0}_{f,j} |_{pr} (T_{ref}) \right]$$
$$+ \sum n_j \Delta \overline{h}_{s,j} |_{react} (T) - \sum n_j \Delta \overline{h}_{s,j} |_{pr} (T)$$
$$= \left[\Delta \overline{h}_C (T_{ref}) \right] + \sum n_j \Delta \overline{h}_{s,j} |_{react} (T) - \sum n_j \Delta \overline{h}_{s,j} |_{pr} (T) ,$$

where $\Delta \overline{h}_C (T_{ref}) = 802580.0$ kJ/kmol has already been calculated in Problem 3.5. The sensible enthalpies of the products thus evaluate to

$$\sum n_j \Delta \overline{h}_{s,j} |_{pr} (500 \text{ K}) = 1 \times \Delta \overline{h}_{s,CO_2} + 2 \times \Delta \overline{h}_{s,H_2O} + 7.52 \times \Delta \overline{h}_{s,N_2}$$
$$= 8303.2 + 2 \times 6928.2 + 7.52 \times 5915.4$$
$$= 66643.4 \text{ kJ/kmol of fuel.}$$

Similarly, the change in fuel sensible heat is evaluated from Appendix B, Eq. B.6 and Table B.4 by integration as $\Delta \overline{h}_{s,fu} = \int_{298}^{500} \overline{c}_p \, dT = \overline{h^0} (500) - \overline{h^0} (298)$. Thus, noting that $\overline{h^0} (500) = -66383.97$ and $\overline{h^0} (298) = -74600$, we have $\Delta \overline{h}_{s,fu} = 8216.03$ kJ/kmol of fuel. Hence,

$$\sum n_j \, \Delta \bar{h}_{s,j} \mid_{react} (500 \text{ K}) = 1 \times \Delta \bar{h}_{s,fu} + 2 \times \Delta \bar{h}_{s,O_2} + 7.52 \times \Delta \bar{h}_{s,N_2}$$
$$= 8216.03 + 2 \times 6126.7 + 7.52 \times 5915.4$$
$$= 64953.24 \text{ kJ/kmol of fuel.}$$

Thus, $\Delta H_C \ (LHV) = (802580 + 64953.24 - 66643.4)/16 = 50055.6 \text{ kJ/kg of}$ CH_4. Thus, $LHV_{500K} < LHV_{298K}$.

Problem 3.8 Estimate the adiabatic flame temperature (AFT) of methane burning in air at STP and $\Phi = 1$.

To determine the AFT, we must equate product and reactant enthalpies in the following reaction:

$$CH_4 + 2 \, (O_2 + 3.76 \, N_2) = CO_2 + 2 \, H_2O + 7.52 \, N_2 \ .$$

Thus,

$$H_{react} = 1 \times (-74600) = -74600 \text{ kJ} \ .$$

$$H_{pr} = 1 \times \left[-393520 + \Delta \bar{h}_s \ (T_{ad}) \right]_{CO_2} + 2 \times \left[-241830 + \Delta \bar{h}_s \ (T_{ad}) \right]_{H_2O}$$
$$+ 7.52 \times \left[0 + \Delta \bar{h}_s \ (T_{ad}) \right]_{N_2}$$

or, from $H_{pr} = H_{react}$,

$$\sum n_j \, \Delta \bar{h}_{s,j} = 1 \times \left[\Delta \bar{h}_s \ (T_{ad}) \right]_{CO_2} + 2 \times \left[\Delta \bar{h}_s \ (T_{ad}) \right]_{H_2O} + 7.52 \times \left[\Delta \bar{h}_s \ (T_{ad}) \right]_{N_2} = 802580 \ .$$

This equation must be solved for T_{ad} by trial and error. Because T_{ad} ranges from 2000 K to 2400 K for most hydrocarbons, let
Trial 1 : $T_{ad} = 2200$ K

$$\sum n_j \, \Delta \bar{h}_{s,j} = 103560.8 + 2 \times 83160.1 + 7.52 \times 63201.3 = 745154.8 < 802580.$$

Trial 2 : $T_{ad} = 2300$ K

$$\sum n_j \, \Delta \bar{h}_{s,j} = 109657 + 2 \times 88428.4 + 7.52 \times 66841.5 = 789161.9 < 802580.$$

Trial 3 : $T_{ad} = 2400$ K

$$\sum n_j \, \Delta \bar{h}_{s,j} = 115774.6 + 2 \times 93747.9 + 7.52 \times 70498 = 833416.9 > 802580.$$

Therefore, T_{ad} must be between 2300 K and 2400 K, or

$$\frac{T_{ad} - 2300}{2400 - 2300} = \frac{802580 - 789161.9}{833416.9 - 789161.9} = 0.303 \, ,$$

or, $T_{ad} = 2300 + (0.303 \times 100) = 2330.32$ K. The tabulated value in Table C.1 is $T_{ad} = 2242$ K. However, using curve-fit correlation B.3 and Table B.2, $T_{ad} = 2239.4$ K. It is important to note these differences.

Exercises

1. Derive Eq. 3.19.
2. Show that if x_j are known, then $\omega_j = (x_j \, M_j)/(\sum x_j \, M_j)$.
3. Derive Eqs. 3.23 and 3.24.
4. Show that the enthalpy of formation of CO_2 at $T \neq T_{ref}$ can be evaluated as

$$\overline{h^0}_{f,CO_2} (T) = \overline{h^0}_{f,CO_2} (T_{ref}) + \left[\Delta \overline{h}_{s,CO_2} - \Delta \overline{h}_{s,C} - \Delta \overline{h}_{s,O_2} \right]_T \, .$$

5. A mixture of CO_2 and N_2 has equal proportions by volume. Determine the mass fractions of the components and the gas constant of the mixture.
6. 1 kg of a mixture of 50% CO_2 and 50% CH_4 on a volume basis is contained in a rigid vessel at 150 °C and 35 bar. Calculate the volume of the mixture using (a) the ideal gas law and (b) the rule of additive volumes.
7. A mixture of CH_4 and $C_2 H_6$ (in equal mole fractions) is compressed reversibly and isothermally from 65 °C and 7 bar to 50 bar in a steady flow process. Calculate the work required per kg of the mixture.
8. An air–water vapor mixture is compressed isentropically from 1 bar, 40 °C, and 50% relative humidity (RH) to 7 bar. Calculate the temperature of the mixture at the end of compression ($cp_{vap} = 1.88$ kJ/kg-K, $cp_{air} = 1.01$ kJ/kg-K).
9. A rigid tank contains 1 kg of H_2 (14 bar and 280 K). A tap is connected to a high-pressure (50 bar) N_2 line at 300 K, and a sufficient quantity is drawn off until the total pressure in the tank reaches 28 bar. The process is adiabatic. Calculate the final temperature in the tank, the mass of N_2 added, and the change in entropy ($cp_{N_2} = 1.045$ kJ/kg-K, $cp_{H_2} = 14.61$ kJ/kg-K).
10. The exhaust gases (600 °C) of a Compression Ignition engine are to be used to drive a gas turbine (isentropic temperature drop $= 100$ °C). The gas composition by volume is CO_2 (8.1%), $H_2 O$ (9.0%), O_2 (7.5%), and N_2 (75.4%). Estimate the pressure ratio and the enthalpy drop in the turbine. Assume that H_2O vapor acts as an ideal gas and allow for temperature dependence of specific heat.
11. For $\Phi = 0.6$, determine associated air–fuel ratios for CH_4, $C_3 H_8$, and $C_{10} H_{22}$.
12. Calculate the heating value (HHV and LHV) for propane ($C_3 H_8$) when burned in stoichiometric proportions at STP.
13. Calculate the heat of reaction of gaseous octane $C_8 H_{18}$ at 550 K.

14. Propane gas is burned with 50% excess air. The reactants enter the combustion chamber at 40 °C and the products leave at 650 °C. Calculate the heat liberated per kg of fuel.

15. Liquid octane is burned with a certain quantity of air such that the adiabatic flame temperature is 1100 K. Calculate the percent excess air if the reactants enter the combustion chamber at 298 K.

16. Determine the enthalpy of formation of methane if its LHV is 50016 kJ/kg.

17. Determine the AFT for constant-pressure combustion of a stoichiometric propane–air mixture, assuming reactants at 298 K and 1 atm and no dissociation of products.

18. Determine the LHV and AFT of biogas (60% CH_4 and 40% CO_2 by volume) at $\Phi = 1$, T $= 400$ K, and p $= 1$ atm.

19. Show that the *heat of reaction at constant volume* $\bar{q}_v$ (kJ/kmol) is related to *heat of reaction at constant pressure* $\bar{q}_p$ (kJ/kmol) as

$$\bar{q}_p - \bar{q}_v = (n_{pr} - n_{react}) R_u T_i \ .$$

Hence, determine $\bar{q}_p - \bar{q}_v$ for methane and propane for $T_i = 298$ K.

20. Determine the constant-volume AFT of propane at STP and $\Phi = 1$. Also determine the final pressure after combustion.

Chapter 4
Chemical Equilibrium

4.1 Progress of a Chemical Reaction

In the previous chapter, we dealt with *postulated* products of a hydrocarbon chemical reaction for $\Phi \leq 1$ and for $\Phi > 1$. The latter case was to be treated specially because it involves postulation of a two-step reaction mechanism (see Eqs. 3.46 and 3.47). The main point is that the postulated product composition was so far *sensitized* only to the value of Φ.

In reality, the product composition does not conform to the one postulated for *any value of* Φ. This is because, in addition to Φ, the product composition is governed by p, T, and the *rate* at which a reaction proceeds. As we have repeatedly mentioned, the central problem of combustion science is to *predict* product compositions accurately under conditions obtaining in practical devices. This inquiry into the most likely product composition is structured with increasing refinements in the following manner.

1. Product formation is determined by Φ only.
2. Product formation is determined by Φ, p, and T only.
3. Product formation is determined by Φ, p, T, and Chemical Kinetics only.
4. Product formation is determined by Φ, p, T, Chemical Kinetics, and fluid mixing.

The first type of product formation has already been discussed. In this chapter, our aim is to determine the *effect of* Φ, p, and T on the product-formation process (second item in the list above). The idea of *chemical equilibrium* enables us to do this. Effects of Chemical kinetics (third item) and fluid-mixing (fourth item) will be taken up in subsequent chapters.

Of course, specification of product composition is possible only if the system was in chemical equilibrium. Secondly, irrespective of the Φ, p, T, or the rate of reaction, the element-balance principle must be obeyed for all times for both the main and the equilibrium reactions. The main questions are: How do reactants, once reacted, approach equilibrium? Do they *coexist* with products under certain p and T? Under what conditions are only *complete* products formed?

© Springer Nature Singapore Pte Ltd. 2020
A. W. Date, *Analytic Combustion*,
https://doi.org/10.1007/978-981-15-1853-9_4

4.2 Dissociation Reaction

Central to the discussion of chemical equilibrium is the notion of *dissociation reactions*. Under appropriate conditions of p and T, these reactions result in *breakup* of *stable* products, such as CO_2, H_2O or N_2, to form either new compounds (H_2, CO, NO and the like), new *radicals* (OH, HCO, or CH_2O, for example), or even some new products in their *atomic state* (e.g., H, O, N). Thus, for $\Phi > 1$, the presence of H_2 and CO via the water-gas reaction (see Eq. 3.47), for example, can be explained in this way.

To illustrate, consider a mixture of CO and O_2 contained in a rigid *adiabatic* vessel. Thus, we have a constant-mass-volume, adiabatic *closed* system. The system is thus also an isolated system. The mixture is now ignited so the pressure and temperature will rise *until* equilibrium is reached. In this *final* state, the p, T, and *composition* of the system are governed not only by the First law but also by the Second law of thermodynamics. Because of the uncertainty associated with this final state, the reaction is written as

$$CO + \frac{1}{2} O_2 \rightarrow CO_2 .\qquad(4.1)$$

The right arrow simply indicates the *general direction* of the reaction, but this does not imply that in the final state, only CO_2 will be present. In fact, if the temperature reached is sufficiently high, a part of CO_2 (say, α) will also *dissociate* back to CO and O_2. Thus, $\alpha\, CO_2 \rightarrow \alpha\, (CO + 0.5\, O_2)$, and the overall reaction may be written as

$$\left[CO + \frac{1}{2} O_2\right]_{reactants} \rightarrow \left[(1-\alpha)\, CO_2 + \alpha\, \left(CO + \frac{1}{2} O_2\right)\right]_{hot\ products.}$$
$$(4.2)$$

When the reaction is written in this manner, it is called a *dissociation reaction* and α is called the *degree of reaction* [10]. Clearly, if $\alpha = 1$, there will be no chemical reaction. If $\alpha = 0$, complete product(s) will be formed. In general, α is not fixed but varies with time, because a chemical reaction takes place over *finite time* with α varying from $\alpha = 1$ to $\alpha = \alpha_{eq}$, where α_{eq} corresponds to the equilibrium state reached. This state will depend on the final equilibrium p and T. Of course, $\alpha_{eq} = 0$ is a special case that denotes complete product formation. Finally, both the forward and backward reactions take place in *stoichiometric* proportions; that is, the number of moles dissociating and recombining satisfy the equation $CO + 0.5\, O_2 = CO_2$.

The idea of a dissociation reaction also opens up another possibility. Suppose we started with only CO_2 in our system and the system temperature and pressure were now raised. Then, a reaction will again be set up that will form CO and O_2, and it will be written as

$$[CO_2]_{reactant} \rightarrow \left[\beta\, CO_2 + (1-\beta)\, \left(CO + \frac{1}{2} O_2\right)\right]_{products} .\qquad(4.3)$$

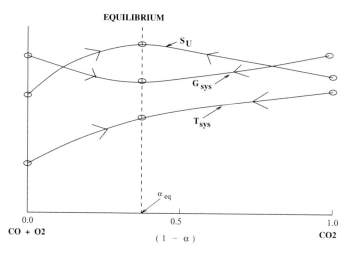

Fig. 4.1 Chemical equilibrium of a constant-volume isolated system

β is the degree of reaction that will equal $1 - \alpha$ of the previous reaction. The equilibrium state for a given p and T will be *identical* irrespective, of whether we started with CO_2 or a mixture of CO and O_2. The exact equilibrium state in either case is determined from the equilibrium condition specified by the Second law of thermodynamics.

4.3 Conditions for Chemical Equilibrium

4.3.1 Condition for a Finite Change

In Chap. 2, the conditions for spontaneous change and equilibrium of a general closed system were derived and given by Eq. 2.77. It is re-stated here:

$$[\Delta G - V \, \Delta p + S \, \Delta T]_{sys} \leq 0 \ . \tag{4.4}$$

The symbol $<$ indicates spontaneous change, whereas the symbol $=$ indicates equilibrium.

In thermodynamics, chemical reactions are studied under *constant temperature and pressure* conditions so $\Delta p = \Delta T = 0$. Further, it is assumed that *temperatures and pressures of both the system and the surroundings remain the same at all times during a reaction*. Thus, $p = p_0$ and $T_{sys} = T_0$ at all times. Equation 4.4 was established for precisely these conditions. Therefore,

$$\Delta G_{sys} \,|_{p_0, T_0} \leq 0 \ . \tag{4.5}$$

This equation is same as Eq. 2.78. Again, when $\Delta G_{sys} < 0$, $\Delta S_U > 0$ and when $\Delta G_{sys} = 0$, $\Delta S_U = 0$. Thus, at equilibrium, S_U will be at a *maximum* but G_{sys} will be at a *minimum*. As such, following a spontaneous reaction, the system will tend to $G = G_{min}$. Qualitatively, the progress of the reaction considered in Eq. 4.2 will therefore be as shown in Fig. 4.1. The figure shows that progress of the system from both left (CO and O_2 as reactants) or right (CO_2 as the initial system) will reach the same equilibrium state.

4.3.2   Consequences for an Infinitesimal Change

From Eq. 4.5 we learn that the Gibbs function for the system will be a minimum at equilibrium. Thus, dropping the subscript *sys* in favor of *mix* and replacing subscripts $(p = p_0, T = T_0)$ in favour of (p, T), we can write that for an infinitesimal change,

$$dG_{mix}\,|_{p,T} = d \sum_j (n_j \, \bar{g}_j)\,|_{p,T} = 0. \tag{4.6}$$

Using Eqs. 2.66 and 3.58, we define $\bar{g}_j$ as

$$\bar{g}_j = \bar{h}_j - T\,\bar{s}_j \tag{4.7}$$

$$= \left[\bar{h}_j - T\,\bar{s}^0_j\,(T)\right] + R_u\,T\,\ln\left(\frac{p_j}{p_{ref}}\right) \tag{4.8}$$

$$= \bar{g}^0_j\,(p_{ref}, T) + R_u\,T\,\ln\left(\frac{p_j}{p_{ref}}\right), \tag{4.9}$$

where, $\bar{g}^0_j\,(p_{ref}, T)$ is the Gibbs function at pressure $p = p_{ref} = 1$ atm.[1] Using Eqs. 3.55 and 3.59, $\bar{g}^0_j\,(p_{ref}, T)$ can be evaluated from

$$\bar{g}^0_j\,(p_{ref}, T) = \bar{h}_j - T\bar{s}^0_j\,(T) = \bar{g}^0_{f,j}\,(T_{ref}) + \Delta\bar{g}_{s,j}(T)$$

where,

$$\bar{g}^0_{f,j}\,(T_{ref}) = \bar{h}^0_{f,j}\,(p_{ref}, T_{ref}) - T\bar{s}^0_{f,j}\,(p_{ref}, T_{ref})$$

$$\Delta\bar{g}_{s,j}(T) = \Delta\bar{h}_{s,j}(T) - T \int_{T_{ref}}^{T} \bar{c}_{p,j}\,\frac{dT}{T}$$

and $\bar{g}^0_{f,j}$ is the *Gibbs function of formation.*

---

[1] Equation 4.9 can also be written in terms of total mixture pressure p as

$$\bar{g}_j = \bar{g}^0_j\,(p_{ref}, T) + R_u\,T\,\left[\ln\left(\frac{p_j}{p}\right) + \ln\left(\frac{p}{p_{ref}}\right)\right]$$

$$= \bar{g}^0_j\,(p_{ref}, T) + R_u\,T\,\left[\ln x_j + \ln\left(\frac{p}{p_{ref}}\right)\right].$$

It is useful to tabulate values of $\overline{g}^0{}_j (p_{ref}, T)$.[2] They are denoted by $\overline{g}^0$ in Appendix A. Reverting to equilibrium condition 4.6, we have

$$dG_{mix} \mid_{p,T} = d \sum_j (n_j \overline{g}_j) \mid_{p,T} = \sum_j \overline{g}_j \, dn_j \mid_{p,T} + \sum_j n_j \, d\overline{g}_j \mid_{p,T} . \quad (4.10)$$

However, from Eq. 4.9 and, for a constant p and T process, the second term on the right hand side of this equation evaluates to

$$\sum_j n_j \, d\overline{g}_j \mid_{p,T} = \sum_j n_j \, d \left[\overline{g}^0{}_j \mid_{p_{ref},T} + R_u \, T \ln \left(\frac{p_j}{p_{ref}} \right) \right]$$

$$= \sum_j n_j \, d \, \overline{g}^0{}_j \mid_{p_{ref},T} + \sum_j n_j \, R_u \, T \, \frac{dp_j}{p_j}$$

$$= 0 + \sum_j \left(\frac{n_{mix} \, R_u \, T}{p} \right) dp_j , \quad \text{as} \quad \frac{n_j}{p_j} = \frac{n_{mix}}{p}$$

$$= \left(\frac{n_{mix} \, R_u \, T}{p} \right) d \sum_j p_j$$

$$= \left(\frac{n_{mix} \, R_u \, T}{p} \right) dp = 0 \quad \text{for p} = \text{constant.} \quad (4.11)$$

Thus, using Eqs. 4.9, 4.10 can be written as

$$dG_{mix} \mid_{p,T} = d \sum_j (n_j \overline{g}_j) \mid_{p,T} = \sum_j \overline{g}_j \, dn_j \mid_{p,T} \quad (4.12)$$

$$= \sum_j \left[\overline{g}^0{}_j (p_{ref}, T) + R_u \, T \ln \left(\frac{p_j}{p_{ref}} \right) \right] dn_j = 0 . \quad (4.13)$$

This equation correlates the manner in which the number of moles n_j of the reactants, and the products change in a constant p and T reaction. It is a very important equation because it enables derivation of an *equilibrium constant*. This constant, in turn, enables us to determine the final composition of the mixture at the given p and T. From the known composition, we can determine *how far* the reaction has proceeded in its march toward completion, or simply the degree of reaction α.

---

[2]These values are the Gibbs function of formation $\overline{g}^0_{f\,j}$ at $T_{ref} = T$, so $\Delta \overline{g}_{s,j} (T) = 0$.

4.4 Equilibrium Constant K_p

4.4.1 Degree of Reaction

For simplicity, consider a general equilibrium reaction involving reactants A_1 and A_2 and products A_3 and A_4 in *stoichiometric* proportions.[3] In terms of number of moles, the reaction is written as[4]

$$n_1 \, A_1 + n_2 \, A_2 \rightleftharpoons n_3 \, A_3 + n_4 \, A_4 \ . \tag{4.14}$$

This reaction is also written in terms of *stoichiometric coefficients* v. Thus,

$$|v_1| \, A_1 + |v_2| \, A_2 \rightleftharpoons |v_3| \, A_3 + |v_4| \, A_4 \ , \tag{4.15}$$

where, by *convention*, $v_j > 0$ or is positive for products and $v_j < 0$ or is negative for reactants. Thus, $n_1 = -v_1, n_2 = -v_2, n_3 = v_3$, and $n_4 = v_4$.

The overall reaction proceeds from left to right such that the reactants *deplete* and products *build up* in time. During this process, the Gibbs function for the mixture decreases, eventually attaining a minimum value at equilibrium. To express how far a reaction has proceeded, we define the *degree of reaction*, now denoted by ϵ (instead of α). It is defined as

$$\epsilon = \frac{n_{j,max} - n_j}{n_{j,max} - n_{j,min}} \ , \tag{4.16}$$

where j stands for reactants and products. Thus, for reactant A_1, for example, $n_{1,max}$ will correspond to the *start* of the reaction and $n_{1,min}$ will correspond to the *end* of the reaction. However, since the initial reactants are in *stoichiometric* proportions, $n_{1,max} = n_{1,i}$ and $n_{1,min} = 0$. Hence,

$$d\epsilon = -\frac{dn_1}{n_{1,max}} = -\frac{dn_1}{n_{1,i}} \ . \tag{4.17}$$

In other words, $dn_1 = -n_{1,i} \, d\epsilon = v_{1,i} \, d\epsilon$. Similarly, for the second reactant, $dn_2 = -n_{2,i} \, d\epsilon = v_{2,i} \, d\epsilon$.

By similar reasoning on the product side, and for A_3, for example, $n_{3,max} = n_{3,f}$ and $n_{3,min} = n_{3,i} = 0$ because initially no products are present. Then

$$d\epsilon = \frac{dn_3}{n_{3,max}} = -\frac{dn_3}{n_{3,f}} \ . \tag{4.18}$$

---

[3]It is not necessary that there be two reactants and two products. There may be one reactant or one product. As we shall see later, in some reactions, there may even be three reactants and/or products. Also, A's may be compounds, or radicals, or even atoms.

[4]In a reaction, $CO + 0.5 \, O_2 \rightleftharpoons CO_2, n_1 = n_{CO} = 1, n_2 = n_{O_2} = 0.5, n_3 = n_{CO_2} = 1$, and $n_4 = 0$.

Hence, $d n_3 = -n_{3,f} d \epsilon = -v_{3,f} d \epsilon$ and, similarly, $d n_4 = -n_{4,f} d \epsilon = -v_{4,f} d \epsilon$. From the above relations and using Eq. 4.12, therefore,

$$dG_{pr} = \bar{g}_{A_3} dn_3 + \bar{g}_{A_4} dn_4 = -\left[\bar{g}_{A_3} v_{3,f} + \bar{g}_{A_4} v_{4,f}\right] d\epsilon . \tag{4.19}$$

$$dG_{react} = \bar{g}_{A_1} dn_1 + \bar{g}_{A_2} dn_2 = \left[\bar{g}_{A_1} v_{1,i} + \bar{g}_{A_2} v_{2,i}\right] d\epsilon . \tag{4.20}$$

4.4.2 Derivation of K_p

For the reaction in Eq. 4.14, $d G_{mix} = d G_{react} + d G_{pr} = 0$ at equilibrium. Thus, using relations in Eqs. 4.19 and 4.20, we have

$$\left[\bar{g}_{A_1} v_{1,i} + \bar{g}_{A_2} v_{2,i} - \bar{g}_{A_3} v_{3,f} - \bar{g}_{A_4} v_{4,f}\right] d\epsilon = 0 , \tag{4.21}$$

or, using Eq. 4.13 to replace $\bar{g}_j$ and noting that $d\epsilon \neq 0$,

$$
\begin{aligned}
& \bar{g^0}_{A_1} v_{1,i} + \bar{g^0}_{A_2} v_{2,i} - \bar{g^0}_{A_3} v_{3,f} - \bar{g^0}_{A_4} v_{4,f} \\
& = -R_u T \left[v_{1,i} \ln \frac{p_{A_1}}{p_{ref}} + v_{2,i} \ln \frac{p_{A_2}}{p_{ref}} - v_{3,f} \ln \frac{p_{A_3}}{p_{ref}} - v_{4,f} \ln \frac{p_{A_4}}{p_{ref}} \right] \\
& = -R_u T \ln \left[\frac{(p_{A_1}/p_{ref})^{v_{1,i}} (p_{A_2}/p_{ref})^{v_{2,i}}}{(p_{A_3}/p_{ref})^{v_{3,f}} (p_{A_4}/p_{ref})^{v_{4,f}}} \right] .
\end{aligned} \tag{4.22}
$$

To simplify further, we define

$$G^0_{react} \equiv \bar{g^0}_{A_1} v_{1,i} + \bar{g^0}_{A_2} v_{2,i} \tag{4.23}$$

$$G^0_{pr} \equiv \bar{g^0}_{A_3} v_{3,f} + \bar{g^0}_{A_4} v_{4,f} . \tag{4.24}$$

Then, from the left hand side of Eq. 4.22, the *finite change*, ΔG^0 can be written as

$$\Delta G^0 = G^0_{pr} - G^0_{react} \tag{4.25}$$

$$= R_u T \ln \left[\frac{(p_{A_1}/p_{ref})^{v_{1,i}} (p_{A_2}/p_{ref})^{v_{2,i}}}{(p_{A_3}/p_{ref})^{v_{3,f}} (p_{A_4}/p_{ref})^{v_{4,f}}} \right] \tag{4.26}$$

$$= -R_u T \ln \left[K_p \right] , \tag{4.27}$$

where

$$K_p = \frac{(p_{A_3}/p_{ref})^{\nu_{3,f}} \, (p_{A_4}/p_{ref})^{\nu_{4,f}}}{(p_{A_1}/p_{ref})^{\nu_{1,i}} \, (p_{A_2}/p_{ref})^{\nu_{2,i}}} \qquad (4.28)$$

$$= \frac{\prod (p_A/p_{ref})^{\nu_{pr}}}{\prod (p_A/p_{ref})^{\nu_{react}}} \qquad (4.29)$$

$$= \exp\left[-\frac{\Delta G^0}{R_u T}\right], \qquad (4.30)$$

where K_p is the *equilibrium constant* and $\prod$ denotes "*product of*". Equation 4.27 is called the *law of mass action*. It represents the equilibrium of a chemical reaction with a constant p and T. The following comments are now in order.

1. If $K_p < 1$ (a *fraction*), then $\ln K_p < 0$ and $\Delta G^0 > 0$ (positive). Hence, from Eq. 4.28 we deduce that *reactants* will be favored and they will *survive* in significant proportions in the equilibrium composition.
2. If $K_p \to 0$, no reaction will take place.
3. If $K_p > 1$, then $\ln K_p > 0$ and $\Delta G^0 < 0$ (negative). Hence, in this case, the *product* proportions will dominate the equilibrium composition.
4. If $K_p \to \infty$, the reaction will be *complete*.
5. As, $x_{A_1} = p_{A_1}/p = (p_{A_1}/p_{ref})\,(p_{ref}/p) = x_{A_1}\,(p_{ref}/p) \ldots$, K_p can be written in terms of the mole fractions of products and reactants. Thus, from Eq. 4.28, it is easy to show that

$$K_p = \left[\frac{(x_{A_3})^{\nu_{3,f}} \, (x_{A_4})^{\nu_{4,f}}}{(x_{A_1})^{\nu_{1,i}} \, (x_{A_2})^{\nu_{2,i}}}\right] \times \left(\frac{p}{p_{ref}}\right)^{(\nu_{3,f}+\nu_{4,f}-\nu_{1,i}-\nu_{2,i})} . \qquad (4.31)$$

6. At constant pressure, $\Delta G^0 = G^0_{pr} - G^0_{react}$ as a function of T can be evaluated from the tabulated values of $\overline{g^0}_{T_{ref}=T}$ in Appendix A. Because, ΔG^0 is a function of T only, from Eqs. 4.28 and 4.30, it follows that K_p is also a function of temperature alone.
7. Therefore, K_p can be evaluated once for all from Eq. 4.30 as a function of T. Table 4.1 gives values of K_p for eight frequently encountered reactions as function of temperature. The water-gas reaction is also included. In some of the reactions, at least one of the participating species is a *radical* or even an *atom*.

The following additional comments concerning K_p are also important.

1. The K_p expression in Eq. 4.30 is derived for the reaction written in Eq. 4.14. If the reaction was written in reverse order, that is

$$n_3\, A_3 + n_4\, A_4 \rightleftharpoons n_1\, A_1 + n_2\, A_2 , \qquad (4.32)$$

then it is easy to show from Eq. 4.29 that K_p (reverse) $= K_p^{-1}$ (forward).
2. From Eq. 2.66, $\Delta G^0 = \Delta H^0 - T \, \Delta S^0$. Therefore, from Eq. 4.30

$$K_p = \exp\left[-\frac{(\Delta H^0 - T\,\Delta S^0)}{R_u\,T}\right]$$

$$\ln K_p = -\frac{(\Delta H^0 - T\,\Delta S^0)}{R_u\,T}$$

$$\frac{d\ln K_p}{d\,(1/T)} = -\frac{\Delta H^0(T, p_{ref})}{R_u} \qquad (4.33)$$

This equation shows that K_p can also be calculated from ΔH^0. Further, $\Delta H^0 = \sum_j (n_j\,\bar{h}_j)_{pr} - \sum_j (n_j\,\bar{h}_j)_{react} = \Delta h_R$. Thus, if K_p increases with temperature, $\Delta H^0 = \Delta h_R$ must be positive or the reaction must be *endothermic*. Conversely, if K_p decreases with temperature, the reaction must be *exothermic*.

3. Equation 4.33 is the called van't Hoff equation. For many reactions, ΔH^0 is nearly constant with temperature. Hence, the equation can be approximated by

$$\ln\left\{\frac{K_p\,(T_2)}{K_p\,(T_1)}\right\} \simeq \frac{\Delta H^0}{R_u}\left(\frac{1}{T_1} - \frac{1}{T_2}\right). \qquad (4.34)$$

Thus, knowing K_p at T_1, for instance, its value at T_2 can be evaluated.

4.5 Problems on Chemical Equilibrium

4.5.1 Single Reaction

Problem 4.1 For the dissociation reaction $CO_2 \rightleftharpoons CO + 0.5\,O_2$, verify the value of K_p given in Table 4.1 at 1500 K and 2500 K and $p = 1$ atm.

From Eq. 4.30, $K_p = \exp\left[-\Delta G^0/(R_u\,T)\right]$ where ΔG^0 will be given by

$$\Delta G^0 = G^0_{pr}\,(T) - G^0_{react}\,(T)$$

$$= \left(\bar{g}^0_{CO} + \frac{1}{2}\bar{g}^0_{O_2}\right) - \left(\bar{g}^0_{CO_2}\right) \quad \rightarrow \quad \text{(see Tables } A.4 \text{ and } A.5)$$

$$= \left(-246108.2 + \frac{1}{2} \times 0\right) - (-396403.9) = 150295.7 \quad \text{at 1500 K}$$

$$= \left(-328691.4 + \frac{1}{2} \times 0\right) - (-396177.7) = 67486.3 \quad \text{at 2500 K.}$$

Therefore,

$$K_p\,(1500K) = \exp\,(-150295.7/8.314/1500) = 5.835 \times 10^{-6} \,,$$

Table 4.1 Values of $\log_{10} K_p$ for single dissociation reaction [49]. Approximate Curve-fit coefficients are given in Appendix B

T (K)	$H_2 \rightleftharpoons 2H$	$O_2 \rightleftharpoons 2O$	$N_2 \rightleftharpoons 2N$	$H_2O \rightleftharpoons$ $H_2 + \frac{1}{2} O_2$
298	−71.228	−81.208	−159.60	−40.048
500	−40.318	−45.880	−92.672	−22.886
1000	−17.292	−19.614	−43.056	−10.062
1500	−9.514	−10.790	−26.434	−5.725
1800	−6.896	−7.836	−20.874	−4.270
2000	−5.582	−6.356	−18.092	−3.540
2200	−4.502	−5.142	−15.810	−2.942
2400	−3.600	−4.130	−13.908	−2.443
2600	−2.836	−3.272	−12.298	−2.021
2800	−2.178	−2.536	−10.914	−1.658
3000	−1.606	−1.898	−9.716	−1.343
3200	−1.106	−1.340	−8.664	−1.067
3500	−0.462	−0.620	−7.312	−0.712
4000	+0.400	+0.340	−5.504	−0.238
4500	+1.074	+1.086	−4.094	+0.133
5000	+1.612	+1.686	−2.962	+0.430
T (K)	$H_2O \rightleftharpoons$ $OH + \frac{1}{2} H_2$	$CO_2 \rightleftharpoons$ $CO + \frac{1}{2} O_2$	$\frac{1}{2} N_2 + \frac{1}{2} O_2 \rightleftharpoons$ NO	$CO_2 + H_2 \rightleftharpoons$ $CO + H_2O$
298	−46.054	−45.066	−15.171	−5.018
500	−26.13	−25.025	−8.783	−2.139
1000	−11.28	−10.221	−4.062	−0.159
1500	−6.284	−5.316	−2.487	+0.409
1800	−4.613	−3.693	−1.962	+0.577
2000	−3.776	−2.884	−1.699	+0.656
2200	−3.091	−2.226	−1.484	+0.716
2400	−2.520	−1.679	−1.305	+0.764
2600	−2.038	−1.219	−1.154	+0.802
2800	−1.624	−0.825	−1.025	+0.833
3000	−1.265	−0.485	−0.913	+0.858
3200	−0.951	−0.189	−0.815	+0.878
3500	−0.547	+0.190	−0.690	+0.902
4000	−0.011	+0.692	−0.524	+0.930
4500	+0.408	+1.079	−0.397	+0.946
5000	+0.741	+1.386	−0.296	+0.956

or $\log_{10} K_p = -5.234$.

Likewise, it can be shown that

$$K_p (2500K) = \exp(-67486.3/8.314/2500) = 0.0389 ,$$

or, $\log_{10} K_p = -1.41$.

In Table 4.1, $\log_{10} K_p (1500K) = -5.316$. This value nearly matches with that calculated earlier. Similarly, at 2500 K, K_p is determined by interpolation from the values given at 2400 K and 2600 K. Thus, $\log_{10} K_p (2500K) = -0.5 (1.679 + 1.210) = -1.4445$. This value also nearly matches with that calculated earlier. In all the following problems, K_p values are therefore evaluated from Table 4.1. Curve-fit coefficients for these values are given in Appendix B (see Eq. B.4 and Table B.3).

Problem 4.2 For the dissociation reaction $CO_2 \rightleftharpoons CO + 0.5\,O_2$, determine (a) the degree of reaction and (b) the product composition at 1500 K and $p = 0.1$, 1.0 and 10 atm.

From Eq. 4.31,

$$K_p = \frac{x_{CO}\,\sqrt{x_{O_2}}}{x_{CO_2}} \left(\frac{p}{p_{ref}}\right)^{0.5} \tag{a1}$$

In terms of degree of reaction, we can write

$$CO_2 \rightarrow \epsilon\,[CO + 0.5\,O_2] + (1 - \epsilon)\,CO_2 .$$

Therefore, $n_{tot,pr} = \epsilon + \epsilon/2 + (1 - \epsilon) = 1 + \epsilon/2$. Hence,

$$x_{CO} = \frac{\epsilon}{1 + \epsilon/2} \qquad x_{O_2} = \frac{\epsilon/2}{1 + \epsilon/2} \qquad x_{CO_2} = \frac{1 - \epsilon}{1 + \epsilon/2} \tag{a2}$$

or, substituting this equation in Eq. a1, and taking $K_p = 4.76 \times 10^{-6}$ from Table 4.1,

$$\frac{\epsilon}{1 - \epsilon}\sqrt{\frac{\epsilon}{2 + \epsilon}} = K_p \left(\frac{p}{p_{ref}}\right)^{-0.5} = 4.7614 \times 10^{-6} \left(\frac{p}{p_{ref}}\right)^{-0.5}$$

This equation must be solved by trial and error for the three values of pressure with $p_{ref} = 1$ atm. The evaluations are given in Table 4.2 which shows that in spite of the 100-fold increase in pressure, the product composition is influenced little and the reactant CO_2 remains almost unreacted at 1500 K. The dissociated amount none the less decreases with increase in pressure.

Problem 4.3 For the dissociation reaction $CO_2 \rightleftharpoons CO + 0.5\,O_2$, determine (a) the degree of reaction and (b) the product composition at $p = 1.0$ atm and $T = 1500$ K, 2000 K and 2500 K.

Table 4.2 Evaluation of ϵ and product composition (Problem 4.2)

p (atm)	ϵ	x_{CO}	x_{O_2}	x_{CO_2}
0.1	7.679×10^{-4}	7.676×10^{-4}	3.838×10^{-4}	0.99885
1.0	3.565×10^{-4}	3.564×10^{-4}	1.782×10^{-4}	0.99946
10.0	1.655×10^{-4}	1.655×10^{-4}	8.274×10^{-5}	0.99975

Table 4.3 Evaluation of ϵ and product composition (Problem 4.3)

T (K)	K_p	ϵ	x_{CO}	x_{O_2}	x_{CO_2}
1500	4.761×10^{-6}	3.565×10^{-4}	3.564×10^{-4}	1.782×10^{-4}	0.99946
2000	1.228×10^{-3}	0.01434	0.01425	0.007123	0.9786
2500	2.901×10^{-2}	0.1119	0.106	0.0530	0.841

The procedure followed here is same as that for the previous problem. It is seen from Table 4.3 that the product composition is now significantly influenced by temperature because of the strong dependence of K_p on temperature.

4.5.2 Two-Step Reactions

Problem 4.4 At very high temperatures, CO_2 dissociation occurs via a two-step mechanism.

$$CO_2 \rightleftharpoons CO + 0.5\,O_2 \tag{I}$$

$$O_2 \rightleftharpoons O + O \tag{II}$$

Calculate the product composition at 5000 K and 1 atm.

Here, we have four species and thus we require four constraints. Two of these constraints are

$$K_{p,I} = \frac{x_{CO}\,\sqrt{x_{O_2}}}{x_{CO_2}} \quad \text{and} \quad K_{p,II} = \frac{x_O\,x_O}{x_{O_2}} \;.$$

Further, because C and O atoms must be conserved, the third constraint is

$$\frac{\text{Number of C atoms}}{\text{Number of O atoms}} = \frac{1}{2} = \frac{x_{CO} + x_{CO_2}}{x_{CO} + 2\,x_{CO_2} + 2\,x_{O_2} + x_O} \;.$$

Finally, the fourth constraint is $x_{CO_2} + x_{CO} + x_{O_2} + x_O = 1$.

From Table 4.2, at 5000 K and 1 atm, $K_{p,I} = 10^{1.386} = 24.322$ and $K_{p,II} = 10^{1.686} = 48.529$. Making substitutions and eliminating mole fractions of O, CO, and CO_2, it can shown that x_{O_2} is governed by

Table 4.4 Iterative solution of Eq. III (Problem 4.4)

l	R/(dR/dy)	y
1	0.233034	0.316228
2	0.012547	0.083194
3	3.6156×10^{-5}	0.070646
4	-9.062×10^{-10}	0.07061

$$2 x_{O_2}^{1.5} + 79.9323 \, x_{O_2} + 338.8 \, x_{O_2}^{0.5} = 24.322 \tag{III}$$

This equation can be solved by the Newton–Raphson technique so that

$$y^{l+1} = y^l - \left[\frac{R}{dR/dy} \right]_{y^l},$$

where $y = \sqrt{x_{O_2}}$, $R = 2 \, y^3 + 79.9323 \, y^2 + 338.8 \, y - 24.322$ is the residual of Eq. III, and, superscript, l, is the iteration number. The iterations are begun with initial guess $x_{O_2} = 0.1$. The results are given in Table 4.4.

Therefore, $x_{O_2} = y^2 = (0.07061)^2 = 0.00498$. Hence, it can be deduced that $x_O = \sqrt{K_{p,11} \times x_{O_2}} = \sqrt{48.529 \times 0.00498} = 0.49189$, $x_{CO} = 2 \, x_{O_2} + x_O = 0.50186$, and $x_{CO_2} = 1 - x_{CO} - x_{O_2} - x_O = 0.00126$. This problem shows that dissociation of CO_2 and O_2 is indeed considerable at high temperature resulting in high mole fractions of O and CO.

Problem 4.5 Consider combustion of $C_{10}H_{22}$ in air with $\Phi = 1.25$. Determine the approximate product composition assuming water-gas reaction whose K_p value is evaluated at T_{wg}. The reactants are at $T = 298$ K and $p = 1$ atm.

In Chap. 3(see Eqs. 3.46 and 3.47), we considered the case of rich mixture. Thus, with $\Phi = 1.25$, $a = (10 + 22/4)/1.25 = 12.4$. Hence, the two-step reaction is written as

$$C_{10}H_{22} + 12.4 \, (O_2 + 3.76 \, N_2) \rightarrow n_{CO_2} \, CO_2 + n_{H_2O} \, H_2O + n_{CO} \, CO \tag{a1}$$
$$+ \, n_{H_2} \, H_2 + 46.624 \, N_2$$

and

$$CO + H_2O \rightleftharpoons CO_2 + H_2 \tag{a2}$$

The total number of product moles is therefore

$$n_{tot,pr} = n_{CO_2} + n_{CO} + n_{H_2O} + n_{H_2} + 46.624.$$

The element balance for reaction (a1) gives

Table 4.5 Product composition for $C_{10}H_{22}$ combustion at $\Phi = 1.25$, partial equilibrium (Problem 4.5)

T_{wg}	K_p	x_{CO_2}	x_{CO}	x_{H_2O}	x_{H_2}	x_{N_2}
2000	0.2208	0.08117	0.06670	0.1377	0.02498	0.6894
2100	0.2060	0.08025	0.06762	0.1386	0.02406	0.6894
2200	0.1923	0.07935	0.06852	0.1395	0.02316	0.6895
2300	0.1819	0.07864	0.06922	0.1402	0.02245	0.6894
2400	0.1722	0.07795	0.06894	0.1409	0.02176	0.6894

$$\text{C balance}: 10 = n_{CO_2} + n_{CO}.$$
$$\text{H balance}: 22 = 2 \times n_{H_2O} + 2 \times n_{H_2}.$$
$$\text{O balance}: 24.8 = 2 \times n_{CO_2} + n_{CO} + n_{H_2O}.$$

From these three equations, it follows that

$$n_{CO} = 10 - n_{CO_2} \quad n_{H_2O} = 14.8 - n_{CO_2} \quad \text{and} \quad n_{H_2} = n_{CO_2} - 3.8.$$
$$n_{tot,pr} = n_{CO_2} + (10 - n_{CO_2}) + (14.8 - n_{CO_2}) + (n_{CO_2} - 3.8) + 46.624 = 67.62.$$

Therefore,

$$x_{CO_2} = \frac{n_{CO_2}}{67.624} \quad x_{CO} = \frac{10 - n_{CO_2}}{67.624} \quad x_{H_2O} = \frac{14.8 - n_{CO_2}}{67.624}.$$
$$x_{H_2} = \frac{n_{CO_2} - 3.8}{67.624} \quad x_{N_2} = \frac{46.624}{67.624} = 0.68946.$$

To close the problem, we use the equilibrium of the water-gas reaction, so

$$K_p = \frac{x_{CO_2}\, x_{H_2}}{x_{H_2O}\, x_{CO}} = \frac{n_{CO_2}\,(n_{CO_2} - 3.8)}{(10 - n_{CO_2})\,(14.8 - n_{CO_2})}.$$

Substituting for mole fractions and rearranging, we have

$$(1 - K_p)\, n_{CO_2}^2 + (24.8\, K_p - 3.8)\, n_{CO_2} - 148\, K_p = 0.$$

Solution of this quadratic equation for n_{CO_2} for a given K_p enables evaluation of n_{CO_2} and hence of all product mole fractions. To evaluate K_p, however, we must choose the appropriate temperature, T_{wg}. At typical combustion temperatures (2000 to 2400 K, say), the product compositions are not very sensitive to the choice of T_{wg}. This is shown in Table 4.5.

Once the product composition is known, it is also possible to evaluate the AFT (see Eq. 3.60) and ΔH_c (see Eq. 3.54).

4.5.3 Multistep Reactions

In the previous problem, equilibrium product composition was derived for a rich mixture using water-gas reaction. For a lean mixture, no such reaction needs to be invoked as was noted in Chap. 3. In reality, however, the product composition is influenced by p and T at all values of Φ. To determine effects of such an influence, a multistep mechanism may be postulated for hydrocarbon reactions. This mechanism is called the *full equilibrium* mechanism, to contrast with the *partial equilibrium mechanism* in which the products for a lean mixture were postulated from the element balance whereas for a rich mixture, they were determined from the water-gas equilibrium reaction. Thus, let the main reaction be represented as

$$C_m H_n + a (O_2 + 3.76 N_2) \rightarrow n_{CO_2} CO_2 + n_{CO} CO$$
$$+ n_{H_2O} H_2O + n_{H_2} H_2 + n_H H + n_{OH} OH$$
$$+ n_{N_2} N_2 + n_{NO} NO + n_N N$$
$$+ + n_{O_2} O_2 + n_O O \ ,$$

where $a = (m + n/4)/\Phi$. Thus, eleven species are permitted in the products. We therefore need eleven constraints to determine their composition. Four of these are provided by the element balance. Thus

$$C \text{ balance} : m = n_{CO_2} + n_{CO}$$
$$H \text{ balance} : n = 2 \times n_{H_2O} + 2 \times n_{H_2} + n_H + n_{OH}$$
$$O \text{ balance} : 2\,a = 2 \times n_{CO_2} + n_{CO} + n_{H_2O} + n_{OH} + n_{NO} + 2 \times n_{O_2} + n_O$$
$$N \text{ balance} : 7.52\,a = 2 \times n_{N_2} + n_{NO} + n_N \ .$$

Further, the mass balance gives

$$(12\,m + n) + a\,(32 + 3.76 \times 28) = 44\,n_{CO_2} + 28\,n_{CO}$$
$$+ 18\,n_{H_2O} + 2\,n_{H_2} + n_H + 17\,n_{OH}$$
$$+ 28\,n_{N_2} + 30\,n_{NO} NO + 14\,n_N$$
$$+ 32\,n_{O_2} + 16\,n_O \ .$$

Dividing both sides of this equation by $n_{tot,pr}$, we have

$$n_{tot} = [(12\,m + n) + a\,(32 + 3.76 \times 28)] / \big[44\,x_{CO_2} + 28\,x_{CO} + 18\,x_{H_2O} + 2\,x_{H_2}$$
$$+ x_H + 17\,x_{OH} + 28\,x_{N_2} + 30\,x_{NO} + 14\,x_N + 32\,x_{O_2} + 16\,x_O\big] \ . \tag{A}$$

Thus, the four element-balance constraints can be expressed as

$$x_{CO_2} + x_{CO} = \frac{m}{n_{tot}} = m^*, \quad (1)$$

$$2 x_{H_2O} + 2 x_{H_2} + x_H + x_{OH} = \frac{n}{n_{tot}} = n^*, \quad (2)$$

$$2 x_{CO_2} + x_{CO} + x_{H_2O} + x_{OH} + x_{NO} + 2 x_{O_2} + x_O = \frac{2a}{n_{tot}} = 2a^*, \quad (3)$$

$$2 x_{N_2} + x_{NO} + x_N = 7.52 a^* \quad (4).$$

Additional seven constraints are derived from the seven reactions given in Table 4.1. Thus,

$$R_1 \; - \; H_2 \rightleftharpoons H + H : x_H = A_1 \sqrt{x_{H_2}}, \quad A_1 = \sqrt{K_{p1}/(p/p_{ref})} \quad (5)$$

$$R_2 \; - \; O_2 \rightleftharpoons O + O : x_O = A_2 \sqrt{x_{O_2}}, \quad A_2 = \sqrt{K_{p2}/(p/p_{ref})} \quad (6)$$

$$R_3 \; - \; N_2 \rightleftharpoons N + N : x_N = A_3 \sqrt{x_{N_2}}, \quad A_3 = \sqrt{K_{p3}/(p/p_{ref})} \quad (7)$$

$$R_4 \; - \; H_2O \rightleftharpoons H_2 + 0.5 O_2 : x_{H_2} = A_4 \frac{x_{H_2O}}{\sqrt{x_{O_2}}}, \quad A_4 = K_{p4}/\sqrt{(p/p_{ref})} \quad (8)$$

$$R_5 \; - \; H_2O \rightleftharpoons 0.5 H_2 + POH : x_{OH} = A_5 \frac{x_{H_2O}}{\sqrt{x_{H_2}}}, \quad A_5 = K_{p5}/\sqrt{(p/p_{ref})} \quad (9)$$

$$R_6 \; - \; CO_2 \rightleftharpoons CO + 0.5 O_2 : x_{CO} = A_6 \frac{x_{CO_2}}{\sqrt{x_{O_2}}}, \quad A_6 = K_{p6}/\sqrt{(p/p_{ref})} \quad (10)$$

$$R_7 \; - \; 0.5 O_2 + 0.5 N_2 \rightleftharpoons NO : x_{NO} = A_7 \sqrt{x_{N_2} x_{O_2}}, \quad A_7 = K_{p7} \quad (11).$$

These eleven equations (along with Eq. (A) for n_{tot}) are now used to determine the eleven mole fractions. Through algebraic manipulations, it is possible to derive three main equations, which must be solved iteratively. These are

$$x_{N_2} + \frac{1}{2} \left[A_7 \sqrt{x_{O_2}} + A_3 \right] \sqrt{x_{N_2}} - 3.76 a^* = 0, \quad (4.35)$$

$$2 \left[\sqrt{x_{O_2}} + A_4 \right] x_{H_2O} + \left[A_1 \sqrt{A_4} x_{O_2}^{0.25} + \frac{A_5}{\sqrt{A_4}} x_{O_2}^{0.75} \right] \sqrt{x_{H_2O}}$$
$$- n^* \sqrt{x_{O_2}} = 0, \quad (4.36)$$

$$\text{and } 2 x_{O_2} + \left[A_7 \sqrt{x_{N_2}} + A_2 \right] \sqrt{x_{O_2}} + \frac{A_5}{\sqrt{A_4}} \sqrt{x_{H_2O}} x_{O_2}^{0.25}$$
$$+ \frac{m^* \sqrt{x_{O_2}}}{A_6 + \sqrt{x_{O_2}}} + \left[x_{H_2O} + m^* - 2 a^* \right] = 0. \quad (4.37)$$

Equations 4.35 and 4.36 are quadratic, whereas Eq. 4.37 is transcendental and therefore must be solved by the Newton–Raphson procedure. Simultaneous solution of the three equations yields x_{N_2}, x_{H_2O}, and x_{O_2}. The remaining eight mole-fractions can therefore be evaluated using Eqs. (1) to (11).

Problem 4.6 Determine product composition for $C_{10}H_{22}$ in the range[5] $0.6 < \Phi < 1.4$ using (a) partial-equilibrium mechanism and (b) full-equilibrium mechanism.

For partial-equilibrium, and $\Phi \leq 1$, the product composition can be readily evaluated using the procedure described in Problem 3.2 of Chap. 3. For $\Phi > 1$, the procedure set out in Problem 4.5 is used with $T_{wg} = 2200$ K and p $= 1$ atm. In Fig. 4.2, these predicted compositions are shown in the top figure using dotted lines. Note that x_{CO_2} rises for $\Phi < 1$ but then falls for $\Phi > 1$ as a result of dissociation via the water-gas reaction. $x_{CO} = 0$ for $\Phi < 1$, as it must be, but then appears for $\Phi > 1$. The same holds for x_{H_2}. x_{O_2} declines steadily with Φ for $\Phi < 1$ and then disappears for the rich mixture. x_{H_2O}, however, steadily increases with Φ, whereas x_{N_2} declines very slowly with Φ.

For full equilibrium, the product compositions are tabulated in Table 4.6 and shown by solid lines in Fig. 4.2. The top figure shows variations of the major components. The figure confirms that the values predicted by the partial equilibrium assumption are in close agreement with those predicted by the full equilibrium assumption. For this reason, the simpler partial equilibrium assumption is invoked for engineering applications. However, the bottom figure shows that the full equilibrium model is also capable of predicting minor species, and their mole fractions are extremely small. That is, they are in ppm proportions.[6] Notice that NO and OH, though small, assume significant proportions so as to be of concern for environmental considerations. Also, near $\Phi = 1$, NO concentrations fall rapidly on the fuel-rich side. This fact is used in IC engines, in which equivalence ratio is maintained near 1.03 to 1.05.

In general, we may say that the partial equilibrium assumption would suffice if we were interested only in energy calculations. However, if environmental considerations are also important, full equilibrium assumption must be invoked.

Problem 4.7 Using the product composition for $C_{10}H_{22}$ combustion shown in Table 4.6, evaluate the AFT for each Φ. Assume reactants at 1 atm and 298 K.

In this problem, $h_{pr}(T_{ad}) = h_{react}(298)$. Hence, from the property tables,

$$\sum n_j \, \Delta\bar{h}_s \, (T_{ad}) = H_{react} \, (298) - \sum n_j \, \overline{h^0}_{f,j}$$
$$= -300900 + 393520 \, n_{CO_2} + 110530 \, n_{CO}$$
$$+ 241830 \, n_{H_2O} - 38990 \, n_{OH} - 218000 \, n_H$$
$$- 90291 \, n_{NO} - 249170 \, n_O \, ,$$

where $n_j = x_j \times n_{tot}$ are known. This equation must be solved by trial and error for each Φ in the manner of Problem 3.8 in Chap. 3. The predicted values of T_{ad} are shown in the last column of Table 4.6. These values are compared with those evaluated using partial equilibrium in Fig. 4.3. For the lean mixture, T_{ad} increases

---

[5]The importance of this range will be appreciated later, when flammability limits are considered in Chap. 8.
[6]ppm denotes *parts per million*. 1 ppm $= 10^6 \, x$ where x is the mole fraction.

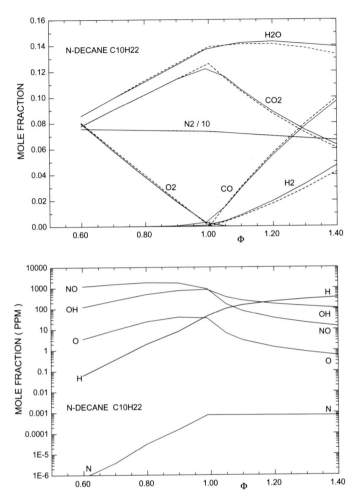

Fig. 4.2 Variation of equilibrium product composition with Φ for $C_{10}H_{22}$ - partial equilibrium (dotted line) - full equilibrium (solid line)

with Φ. The maximum T_{ad} occurs at $Φ \simeq 1.01$. In the rich mixture, T_{ad} decreases with increase in Φ. This is a typical behavior when full equilibrium products are used to evaluate AFT.

Similarly, T_{ad} can also be evaluated for the product compositions determined for partial equilibrium. These evaluations are also shown in Fig. 4.3 by dotted line. The behavior is similar, but $T_{ad,max}$ now occurs exactly at $Φ = 1$.

Table 4.6 Product composition and AFT for $C_{10}H_{22}$ combustion - full equilibrium (Problem 4.6)

Φ	x_{CO_2}	x_{CO}	x_{H_2O}	x_{H_2}	x_{O_2}	x_{N_2}
0.600E+00	0.779E−01	0.463E−05	0.857E−01	0.174E−05	0.799E−01	0.757
0.700E+00	0.903E−01	0.217E−04	0.992E−01	0.717E−05	0.592E−01	0.751
0.800E+00	0.102E+00	0.106E−03	0.113E+00	0.307E−04	0.388E−01	0.746
0.900E+00	0.114E+00	0.427E−03	0.126E+00	0.114E−03	0.189E−01	0.741
0.990E+00	0.123E+00	0.289E−02	0.137E+00	0.713E−03	0.307E−02	0.739
0.101E+01	0.123E+00	0.528E−02	0.139E+00	0.132E−02	0.921E−03	0.738
0.102E+01	0.121E+00	0.743E−02	0.139E+00	0.189E−02	0.452E−03	0.733
0.103E+01	0.119E+00	0.100E−01	0.139E+00	0.258E−02	0.241E−03	0.730
0.104E+01	0.117E+00	0.128E−01	0.139E+00	0.336E−02	0.143E−03	0.728
0.105E+01	0.115E+00	0.156E−01	0.139E+00	0.418E−02	0.928E−04	0.726
0.110E+01	0.106E+00	0.296E−01	0.140E+00	0.866E−02	0.217E−04	0.716
0.120E+01	0.886E−01	0.552E−01	0.139E+00	0.191E−01	0.439E−05	0.698
0.130E+01	0.745E−01	0.775E−01	0.136E+00	0.312E−01	0.158E−05	0.681
0.140E+01	0.629E−01	0.968E−01	0.131E+00	0.445E−01	0.719E−06	0.665
Φ	O (PPM)	H (PPM)	N (PPM)	OH (PPM)	NO (PPM)	T_{ad} (K)
0.600E+00	0.354E+01	0.630E−01	0.446E−06	0.122E+03	0.120E+04	1710.40
0.700E+00	0.946E+01	0.348E+00	0.375E−05	0.251E+03	0.154E+04	1896.27
0.800E+00	0.238E+02	0.198E+01	0.316E−04	0.495E+03	0.186E+04	2072.00
0.900E+00	0.390E+02	0.809E+01	0.156E−03	0.753E+03	0.175E+04	2238.60
0.100E+01	0.368E+02	0.432E+02	0.773E−03	0.859E+03	0.952E+03	2366.79
0.101E+01	0.201E+02	0.588E+02	0.773E−03	0.641E+03	0.521E+03	2381.25
0.102E+01	0.141E+02	0.703E+02	0.770E−03	0.536E+03	0.364E+03	2380.05
0.103E+01	0.103E+02	0.823E+02	0.769E−03	0.458E+03	0.265E+03	2374.91
0.104E+01	0.794E+01	0.938E+02	0.768E−03	0.403E+03	0.204E+03	2367.87
0.105E+01	0.639E+01	0.105E+03	0.766E−03	0.362E+03	0.164E+03	2359.96
0.110E+01	0.309E+01	0.151E+03	0.761E−03	0.252E+03	0.788E+02	2317.03
0.120E+01	0.139E+01	0.224E+03	0.752E−03	0.168E+03	0.350E+02	2230.19
0.130E+01	0.834E+00	0.286E+03	0.742E−03	0.129E+03	0.207E+02	2146.11
0.140E+01	0.563E+00	0.351E+03	0.734E−03	0.107E+03	0.139E+02	2065.12

4.5.4 Constant-Volume Combustion

So far we have considered constant-pressure combustion as in a steady-flow device. However, in constant volume combustion, both pressure and temperature change and it is of interest to determine equilibrium product composition. Thus, we consider the following problem.

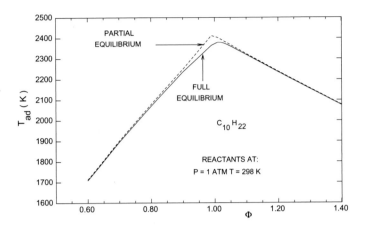

Fig. 4.3 Variation of adiabatic flame temperature of $C_{10}H_{22}$ with Φ

Problem 4.8 Determine the product composition following adiabatic constant-volume combustion of CO in *moist* air ($x_{H_2O} = 0.02$). The initial temperature and pressure are 1000 K and 1 atm. Assume partial equilibrium products with H_2O and N_2 taking no part in reaction.

To solve the problem, we first determine the initial reactant moles.

$$\text{Reactants} = CO + \frac{0.5}{\Phi} (O_2 + 3.76\, N_2) + n_{H_2O}\, H_2O$$

Because the initial mole fraction of H_2O is given, it is easy to deduce that,

$$n_{H_2O} = \left(\frac{x_{H_2O}}{1 - x_{H_2O}}\right) \times \left(1 + \frac{2.38}{\Phi}\right) = 0.020408 \left(1 + \frac{2.38}{\Phi}\right).$$

Further, H_2O is assumed to take no part in reaction, and hence, the postulated products are

$$\text{Products} \rightarrow n_{CO_2}\, CO_2 + n_{CO}\, CO + n_{O_2}\, O_2$$
$$+ n_{H_2O}\, H_2O + \frac{0.5 \times 3.76}{\Phi}\, N_2 \qquad (4.38)$$

Thus, the element balance gives product mols as,

$$n_{CO} = 1 - n_{CO_2} ,$$

$$n_{O_2} = \frac{0.5}{\Phi} - 0.5\, n_{CO_2} ,$$

$$n_{tot,i} = 1 + \frac{2.38}{\Phi} + n_{H_2O}$$

$$n_{tot,pr} = n_{tot,i} - 0.5\, n_{CO_2} , \tag{4.39}$$

where subscripts i and pr refer to reactants and products, respectively, and n_{CO_2} is determined from equilibrium relation for CO-oxidation reaction as

$$K_p = \frac{x_{CO_2}}{x_{CO}\sqrt{x_{O_2}}} \left(\frac{p_{pr}}{p_{ref}}\right)^{-0.5} = \frac{n_{CO_2}}{n_{CO}} \sqrt{\frac{n_{tot,pr}}{n_{O_2}}} \left(\frac{p_{pr}}{p_{ref}}\right)^{-0.5} , \tag{4.40}$$

where p_{pr} is product pressure. It is related to product temperature T_{pr} via the ideal gas law. Thus,

$$\rho = \left(\frac{p}{R_{mix}\, T}\right)_{pr} = \left(\frac{p}{R_{mix}\, T}\right)_i . \tag{4.41}$$

R_{mix} changes because number of moles change from $n_{tot,i}$ to $n_{tot,pr}$ in this reaction. Therefore, the product composition must be determined by iteration. The procedure is as follows.

1. For Given p_i and T_i, determine $n_{tot,i}$, $M_{mix,i}$, and $R_{mix,i}$.
2. Assume n_{CO_2} and T_{pr}.
3. Hence, evaluate $n_{tot,pr}$, n_{O_2}, and n_{CO} from Eq. 4.39. n_{H_2O} and n_{N_2} remain unchanged. Hence, evaluate $M_{mix,pr}$, and $R_{mix,pr}$.
4. Determine p_{pr} from Eq. 4.41, and K_p (T_{pr}) for CO_2 dissociation from Appendix B (Eq. B.4 and Table B.3).
5. Determine n_{CO_2} (new) from Eq. 4.40 by the Newton–Raphson procedure, and hence, determine $n_{tot,pr}$ and product composition again.
6. Evaluate absolute fractional changes in n_{CO_2} and ensure that these are less than a convergence criterion.
7. If the criterion is not satisfied, update T_{pr} such that internal energy is conserved (that is, $U_i = U_{pr}$ for constant-volume adiabatic oxidation), and return to step 3 with new values n_{CO_2} and T_{pr}.
8. Upon convergence, evaluate $x_{j,pr}$, and hence, $\omega_{j,pr}$.

Table 4.7 shows the computed values for three values of equivalence ratio Φ. The reader should compare values for $\Phi = 0.5$ with those estimated from kinetic considerations in Chap. 7 (Problem 7.4, Table 7.5).

Table 4.7 Constant-volume combustion product composition (Problem 4.8)

Property	$\Phi = 0.25$	$\Phi = 0.5$	$\Phi = 1.0$
T_{pr} (K)	1941.21	2516.12	2838.13
p_{pr} (bar)	1.851	2.318	2.560
K_p	1269.5	25.24	5.706
$M_{mix,pr}$	29.937	30.91	31.46
$n_{tot,i}$	10.7347	5.877	3.4489
$n_{tot,pr}$	10.2354	5.415	3.1111
ω_{CO_2}	0.1433	0.2428	0.3037
ω_{CO}	0.0001355	0.0127	0.09278
ω_{O_2}	0.1567	0.1028	0.0521
ω_{H_2O}	0.0126	0.0126	0.0126
ω_{N_2}	0.6871	0.629	0.5378
$(U_{pr} - U_i)/U_i$	0.0014	0.0045	0.006

Exercises

1. Show that Gibbs function of formation of CO_2 at $T \neq T_{ref}$ can be evaluated as

$$\overline{g}^0{}_{f,CO_2}(T) = \overline{h}^0{}_{f,CO_2}(T_{ref}) + \left[\Delta\overline{h}_{s,CO_2} - \Delta\overline{h}_{s,C} - \Delta\overline{h}_{s,O_2}\right]_T$$
$$- T\left[S^0_{CO_2} - S^0_C - S^0_{O_2}\right]_T$$

2. Derive K_p for the reverse reaction in Eq. 4.32 from the first principles and show that it equals K_p^{-1} for the forward reaction.

3. Derive Eq. 4.33. Now, consider dissociation reaction $CO_2 \rightleftharpoons CO + 0.5\,O_2$ and evaluate $K_p(T_1)$ at $T_1 = 2000$ K from Table 4.1. Hence, evaluate $K_p(T_2)$ at $T_2 = 2105$ K using Eq. 4.33. Compare this value with that interpolated between 2000 K and 2200 K from Table 4.1.

4. Evaluate the product composition for $C_{10}H_{22}$ combustion using *partial equilibrium* when reactants are at 1 atm and 298 K. Use these compositions to evaluate AFT for each Φ. For $\Phi > 1$, evaluate K_p at $T_{wg} = 2200$ K. Compare these values with those shown in Table 4.6 and Fig. 4.3.

5. Repeat the above problem for $p = 5$ and 10 atm using *full equilibrium*. Compare product compositions and comment on the result.

6. Repeat Problems 4.6 and 4.7 in the text for propane (C_3H_8) combustion.

7. 1 mole of CO is mixed with 1 mole of water vapor at 298 K and 1 atm pressure. The mixture is heated to 1800 K at constant pressure. Calculate the heat required and the final composition of the mixture.

8. Natural gas fired boiler operates with excess air such that O_2 concentration in the flue gas *after* removal of moisture is 2 percent by volume. The flue gas temperature is 700 K and the fuel and air enter the boiler at 298 K. Determine (a) Air-Fuel ratio and (b) the thermal efficiency of the boiler.

9. Calculate the equilibrium composition of reaction $H_2 + \frac{1}{2} O_2 \rightleftharpoons H_2 O$ when the ratio of number of moles of elemental hydrogen and elemental oxygen is unity. The temperature is 2000 K and $p = 1$ atm.

10. A 15 percent rich mixture of propane and air is supplied to a gas engine at 1 atm and 298 K. The compression ratio is 7:1 and the compression stroke is adiabatic. Combustion takes place at constant volume; the products of combustion contain CO_2, H_2O, CO, H_2, N2, and small traces of O_2 at 2100 K. Determine the product composition. (Hint: To explain the presence of H_2, CO and O_2 in the products, the reaction mechanism comprises the water-gas shift reaction and CO_2 dissociation reaction.)

11. Develop a computer program for a fuel-rich (STP and $\Phi > 1$) combustion of any hydrocarbon $C_m H_n$ allowing for water-gas shift reaction (that is, partial equilibrium assumption) whose K_p value is evaluated using curve-fit coefficients given in Eq. B.4 and Table B.3 at prescribed temperature T_{wg} (in the range 2000 to 2400 K). The program should evaluate product composition, AFT and ΔH_c for each Φ.

12. Repeat Problem 4.8 in the text for $\Phi = 0.5$ assuming heat loss $q_{loss} = F u_i$ where $0 \leq F \leq 0.5$. Plot variation of final T_{pr}, p_{pr}, K_p, and $\omega_{j,pr}$ as a function of F.

13. Evaluate product composition, AFT, and final pressure for constant-volume combustion of C_3H_8 initially at STP and $\Phi = 1.0$. Assume Full equilibrium. Compare your results with exercise Prob 20 in Chap. 3.

Chapter 5
Chemical Kinetics

5.1 Importance of Chemical Kinetics

In the previous chapter, we analyzed chemical reactions in the manner of thermodynamics—that is, by considering the *before* and *after* of a chemical reaction. This analysis did not consider the *rate* of change of states during the process. Nonetheless, thermodynamic analysis coupled with the chemical equilibrium considerations enabled us to evaluate useful quantities, such as the (a) equilibrium composition of the *postulated* products, (b) heat of reaction or the calorific value of a fuel, and (c) adiabatic flame temperature.

In this chapter, we introduce the empirical science of *chemical kinetics* whose precise purpose is to determine the *rate of a chemical reaction*—in other words, the time rates of *depletion of reactants* and of *formation of products*. In this sense, the departure from chemical equilibrium thermodynamics to chemical kinetics maybe viewed as analogous to the familiar departure made from thermodynamics to the empirical science of heat transfer. The latter, again, deals with *rates of heat transfer* across a finite temperature difference.

The science of heat transfer introduces rate quantities such as the *heat transfer coefficient* α, having units W / m^2-K and the transport property *thermal conductivity* k, having units W/m-K. Both these quantities are used in the design of practical equipment. Thus, with the knowledge of α (which may be obtained from a correlation, see for example, Ref. [51]), we are able to *size* a heat exchanger by estimating the heat transfer surface area, determine the time required for a cooling process, or ensure that the temperature of solid elements in a furnace or a nuclear reactor is always well below safe limits. Sizing (or, shaping), safety and economics are the three main considerations in the design of thermo-fluid equipment. The science of heat transfer, often coupled with the science of fluid mechanics helps us to meet these considerations.

The science of chemical kinetics has the same practical purpose in relation to combustion equipment. Thus, the knowledge of chemical reaction rates enable sizing of a combustion chamber so the required heat is released within the chamber and the

© Springer Nature Singapore Pte Ltd. 2020
A. W. Date, *Analytic Combustion*,
https://doi.org/10.1007/978-981-15-1853-9_5

products of combustion will have the *desired* composition and temperature. Sizing is important to ensure *compactness*, as in an aircraft engine, or for economic design. Knowledge of product composition and temperature are important because they may affect downstream processes, or simply for environmental considerations that require that pollutant (NO_x, SO_2, CO, CO_2, O_3 etc) concentrations are within permissible limits.

The purpose of chemical kinetics is thus to determine the rates of reactant depletion and product formation and the quantities on which the rates depend. The chemical kinetic rates are specified in *volumetric* terms as kg/m^3-s or $kmol/m^3$-s and are denoted by the symbol R. The empirical science of chemical kinetics derives greatly from the *kinetic theory of gases* and from *chemistry*. Thus, the scope of science is very vast. For our purpose of assisting in the design of practical combustion equipment, however, we shall touch on kinetic theory and chemistry only in the bare necessities. Just as we determine the heat transfer coefficient from a correlation, the value of R will be determined from a *reaction rate formula* that will specify the appropriate quantities on which R depends.

Of course, the *true* product composition in a practical combustion equipment will also be influenced by *fluid mixing*. This matter will be taken up in subsequent chapters. For the purposes of the present chapter, though, it will be assumed that fluid mixing is perfect.

5.2 Reformed View of a Reaction

When a fuel is burned, the *net-effect* reaction can be expressed as

$$1 \text{ mole of fuel} + a_{stoic} \text{ moles of oxidant} \rightarrow \text{ products.} \qquad (5.1)$$

This net-effect reaction is called the *global* reaction, in which the symbol $\rightarrow$ signifies the direction of the reaction, implying that the reactants fuel and oxygen in oxidant air *combine* to yield products. Thus, if the fuel was ethane (C_2H_6), the reaction informs that 1 mole of ethane will *combine* with $a_{stoich} = (2 + 6/4) = 3.5$ moles of O_2 in the oxidant air to form products.

In contrast, however, the science of chemistry informs us that *no encounter* between one ethane molecule and 3.5 oxygen molecules really ever takes place. Then how does a reaction get executed at all? The answer is that the reaction is executed through *intermediates*. Each atom from the reactant molecule enters into one or more intermediate *compounds*, as a rule, before taking its final place in a product molecule. The intermediate compounds are of two types.

Stable: $CH_4, C_2H_2, CO, CO_2, H_2O, H_2$, etc.

Unstable: H, OH, O, C, C_2, CH, N, etc.

Here, *unstable* means that the species is *incapable of existing in equilibrium at normal temperature and pressure* in significant proportions.[1] Thus, if our interest was only in realizing the heat released, then knowledge of the *stable* products having significant proportions suffices. However, if our interest was in knowing pollutant formation or in determining the *colors* acquired by a flame then knowledge of proportions of unstable species becomes important.

The great majority of molecular encounters that result in a chemical reaction involve at least one radical species. Such reactions are called *elementary* reactions or *reaction steps*. Typical ones are:

$$O_2 + H \rightleftharpoons OH + O \, ,$$

$$H_2 + OH \rightleftharpoons H + H_2O \, ,$$

$$H_2 + O \rightleftharpoons 2OH \, ,$$

$$\text{and } O_2 + N \rightleftharpoons NO + O \, .$$

Large molecules, such as ethane, may react with radicals directly or they may break (or dissociate) into smaller molecules such as CH_4, C_2H_2, and the like. The global reaction is meant to be executed through many such reaction steps.

Specification of the global reactions along with the many elementary reactions (that define the path of the global reaction) constitutes the *reaction mechanism*. The methane (CH_4) reaction mechanism, for example, is the most extensively studied. The global reaction is

$$CH_4 + 2\,(O_2 + 3.76\,N_2) \rightarrow CO_2, H_2O, \ldots \text{ others.}$$

For engineering purposes, twenty-two elementary reactions given in Table 5.1 are considered adequate. The first sixteen reactions involve elements C, H, and O.[2] The last six reactions involve elements N, H, and O. Thus, there are twenty-two elementary reactions and sixteen species. This is the simplest reaction mechanism for CH_4 combustion. The more elaborate mechanism involves as many as 149 elementary reactions and thirty-three species [40].

The overall implication is that although the intermediate species play a vital role in combustion reactions, they do not by themselves attain concentrations of the same order of magnitude as the main reactants (CH_4 and O_2) and the main products (CO_2, H_2O). In a sense, this mechanism is similar to the one postulated for the *full-equilibrium* calculation of hydrocarbon reactions in the previous chapter. The difference is that, whereas in that chapter the proportions of minor species were

---

[1] We have already been informed about the smallness of proportions through the chemical equilibrium analysis of the previous chapter. In fact, the proportions were found in ppm or with $w_j \simeq 10^{-6}$.

[2] Species M in these reactions stands for *inert* species, whose presence is necessary to enable the reaction to take place.

Table 5.1 Elementary reactions in CH_4 reaction mechanism

No.	Reaction
1	$CH_4 + M \rightleftharpoons CH_3 + H + M$
2	$CH_4 + O \rightleftharpoons CH_3 + OH$
3	$CH_4 + H \rightleftharpoons CH_3 + H_2$
4	$CH_4 + OH \rightleftharpoons CH_3 + H_2O$
5	$CH_3 + O \rightleftharpoons HCHO + H$
6	$CH_3 + O_2 \rightleftharpoons HCHO + OH$
7	$CH_3 + O_2 \rightleftharpoons HCO + H_2O$
8	$HCO + OH \rightleftharpoons CO + H_2O$
9	$HCHO + OH \rightleftharpoons H + CO + H_2O$
10	$CO + OH \rightleftharpoons CO_2 + H$
11	$H + O_2 \rightleftharpoons OH + O$
12	$O + H_2O \rightleftharpoons 2\,OH$
13	$H + H_2O \rightleftharpoons OH + H_2$
14	$H_2O + M \rightleftharpoons H + OH + M$
15	$HCO + M \rightleftharpoons H + CO + M$
16	$O + H_2 \rightleftharpoons H + OH$
17	$O + N_2 \rightleftharpoons NO + N$
18	$N + O_2 \rightleftharpoons NO + O$
19	$N_2 + O_2 \rightleftharpoons N_2O + O$
20	$N_2O + O \rightleftharpoons 2\,NO$
21	$N + OH \rightleftharpoons NO + H$
22	$N_2 + OH \rightleftharpoons N_2O + H$

evaluated from equilibrium considerations, now they will be evaluated from *rate* or kinetic considerations.

5.3   Reaction Rate Formula

5.3.1   Types of Elementary Reactions

Elementary reactions are of three types:

1. Collision-controlled bimolecular;
2. Decomposition-controlled or dissociation-controlled; and
3. Trimolecular.

Most elementary reactions are of the first type. In this type, molecules A and B *collide* and, *if their energy levels and orientations* are right, they react *immediately*. Thus, the collision-controlled bimolecular reaction is written as

$$A + B \rightarrow C + D \, .$$

In the dissociation reaction, molecules of A collide with *inert* molecules M and, *if their energy levels and orientations* are right, they decompose *immediately*. Thus,

$$A + M \rightarrow C + D + M \, .$$

Although the reaction between A and M is written as bimolecular, in effect it is only *unimolecular* $A \rightarrow C + D$, because M appears on both the reactant and the product sides. The reaction represents the *dissociation* of A into C and D. Dissociation reactions are usually unimolecular. In both the uni- and bimolecular reaction types, *immediately* means *in a time that is much shorter than the average time between successive molecular collisions.*

Finally, the *trimolecular* reactions are written as

$$A + B + M \rightarrow C + M \, .$$

These are usually *reverse* reactions of unimolecular dissociation reactions. They are also usually very slow but are important for forming stable compounds. These descriptions can be verified from the reactions given in Table 5.1. The reverse reactions 12 and 20 are thus unimolecular, whereas the reverse reaction 14 is trimolecular forming a stable compound H_2O. All others are bimolecular.

In the next subsection, the rate-formula is derived for the commonly occurring bimolecular reaction using some results from the kinetic theory of gases.

5.3.2 Rate Formula for $A + B \rightarrow C + D$

In this bimolecular reaction, let the concentration of A be given by

$$[A] = \frac{\rho_A}{M_A} \qquad \frac{\text{kmol}}{\text{m}^3} \, . \tag{5.2}$$

Then, the rate of reaction of A, [R_A] is given by

$$[R_A] = - \frac{d \, [A]}{dt} \qquad \frac{\text{kmol}}{\text{m}^3\text{-s}} \, , \tag{5.3}$$

or, in mass units,

$$R_A = M_A \, [R_A] \qquad \frac{\text{kg}}{\text{m}^3\text{-s}} \, . \tag{5.4}$$

The negative sign in Eq. 5.3 indicates that reactant A is being *depleted*.

From kinetic theory of gases, the average molecular velocity is given by

$$\bar{u}_{mol} = \sqrt{\frac{8}{\pi} \frac{R_u}{M} T} \quad \frac{m}{s} .$$ (5.5)

Our interest now is to estimate the probable number of collisions between molecules of A and B per unit volume of the gas per unit time. This number of collisions (or, collision frequency) is denoted by Z (number of collisions / m^3-s). To estimate[3] this number, we proceed as follows.

Let one molecule of A, during its flight, sweep out a volume $\dot{V}_{swept}$ in unit time. It can, therefore, be expressed as

$$\dot{V}_{swept,A} = \frac{\pi}{4} d_A^2 \bar{u}_{mol} = \frac{\pi}{4} d_A^2 \sqrt{\frac{8}{\pi} \frac{R_u}{M_A} T} \quad \frac{m^3}{s} ,$$ (5.6)

where d_A is the *hard-sphere* diameter of molecule A. Therefore, the rate ($\dot{N}_B$) of number of B-molecules *likely to be encountered* by one molecule of A can be estimated as

$$\dot{N}_B = n_B^* \dot{V}_{swept,A} = n_B^* \frac{\pi}{4} d_A^2 \sqrt{\frac{8}{\pi} \frac{R_u}{M_A} T} \quad \frac{\text{molecules}}{s} ,$$ (5.7)

where n_B^* (molecules/m^3) represents number of molecules of B per unit volume. Therefore, the total number of collisions between A and B can be estimated as

$$Z \propto n_A^* \times \dot{N}_B \qquad \frac{\text{collisions}}{m^3\text{-s}}$$ (5.8)

$$\propto n_A^* n_B^* \frac{\pi}{4} d_A^2 \sqrt{\frac{8}{\pi} \frac{R_u}{M_A} T} \quad \frac{\text{molecules of A}}{m^3\text{-s}} ,$$ (5.9)

where n_A^* represents the number of molecules of A per unit volume and, therefore, also equals the number of collisions. The actual rate, however, will also be influenced by d_B and M_B. Therefore, in general, the rate of collisions is calculated from the formula

$$Z = \frac{n_A^* n_B^*}{S_{AB}} \pi d_{AB}^2 \sqrt{\frac{8}{\pi} \frac{R_u}{M_{AB}} T} \quad \frac{\text{molecules of A}}{m^3\text{-s}} ,$$ (5.10)

where the mean molecular diameter d_{AB} and mean molecular weight M_{AB} are given by

$$d_{AB} = \frac{1}{2}(d_A + d_B) \quad M_{AB} = \frac{M_A M_B}{M_A + M_B} .$$ (5.11)

---

[3]The collision between molecules is a probabilistic phenomenon, hence the use of the word *estimate* rather than *determine*.

S_{AB} is called the *symmetry number*. It is given by

$$S_{AB} = 2 \quad \text{if} \quad A = B$$
$$= 1 \quad \text{if} \quad A \neq B . \tag{5.12}$$

The symmetry number simply acknowledges that for encounters between molecules of the same gas ($A = B$), the number of collisions must be half the number of collisions as when A and B molecules are of different gas species.[4]

The reaction rate $[R_A]$ must be *proportional* to Z. To be dimensionally correct, the rate formula must take the form

---

[4]Note following two points in respect of Eq. 5.10

1. The molecular diameters are a function of temperature. This can be inferred from the formula for viscosity of a gas given by the kinetic theory.

$$\mu_g = \frac{5}{16} \left[\frac{M_g \, R_u \, T}{\pi} \right]^{0.5} \frac{1}{N_{avo} \, d_g^2} \quad \frac{\text{N-s}}{\text{m}^2} . \tag{5.13}$$

It is well known from experimentally measured data that $\mu_g \propto T^m$ where $m > 0.5$. Thus, for air, for example, $m \simeq 0.62$. From Eq. 5.13, it is apparent that this dependence ($\mu_g \sim T$) is possible only if d_g is a function of *temperature*. In fact, d_g must *decrease* with an increase in temperature.

2. The value of number of molecules per unit volume n^* can be recovered from the ideal gas relation $p \, v = R_g \, T$ as follows:

$$p = \frac{R_u \, T}{v_g \, M_g} = \left[\frac{R_u}{N_{avo}} \right] \left[\frac{N_{avo}}{v_g \, M_g} \right] T \tag{5.14}$$
$$= \kappa_B \, n_g^* \, T ,$$

where $N_{avo} = 6.022 \times 10^{26}$ molecules/kmol is the Avogadro's number and $\kappa_B = 8314/6.022 \times 10^{26} = 1.38 \times 10^{-23}$ J/molecules-K (Boltzman's constant). Thus, knowing p and T, n^* can be determined. Incidentally, n^* can also be written as

$$n_g^* = \frac{N_{avo}}{v_g \, M_g} = \frac{\rho_g}{M_g} \, N_{avo} = [g] \, N_{avo} . \tag{5.15}$$

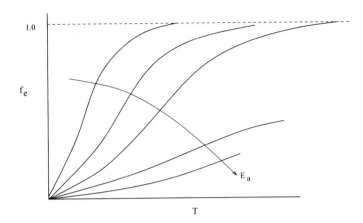

Fig. 5.1 Variation of the energy factor f_e with temperature

$$[R_A] \propto \frac{Z}{N_{avo}} \left[\frac{\text{no of molecules of A}}{\text{m}^3\text{-s}}\right] \left[\frac{\text{kmol of A}}{\text{no of molecules of A}}\right]$$

$$= (f_s \ f_e) \frac{Z}{N_{avo}} \frac{\text{kmol}}{\text{m}^3\text{-s}} \ , \tag{5.16}$$

where the constant of proportionality is taken as $(f_s \ f_e)$. f_s is called the *steric or the orientation factor* and f_e is called the *energy factor*. It becomes necessary to introduce these dimensionless factors because not all collisions between A and B will produce a reaction; only a fraction of them will. Both f_e and f_s are thus fractions.

Energy Factor f_e

In kinetic theory, the energy factor f_e is associated with the idea that for a chemical reaction to take place, the collision energy between molecules of A and B *must exceed* a threshold value E_a, which is also known as the *activation energy* . Because the molecular velocities increase with temperature (see Eq. 5.5), the collision energy also must increase with temperature and, therefore, f_e must increase with temperature. These ideas are embodied in the following mathematical expression:

$$f_e = \exp\left(-\frac{E_a}{R_u T}\right) \ , \tag{5.17}$$

where the activation energy E_a is a function of reactant molecules A and B under consideration. Its value is thus different for different reactants. The variation of f_e with T is shown in Fig. 5.1 with E_a as a parameter.

When E_a is small or T is very large, $f_e \rightarrow 1$, indicating that all collisions have exceeded the threshold value. When E_a is large or T is small, however, $f_e \rightarrow 0$. As such, practically none of the collisions has exceeded the threshold value and the probability of a reaction is very small.

Orientation Factor f_s

The orientation factor f_s embodies the idea that it is not sufficient that the collision energy levels are high (or $> E_a$). In addition, the orientation of the molecules must also be correct. In other words, the molecules must collide with each other in *sensitive spots* for a reaction to occur at all. Thus, f_s is obviously a probabilistic factor.[5]

Substituting expressions for Z and f_e and noting the definition in Eq. 5.15; Eq. 5.16 can be written as

$$[R_A] = -\frac{d\,[A]}{dt} = k_{bi}\,(T)\,[A]\,[B] \qquad \frac{\text{kmol of A}}{\text{m}^3\text{-s}}\,, \qquad (5.18)$$

where

$$k_{bi}\,(T) = \left[f_s\,\frac{N_{avo}}{S_{AB}}\,\pi\,d_{AB}^2\,\sqrt{\frac{8}{\pi}\,\frac{R_u}{M_{AB}}\,T}\,\right]\,\exp\left(-\frac{E_a}{R_u\,T}\right) \qquad (5.19)$$

$$= A\,T^b\,\exp\left(-\frac{E_a}{R_u\,T}\right)\,\frac{\text{m}^3}{s\text{-kmol of A}}\,. \qquad (5.20)$$

The values of A, b and E_a are typically determined from experimental data.[6] k_{bi} (T) is called the *rate coefficient for the bimolecular reaction* and A is called the preexponential factor. Typical values of k_{bi} are given in Table 5.2. The table shows that k_{bi} for the forward and backward reactions are different. As such, the species concentrations will be influenced by the *net-effect* of magnitudes of k_{bi} during forward and backward reactions.

To express the rate in *mass units*, Eq. 5.20 must be multiplied by the molecular weight M_A (see Eq. 5.4). Thus,

$$R_A = M_A\,[R_A] = -M_A\,\frac{d\,[A]}{dt} = -\frac{d\,\rho_A}{dt}$$

$$= M_A\,k_{bi}\,[A]\,[B]$$

$$= M_A\,k_{bi} \times \frac{\rho_A}{M_A} \times \frac{\rho_B}{M_B}$$

$$= \frac{k_{bi}}{M_B}\,\rho_A\,\rho_B = \frac{k_{bi}}{M_B} \times \frac{\rho_A}{\rho_{mix}} \times \frac{\rho_B}{\rho_{mix}} \times \rho_{mix}^2$$

$$= \frac{k_{bi}}{M_B}\,\rho_{mix}^2\,\omega_A\,\omega_B\,, \qquad (5.21)$$

where the mixture density is given by $\rho_{mix} = (p\,M)_{mix}/(R_u\,T)$.

---

[5]The reader may imagine an encounter between two standard soccer balls, which have white and black patches on them. If the balls are thrown at each other, then clearly the event that two black patches (considered sensitive) will hit each other is probabilistic.

[6]The value of E_a (which is a function of reactant molecules A and B) can also be determined from advanced theories that postulate a change in the structure of molecules during breaking and forming of chemical bonds.

Table 5.2 Rate coefficient k_{bi} for bimolecular Reaction (Eq. 5.20). Temperature range: 300–2500 K — E_a (J/kmol) [122]

No.	Reaction	A	b	E_a
1	$H + O_2 \rightarrow OH + O$	1.2×10^{14}	-0.91	6.91×10^4
2	$OH + O \rightarrow O_2 + H$	1.8×10^{10}	0.0	0.0
3	$OH + H_2 \rightarrow H + H_2O$	10^5	1.6	1.38×10^4
4	$H + H_2O \rightarrow OH + H_2$	4.6×10^5	1.6	7.77×10^4

Two observations are important in respect of the reaction rate formula.

1. Because 1 mole of A combines with one mole of B,

$$[R_B] = - \frac{d\,[B]}{dt} = [R_A] = k_{bi}\,(T)\,[A]\,[B] \ . \tag{5.22}$$

2. Consider the first two reactions in Table 5.2. For this reaction pair, the net reaction rate of species H, will be given by the algebraic sum of forward and backward reaction rates.

$$[R_H]_{net} = [R_H]_{forward} + [R_H]_{backward}$$
$$= -\,1.2 \times 10^{14}\,T^{-0.91}\,\exp\left(-\,\frac{6.91 \times 10^4}{8314 \times T}\right)\,[H]\,[O_2]$$
$$+\,1.8 \times 10^{10}\,[OH]\,[O] \ . \tag{5.23}$$

$[R_H]_{backward}$ is positive because H appears as a product in reaction 2 and, hence, it is *generated*. Also, it is a function of concentrations of reactants OH and O in this reaction.

5.3.3   Tri- and Unimolecular Reactions

Following the logic of expressing the reaction rate of a bimolecular reaction, the rate of a trimolecular reaction, $A + B + M \rightarrow C + M$ is given by

$$[R_A] = k_{tri}\,(T)\,[A]\,[B]\,[M] \ . \tag{5.24}$$

In a similar fashion, for unimolecular reaction $A \rightarrow B + C$, the rate is given by

$$[R_A] = k_{uni}\,(T)\,[A] \ . \tag{5.25}$$

If the unimolecular reaction takes place in presence of an inert species, as in $A + M \rightarrow B + C + M$, then

Table 5.3   Rate coefficient k_{tri} (Eq. 5.24) and k_{ini} (Eq. 5.25) $-$ E_a (J/kmol)

No.	Reaction	A	b	E_a	< $T(K)$ <
1	$H + H + M \rightarrow H_2 + M$				
	(M = Ar)	6.4×10^{14}	-1.0	0.0	300–5000
	(M = H)	9.7×10^{13}	-0.6	0.0	100–5000
2	$H + OH + M \rightarrow H_2O + M$				
	(M = H_2O)	1.4×10^{17}	-2.0	0.0	1000–3000
3	$O + O + M \rightarrow O_2 + M$				
	(M = Ar)	1.0×10^{11}	-1.0	0.0	300–5000
4	$H_2O + M \rightarrow H + OH + M$				
	(M = H_2O)	1.6×10^4	0	4.78×10^8	2000–3000
5	$O_2 + M \rightarrow O + O + M$				
	Low Pr (M = Ar)	1.2×10^{11}	0.0	4.51×10^8	2000–3000
	High Pr (M = Ar)	1.5×10^6	1.14	0.0	300–2500

$$[R_A] = k_{uni} (T) [A] [M] \ . \tag{5.26}$$

Table 5.3 gives values of coefficients for a few uni- and trimolecular reactions.

5.3.4   Relation Between Rate Coefficient and K_p

Consider the elementary reaction

$$A + B \overset{k_f}{\underset{k_b}{\rightleftharpoons}} C + D$$

in which k_f and k_b are forward and backward reaction coefficients. As we noted, the rate coefficients are determined experimentally. However, usually it is difficult to determine both k_f and k_b with equal certainty. Therefore, the more reliable coefficient (k_f or k_b) is taken as given, and the other is evaluated from the knowledge of the *equilibrium constant* introduced in the previous chapter. To demonstrate how this is done, we write

$$\frac{d [A]}{dt} \Big|_{net} = \frac{d [A]}{dt} \Big|_{forward} + \frac{d [A]}{dt} \Big|_{backward}$$
$$= - k_f [A] [B] + k_b [C] [D] \ . \tag{5.27}$$

When equilibrium is reached, however,

$$\frac{d\,[A]}{dt}\bigg|_{net} = 0 \,.$$

(5.28)

Hence, it follows that at equilibrium,

$$\frac{k_f\,(T)}{k_b\,(T)} = \frac{[C]\,[D]}{[A]\,[B]}\,.$$

(5.29)

Because only one mole of each species is involved in the reaction (that is, ν for each species is 1), the equilibrium constant K_p will be given by[7]

$$K_p = \frac{p_C\,p_D}{p_A\,p_B} = \frac{[C]\,[D]}{[A]\,[B]}$$

$$= \frac{k_f}{k_b}\,.$$

(5.30)

Thus, as K_p can be determined for the reaction from the tabulated Gibbs functions, k_f or k_b can be determined as desired. Two problems below will illustrate how this is accomplished.

Problem 5.1 In the reaction $O + N_2 \rightleftharpoons NO + N$, the forward rate coefficient is given by

$$k_f = 1.82 \times 10^{11}\,\exp\left(-\frac{38370}{T}\right)\quad\frac{m^3}{kmol\text{-}s}$$

Calculate k_b at 1500 K, 2000 K and 2500 K.

We first calculate K_P from Eq. 4.30 for the three temperatures. Thus, using property tables, at 1500 K

$$\Delta G^0_{1500} = \left[\overline{g}^0_{NO} + \overline{g}^0_N - \overline{g}^0_O - \overline{g}^0_{N_2}\right]_{1500}$$

$$= [71491.8 + 379577.8 - 154790.4 - 0]$$

$$= 296279.2\quad\frac{kJ}{kmol}\,.$$

Therefore,

$$K_{p,1500} = \exp\left[-\frac{296279.2}{8.314 \times 1500}\right] = 4.811 \times 10^{-11}\,.$$

Similar evaluations are carried out at other temperatures. Thus, evaluating k_f from the given formula, k_b is determined from Eq. 5.30. These evaluations are shown in Table 5.4.

It is also possible to *correlate* k_b variations with temperature. Thus, let

---

[7]From Eq. 3.18, $[j] = \rho_j/M_j = p_j/(R_u\,T)$.

Table 5.4 Evaluations for Problem 5.1

T (K)	ΔG^0	K_p	k_f	k_b
1500	296279.2	4.811×10^{-11}	1.415	2.995×10^{10}
2000	289733.4	2.708×10^{-8}	847.4	3.129×10^{10}
2500	283313	1.203×10^{-6}	3.931×10^4	3.267×10^{10}

Table 5.5 Evaluations for Problem 5.2

T (K)	ΔG^0	K_p	k_f	k_b
1500	153206	4.62×10^{-6}	5.344×10^3	1.157×10^9
2000	159654	6.763×10^{-5}	2.29×10^5	3.38×10^9
2500	165969	3.405×10^{-4}	2.296×10^6	6.74×10^9

$$k_b = a \, T^b \, \exp\left(-\frac{c}{T}\right) \, ,$$

or

$$\ln k_b = \ln a + b \ln T - \frac{c}{T} \, ,$$

where a, b and c can be determined from three values of k_b at three values of T in Table 5.4. The result is

$$k_b = 7.922 \times 10^{10} \, T^{-0.08062} \, \exp\left(-\frac{629.85}{T}\right)$$

Noting the rather small value of b = −0.08062, the correlation can be written in alternative form [122]

$$k_b = 3.8 \times 10^{10} \, \exp\left(-\frac{425}{T}\right) \quad \frac{m^3}{kmol\text{-}s} \, .$$

Finally, if the reaction were written in reverse order as $NO + N \rightleftharpoons O + N_2$ then $k_{f,reverse} = k_{b,forward}$ and $k_{b,reverse} = k_{f,forward}$.

Problem 5.2 In the reaction $NO + O \rightleftharpoons N + O_2$, the forward rate coefficient is given by

$$k_f = 3.8 \times 10^6 \, T \, \exp\left(-\frac{20820}{T}\right) \quad \frac{m^3}{kmol\text{-}s} \, .$$

Calculate k_b at 1500 K, 2000 K and 2500 K.

This problem is again solved in the manner of the previous one. The results are tabulated in Table 5.5. Correlating the k_b-data,

$$k_b = 8.5176 \times 10^7 \, T^{0.8176} \, \exp\left(-\frac{5068.44}{T}\right)$$

This coefficient can also be expressed as [122]

$$k_b = 1.8 \times 10^7 \, T \exp\left(-\frac{4680}{T}\right) \quad \frac{m^3}{kmol\text{-}s} \, .$$

5.4 Construction of Global Reaction Rate

5.4.1 Useful Approximations

We have already recognized that the global reaction in Eq. 5.1 comprises several elementary reactions. Therefore, the fuel consumption rate,[8] $[R_{fu}] = -d\,[C_m H_n]/dt$, will depend on k_f and k_b values of a large number of elementary reactions. Determination of $[R_{fu}]$ in this manner poses a formidable problem. Instead, what we would ideally like is that the global reaction rate should be expressible in the simple form

$$[R_{fu}] = k_G \, [fu]^x \, [O_2]^y \, , \tag{5.31}$$

where k_G is the *global rate coefficient*. This rate will, of course, depend on k_f and k_b values of the elementary reactions.

To illustrate the elaborate procedure for determining k_G, we first take a rather simple global reaction with only two elementary reactions and three species [124]. Thus, consider reactions $A_1 \rightarrow A_2$ with rate coefficient k_{12} and $A_2 \rightarrow A_3$ with rate coefficient k_{23}. Then

$$[R_{A_1}] = \frac{d}{dt} \, [A_1] = -k_{12} \, [A_1] \tag{5.32}$$

$$[R_{A_2}] = \frac{d}{dt} \, [A_2] = k_{12} \, [A_1] - k_{23} \, [A_2] \tag{5.33}$$

$$[R_{A_3}] = \frac{d}{dt} \, [A_3] = k_{23} \, [A_2] \, . \tag{5.34}$$

If it is assumed that at t = 0, A_1 is the only species present, then initial conditions are, say, $[A_1]_{t=0} = [A_1]_0$, $[A_2]_{t=0} = [A_3]_{t=0} = 0$. The analytic solution to the system of equations can be derived as

$$\frac{[A_1]}{[A_1]_0} = \exp\left(-k_{12} \, t\right) \tag{5.35}$$

$$\frac{[A_2]}{[A_1]_0} = \frac{r_k \, \{\exp\left(-k_{23} \, t\right) - \exp\left(-k_{12} \, t\right)\}}{r_k - 1} \tag{5.36}$$

$$\frac{[A_3]}{[A_1]_0} = 1 - \frac{\{r_k \, \exp\left(-k_{23} \, t\right) - \exp\left(-k_{12} \, t\right)\}}{r_k - 1} \, , \tag{5.37}$$

---

[8]Henceforth, the subscript fu will denote fuel $C_m H_n$.

Fig. 5.2 Time variations of species A_1, A_2 and A_3 for the reaction $A_1 \rightarrow A_2 \rightarrow A_3$

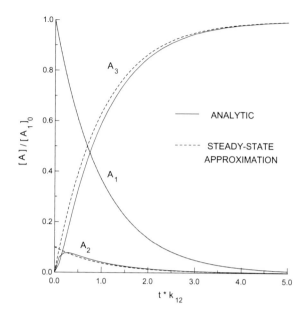

where $r_k = k_{12}/k_{23}$. Such analytic solutions are possible only when the numbers of reactions and species are small.

Steady-State Approximation

Consider a case in which A_2 is highly reactive. That is, $k_{23} >> k_{12}$ or $r_k \rightarrow 0$. For this highly reactive case, the rate of consumption of A_2 will almost equal rate of its production. As such, Eq. 5.33 can be written as

$$\frac{d}{dt}[A_2] = k_{12}[A_1] - k_{23}[A_2] = 0 . \qquad (5.38)$$

This is known as the *steady-state* approximation, which is frequently invoked to simplify solutions in a multistep reaction mechanism. Using Eqs. 5.32 and 5.38, it is a straightforward matter to determine the product composition from Eq. 5.34 as

$$\frac{[A_3]}{[A_1]_0} = 1 - \exp(-k_{12} t) \qquad (5.39)$$

$$\frac{[A_2]}{[A_1]_0} = r_k \exp(-k_{12} t) . \qquad (5.40)$$

Equation 5.35 for A_1 remains unchanged. Figure 5.2 shows the analytic solutions (Eqs. 5.35, 5.36 and 5.37), as well as solutions obtained under the steady-state approximation for the case in which $k_{23} = 10$, $k_{12} = 1$, and $r_k = 0.1$. It is seen that except for the initial times, the approximate solutions are in close agreement with the exact ones.

Near-Equilibrium Approximation

Like the steady-state approximation, the *near-equilibrium* approximation is also invoked to simplify the reaction mechanism. This approximation is invoked at high temperatures (typically >1600 K). Concentrations of radical (minor) species are to be expressed in terms of concentrations of the more stable species. Thus, consider the following three reactions involving three stable (with superscript *) and three radical species.

$$A + B^* \underset{k_{1b}}{\overset{k_{1f}}{\rightleftharpoons}} C + D \, , \tag{5.41}$$

$$D + E^* \underset{k_{2b}}{\overset{k_{2f}}{\rightleftharpoons}} C + A \, , \tag{5.42}$$

$$\text{and } C + E^* \underset{k_{3b}}{\overset{k_{3f}}{\rightleftharpoons}} F^* + A \, . \tag{5.43}$$

Then, at equilibrium,

$$k_{1f} [A] [B^*] = k_{1b} [C] [D] \, , \tag{5.44}$$
$$k_{2f} [D] [E^*] = k_{2b} [C] [A] \, , \tag{5.45}$$
$$\text{and } k_{3f} [C] [E^*] = k_{3b} [F^*] [A] \, . \tag{5.46}$$

These equations can be solved simultaneously to yield concentrations of the minor species A, C, and D.

$$[A] = \left\{ \frac{k_{1f} \, k_{2f} \, k_{3f}^2}{k_{1b} \, k_{2b} \, k_{3b}^2} \right\}^{0.5} [B^*]^{0.5} [E^*]^{1.5} [F^*]^{-1} \tag{5.47}$$

$$[C] = \left\{ \frac{k_{1f} \, k_{3f}}{k_{1b} \, k_{3b}} \right\} [B^*] [E^*] [F^*]^{-1} \tag{5.48}$$

$$[D] = \left\{ \frac{k_{1f} \, k_{2f}}{k_{1b} \, k_{2b}} \right\}^{0.5} [B^*]^{0.5} [E^*]^{0.5} \, . \tag{5.49}$$

Thus, the concentrations of the minor species (which are difficult to measure) have been expressed in terms of the stable species and the quantities in curly brackets represent effective rate coefficients as functions of temperature. In the subsection that follows, we consider the application of the steady-state and near-equilibrium assumptions to the reaction involving dissociation of N_2 and O_2 and determine global rate k_G in a simplified manner.

5.4.2 *Zeldovich Mechanism of NO Formation*

Consider the global reaction involving *shock-heating of air*, comprising N_2 and O_2 so as to form NO.

$$N_2 + O_2 \rightarrow 2\,NO \tag{5.50}$$

This NO-formation reaction is executed through three elementary reactions, called the Zeldovich mechanism.

$$N_2 + O \underset{k_{1b}}{\overset{k_{1f}}{\rightleftharpoons}} NO + N \tag{5.51}$$

$$N + O_2 \underset{k_{2b}}{\overset{k_{2f}}{\rightleftharpoons}} NO + O \tag{5.52}$$

$$O_2 \underset{k_{3b}}{\overset{k_{3f}}{\rightleftharpoons}} O + O \tag{5.53}$$

In these reactions, N and O are *intermediate*, or minor, species. From the two problems solved, it is clear that reaction 5.51 is same as that in Problem 5.1 but reaction 5.52 is stated in a reverse order of that stated in Problem 5.2. Therefore, to recapitulate,

$$k_{1f} = 1.82 \times 10^{11} \exp\left(-\frac{38370}{T}\right) \qquad k_{1b} = 3.8 \times 10^{10} \exp\left(-\frac{425}{T}\right)$$

$$k_{2f} = 1.8 \times 10^{7}\, T \exp\left(-\frac{4680}{T}\right) \qquad k_{2b} = 3.8 \times 10^{6}\, T \exp\left(-\frac{20820}{T}\right).$$

For the third reaction, refer to Table 5.3. Then

$$k_{3f} = 1.2 \times 10^{11} \exp\left(-\frac{4.51 \times 10^8}{8314 \times T}\right) \qquad k_{3b} = \frac{10^{11}}{T}.$$

Full Mechanism

Our interest is to determine the NO formation rate—that is, $d\,x_{NO}/dt$. This rate is determined by solving the following five equations simultaneously:

$$\frac{d}{dt}\,[NO] = k_{1f}\,[N_2]\,[O] - k_{1b}\,[NO]\,[N]$$
$$+\, k_{2f}\,[N]\,[O_2] - k_{2b}\,[NO]\,[O] \tag{5.54}$$

$$\frac{d}{dt} [N] = k_{1f} [N_2] [O] - k_{1b} [NO] [N]$$
$$- k_{2f} [N] [O_2] + k_{2b} [NO] [O] \tag{5.55}$$

$$\frac{d}{dt} [N_2] = -k_{1f} [N_2] [O] + k_{1b} [NO] [N] \tag{5.56}$$

$$\frac{d}{dt} [O] = -k_{1f} [N_2] [O] + k_{1b} [NO] [N]$$
$$+ k_{2f} [N] [O_2] - k_{2b} [NO] [O]$$
$$+ k_{3f} [O_2] [Ar] - k_{3b} [O] [O] [Ar] \tag{5.57}$$

$$\frac{d}{dt} [O_2] = -k_{2f} [N] [O_2] + k_{2b} [NO] [O]$$
$$- k_{3f} [O_2] [Ar] + k_{3b} [O] [O] [Ar] \ . \tag{5.58}$$

The preceding equations[9] must be solved for a *fixed temperature and pressure* to determine variation of each [j] with time. We take T = 2500 K and p = 3 atm, for example. Then, the equations are solved in a fully implicit manner with initial conditions (t = 0) $x_{O_2} = 0.21$, $x_{N_2} = 0.78$, and $x_{Ar} = 0.01$, or using Eq. 3.18, $[j] = x_j \times p/(R_u T)$. The concentrations of all other species are initially zero. The computed results are shown in Fig. 5.3.

The top figure in Fig. 5.3 shows the variations of $d\, x_{NO}/dt$, $d\, x_{O_2}/dt$, and $d\, x_{N_2}/dt$. The rates of O and N are not shown because they are extremely small. The figure shows that NO is formed at the expense of N_2 and O_2. All rates become negligible after approximately 300 ms. The bottom figure shows variations of x_{N_2}, x_{O_2}, x_O, x_{NO}, and x_N with time evaluated with two time steps, $\Delta t = 10^{-5}$ and

---

[9]In general, reaction mechanisms involve many reactions and several species. Therefore, it becomes necessary to write the resulting equations more compactly. This is done as follows: Let subscript j refer to the species and i to the reaction. Then, the *production rate* of each species can be written as

$$\frac{d\, x_j}{dt} = \sum_{i=1}^{NR} \nu_{ji}\, q_j \tag{5.59}$$

$$\nu_{ji} = (\nu_{pr} - \nu_r)_{ji} \tag{5.60}$$

$$q_j = k_{fi} \prod_{j=1}^{NS} x_j^{\nu_{r,ji}} - k_{bi} \prod_{j=1}^{NS} x_j^{\nu_{pr,ji}} \ , \tag{5.61}$$

where NR represents the number of reactions, NS represents the number of species, q_j is called the *rate progress variable* and $\nu_{r,ji}$ and $\nu_{pr,ji}$ are *stoichiometric coefficients* on the reactant and product sides, respectively. Their value may be zero in some reactions and 1 in others. k_{fi} and k_{bi} are forward and backward reaction rates of reaction i. The coefficients are related by

$$\sum_{j=1}^{NS} \nu_{r,ji}\, x_j = \sum_{j=1}^{NS} \nu_{pr,ji}\, x_j \qquad i = 1, 2, \dots, NR \ . \tag{5.62}$$

Fig. 5.3 Rate of formation of NO (top figure), mole fractions (bottom figure), Implicit calculation at $T = 2500$ K and $p = 3$ atm

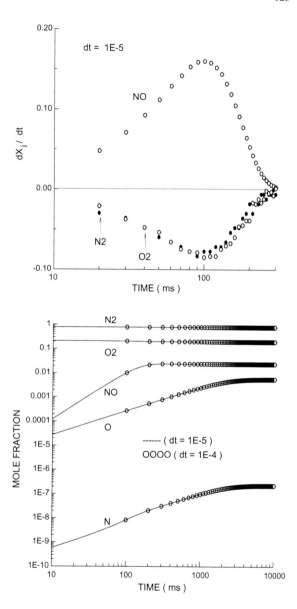

$\Delta t = 10^{-4}$. The figure confirms that the calculated mole fractions are not influenced by the choice of time step and that a steady state is indeed reached. The x_{NO} reaches near-steady state in about 250 ms, whereas x_O and x_N take longer to reach steady state. The bottom figure shows that N_2 and O_2 mole fractions remain very nearly at their equilibrium values. The composition after 500 ms is: $x_{NO} = 0.0234$, $x_N = 1.66 \times 10^{-7}$, $x_O = 4.18 \times 10^{-4}$, $x_{N_2} = 0.768$, and $x_{O_2} = 0.198$. The rates of NO and N are: $dx_{NO}/dt = 1.12 \times 10^{-3}$ s^{-1} and $dx_N/dt = 1.12 \times 10^{-8}$ s^{-1}.

Table 5.6 Rate coefficients for Zeldovich mechanism

T (K)	k_{1f}	k_{1b}	k_{2f}	k_{2b}
1500	1.41	2.86×10^{10}	1.15×10^{9}	5.34×10^{3}
2000	847.5	3.07×10^{10}	3.38×10^{9}	2.29×10^{5}
2500	3.93×10^{4}	3.20×10^{10}	6.73×10^{9}	2.30×10^{6}

Reduced Mechanism—Equilibrium Considerations

From the preceding results, the full mechanism can be simplified. Thus, we construct a *reduced mechanism* by observing the magnitudes of the reaction rate coefficients in different reactions. Table 5.6 shows the magnitudes of forward and backward coefficients for the three reactions. The table shows that $k_{1f} \ll k_{1b}$. Therefore, the *N-formation rate* in reaction 5.51 will be controlled by the *slower* rate k_{1f}. Similarly, in reaction 5.52, the *N-destruction rate* will be controlled by k_{2f}, because $k_{2b} < k_{2f}$. Therefore, effectively, formation rates of NO (Eq. 5.54) and N (Eq. 5.55) can be taken as

$$\frac{d}{dt} [N] = k_{1f} [N_2] [O] - k_{2f} [N] [O_2] \tag{5.63}$$

$$\text{and } \frac{d}{dt} [NO] = k_{1f} [N_2] [O] + k_{2f} [N] [O_2] . \tag{5.64}$$

Further, at high temperature, we shall choose to invoke the *near-equilibrium* assumption, so the concentration of O will depend on the equilibrium constant K_{p3} and stable species O_2:

$$[O] = \left[[O_2] \frac{K_{p3} \, p_{ref}}{R_u \, T} \right]^{0.5} . \tag{5.65}$$

The reduced mechanism will thus comprise Eqs. 5.63, 5.64, and the O-concentration is evaluated from Equ. 5.65. N_2 and O_2 can be assumed to retain their initial values. The results obtained with the reduced mechanism at steady state after 50 ms are: $x_{NO} = 0.0246$, and $x_N = 1.49 \times 10^{-7}$ and the rates are: $dx_{NO}/dt = 7.45 \times 10^{-4}$ s^{-1} and $dx_N/dt \simeq 1.0 \times 10^{-9}$ s^{-1}. These values compare reasonably well with the full mechanism.

Calculations with both the full and the reduced mechanisms must be performed for different combinations of p and T. The NO-formation rate can then be correlated with equilibrium values of N_2 and O_2 to arrive at the global rate expression for the formation of NO. This task can be simplified by making the *steady-state assumption*.

Reduced Mechanism—Steady-State Assumption

Observations from both the full and the reduced-equilibrium mechanisms suggest that we can make a *steady state* assumption with respect to Eq. 5.63 or, set $d \, x_N/dt \simeq 0$. Thus,

$$[N_{ss}] = \frac{k_{1f} \ [N_2] \ [O]}{k_{2f} \ [O_2]} \ , \tag{5.66}$$

where subscript ss denotes steady-state. Substituting this equation in Eq. 5.64 and using Eq. 5.65, therefore,

$$\frac{d}{dt} \ [NO] = 2 \, k_{1f} \ [N_2] \ [O] = 2 \, k_{1f} \left[\frac{K_{p3} \, p_{ref}}{R_u \, T} \right]^{0.5} [N_2]_{eq} \ [O_2]_{eq}^{0.5} \ , \tag{5.67}$$

or,

$$[R_{NO}] = k_G \ [N_2]_{eq} \ [O_2]_{eq}^{0.5} \ , \tag{5.68}$$

where

$$k_G = 2 \, k_{1f} \left[\frac{K_{p3} \, p_{ref}}{R_u \, T} \right]^{0.5}. \tag{5.69}$$

In these equations, the subscript eq denotes equilibrium values that reflect the effect of *system pressure* p. Eqs. 5.67 or 5.68, of course, evaluate only the initial (or short-time) NO-formation rate. These equations are found to fit the experimental data very well. The N_2 oxidation discussed here has shown typical simplifications that are made to evaluate the global reaction coefficient k_G in all reactions of engineering interest.

5.4.3   Quasi-global Mechanism

As noted earlier, the global hydrocarbon reaction is executed through a large number of elementary reactions. Applying the full mechanism is, therefore, cumbersome. Hence, a reduced mechanism is constructed employing steady-state and near-equilibrium approximations with respect to some of the elementary reactions. Such mechanisms are described in Ref. [91].

Such a detailed study of chemical reactions suggests many simplifying paths to aid engineering calculations. One such path is that of the *quasi-global mechanism*. In this mechanism, the main global reaction rate is evaluated from reaction rates of several quasi-global reactions involving only the major stable species. Thus, Hautmann et al. [43] describe a four-step quasi-global mechanism that can be applied to a $C_n \, H_{2n+2}$ fuel (with $n > 2$). The reaction steps are

$$C_n H_{2n+2} \rightarrow \frac{n}{2} \, C_2 \, H_4 + H_2 \tag{5.70}$$

$$C_2 H_4 + O_2 \rightarrow 2 \, CO + 2 \, H_2 \tag{5.71}$$

$$CO + \frac{1}{2} \, O_2 \rightarrow CO_2 \tag{5.72}$$

$$H_2 + \frac{1}{2} \, O_2 \rightarrow H_2O \ , \tag{5.73}$$

Table 5.7 Rate constants for $C_n H_{2n+2}$ combustion

Rate Eq.	(5.74)	(5.75)	(5.76)	(5.77)
x	14.39	12.57	12.35	11.39
E_a/R_u (K)	24962	25164	20131	20634
a	0.5	0.9	1.0	0.85
b	1.07	1.18	0.25	1.42
c	0.4	− 0.37	0.5	− 0.56

where the reaction rates in kmol/m^3-s are expressed as[10]

$$[R_{fu}] = -10^x \exp\left(-\frac{E_a}{R_u T}\right) [fu]^a \ [O_2]^b \ [C_2H_4]^c \qquad (5.74)$$

$$[R_{C_2H_4}] = -10^x \exp\left(-\frac{E_a}{R_u T}\right) [C_2H_4]^a \ [O_2]^b \ [fu]^c \qquad (5.75)$$

$$[R_{CO}] = -10^x \exp\left(-\frac{E_a}{R_u T}\right) [CO]^a \ [O_2]^b \ [H_2O]^c \qquad (5.76)$$

$$[R_{H_2}] = -10^x \exp\left(-\frac{E_a}{R_u T}\right) [H_2]^a \ [O_2]^b \ [C_2H_4]^c \ . \qquad (5.77)$$

The values of x, a, b, c, and E_a are given in Table 5.7. To account for the effect of the equivalence ratio Φ, Eq. 5.76 is multiplied by $7.93 \times \exp(-2.48\,\Phi)$. The constants are valid for $0.12 < \Phi < 2$, $1 < p(atm) < 9$ and $960 < T(K) < 1540$ [19].

5.5 Global Rates for Hydrocarbon Fuels

Global rates for hydrocarbon fuels have been determined in the manner described in the previous section by considering multi-step reaction mechanisms. The global reaction and the reaction rate are expressed as

$$C_m H_n + \left(m + \frac{n}{4}\right) O_2 \rightarrow m\, CO_2 + \frac{n}{2}\, H_2O \ . \qquad (5.78)$$

$$[R_{C_m H_n}] = -\frac{d}{dt}[C_m H_n]$$

$$= A \exp\left(-\frac{E_a}{R_u T}\right) [C_m H_n]^x \ [O_2]^y \quad \frac{kmol}{m^3\text{-s}} \ . \qquad (5.79)$$

---

[10]$[fu] \equiv [C_n H_{2n+2}]$.

Table 5.8 Rate constants in Eq. 5.79

Fuel	A	E_a/R_u (K)	x	y
CH_4	1.3×10^9	24358	-0.3	1.3
C_2H_6	6.1857×10^9	15098	0.1	1.65
C_3H_8	4.836×10^9	15098	0.1	1.65
C_6H_{14}	3.205×10^9	15098	0.25	1.5
C_7H_{16}	2.868×10^9	15098	0.25	1.5
C_8H_{18}	4.049×10^{10}	20131	0.25	1.5
$C_{10}H_{22}$	2.137×10^9	15098	0.25	1.5
C_6H_6	1.125×10^9	15098	-0.1	1.85
C_2H_4	1.125×10^{10}	15098	0.1	1.65
C_3H_6	2.362×10^{10}	15098	-0.1	1.85
C_2H_2	3.665×10^{10}	15098	0.5	1.25

where the concentrations are expressed in $kmol/m^3$. The values of A, E_a, x and y as determined by Westbrook and Dryer [125], are tabulated in Table 5.8 for a variety of hydrocarbon fuels.[11] Equation 5.79 expresses the reaction rate in the form of Eq. 5.31 as was desired. In terms of mass units, the rate is expressed as

$$R_{fu} = M_{fu} \left[R_{fu} \right]$$
$$= A^* \exp\left(-\frac{E_a}{R_u T} \right) \rho_{mix}^{(x+y)} \, \omega_{fu}^x \, \omega_{O_2}^y \quad \frac{kg}{m^3\text{-s}} \tag{5.80}$$
$$A^* = A \, M_{fu}^{(1-x)} \, M_{O_2}^{-y}$$

It is possible to verify the accuracy of the global reaction rates on the basis of the quasi-global reaction rates. The use of the latter will be demonstrated in Chap. 11.

Exercises

1. Estimate the molecular velocity of O_2 and H_2 at 500 °C.
2. Estimate the collision frequency Z of O_2 and H_2 at 500 °C and p = 10.0 bar.
3. Using Table 5.2, evaluate the steric factor f_s for the reaction $H + O_2 \rightarrow OH + O$ at 2500 K. Given: $d_H = 2.708$ A and $d_{O_2} = 3.067$ A.
4. Consider the reaction $N + OH \leftrightarrow NO + H$. If the forward reaction rate is $7.1 \times 10^{10} \exp(-450/T)$ (m^3/kmol-s), what is the reverse reaction rate at 1500 K, 2000 K, and 2500 K ? Hence, develop a relationship $k_b = a \, T^b \exp(-c/T)$ and determine a, b, and c.

---

[11]Westbrook and Dwyer [125] expressed reaction rates and concentrations in units of $gmol, cm^3, s$. Here, we use units of $kmol, m^3, s$.

5. For the water-gas shift reaction, $CO + H_2O \rightleftharpoons CO_2 + H_2$, Jones and Lin-steadt [56] recommend that $k_f = 0.275 \times 10^{10} \exp(-10.065/T)$. Determine the expression for $k_b(T)$. Hence evaluate K_p at 2000 K. Compare this value with that given in Table 4.1.

6. Derive Eqs. 5.65 and 5.68.

7. Represent Eqs. 5.54 to 5.58 in the manner of Eq. 5.59 and identify $\nu_{r,ji}$, $\nu_{pr,ji}$, and q_j for each species and reaction.

8. Consider the thermal Zeldovich mechanism for formation of NO. What is the time required for formation of 50 ppm NO ? Given: $T = 2100$ K, $\rho_{mix} = 0.167$ m^3/kg, $M_{mix} = 28.778$, $x_O = 7.6 \times 10^{-5}$, $x_{O_2} = 3.025 \times 10^{-3}$, $x_{N_2} = 0.726$. Also check if the assumption of negligible reverse reactions is valid, and determine [N_{ss}].

9. The extended Zeldovich mechanism for NO formation is given by

$$N_2 + O \underset{k_{1b}}{\overset{k_{1f}}{\rightleftharpoons}} NO + N$$

$$N + O_2 \underset{k_{2b}}{\overset{k_{2f}}{\rightleftharpoons}} NO + O$$

$$N + OH \underset{k_{3b}}{\overset{k_{3f}}{\rightleftharpoons}} NO + H$$

In this mechanism, at moderately high temperatures, O_2, N_2, OH, O, and H are assumed to be in equilibrium, whereas concentration of N is derived from steady state assumption. Hence show that

$$[N]_{ss} = \frac{k_{1f} [O]_{eq} [N_2]_{eq} + k_{2b} [O]_{eq} [NO] + k_{3b} [NO] [H]_{eq}}{k_{1b} [NO] + k_{2f} [O_2]_{eq} + k_{3f} [OH]_{eq}}$$

and

$$\frac{d [NO]}{dt} = k_{1f} [O]_{eq} [N_2]_{eq} - k_{1b} [NO] [N]_{ss} + k_{2f} [N]_{ss} [O_2]_{eq}$$
$$- k_{2b} [O]_{eq} [NO] + k_{3f} [N]_{ss} [OH]_{eq} - k_{3b} [NO] [H]_{eq}$$

10. Consider the *adiabatic* combustion of methane at $\Phi = 0.8$ initially at 298 K and 1 atm.

   (a) Determine the AFT and equilibrium composition of products assuming *full equilibrium*.

   (b) Using the result of the previous problem, determine the variation of NO-formation with time (assume $T = T_{ad}$ for all times) by numerical integration. Compare the steady state equilibrium NO value with that estimated from the full equilibrium assumption and comment on the result. Assume

$$k_{3f} = 7.1 \times 10^{10} \exp\left(-\frac{450}{T}\right) \quad k_{3b} = 1.7 \times 10^{11} \exp\left(-\frac{24560}{T}\right).$$

11. Repeat the preceding problem for $\Phi = 1.2$ and comment on the result.
12. In a stationary gas turbine plant, n-decane is burned with *air* with $\Phi = 1$ in the primary combustion zone. The combustion products are at $p = 14$ atm, $T = 2300$ K, and $M_{mix} = 28.47$. The mean residence time of combustion products is 7 ms and the equilibrium concentrations are: $x_O = 7.93 \times 10^{-5}$, $x_{O_2} = 3.62 \times 10^{-3}$, $x_{NO} = 2.09 \times 10^{-3}$, $x_{N_2} = 0.7295$. Assuming the validity of Zeldovich mechanism, calculate the ratio of kinetically controlled and equilibrium concentrations of NO. Also calculate $[N]_{ss}$ and check if the reverse reaction can be neglected.
13. Prove the correctness of Eq. 5.80.
14. Assuming a global one-step reaction for Heptane C_7H_{16} at $\Phi = 1$, $p = 1$ atm, and $T = 1500$ K, calculate variation of $[fu]/[fu]_i$, and $[CO_2]/[CO_2]_f$ as a function of time where subscript i refers to initial and f to final. Hence, estimate the time (ms) required for 90% conversion of Heptane to products. Repeat the problem for C_3H_8 and C_6H_6 and comment on the result.

Chapter 6
Derivation of Transport Equations

6.1 Introduction

Our interest now is to study chemical reactions in the presence of fluid motion so that fluid mixing effects can be incorporated in the determination of the product composition. In the study of transport phenomena in moving fluids, the fundamental laws of motion (conservation of mass and Newton's second law) and energy (first law of thermodynamics) is applied to an elemental fluid known as the *control volume* (CV).

The CV may be defined as an imaginary region in space, across the boundaries of which matter, energy, and momentum may flow. It is a region *within* which a source or a sink of the same quantities may prevail. Further, it is a region on which external forces may act.

In general, a CV may be large or infinitesimally small. However, consistent with the idea of a *differential* in a continuum, an infinitesimally small CV is considered. Thus, when the laws are to be expressed through differential equations, the CV is located *within* a moving fluid. In the *Eulerian approach*, the CV is assumed *fixed* in space and the fluid is assumed to flow through and around the CV. The advantage of this approach is that any measurements made using stationary instruments (velocity, temperature, mass fraction, etc) can be compared directly with the solutions of differential equations.

The fundamental laws define *total flows* of mass, momentum, and energy not only in terms of *magnitude*, but also in terms of *direction*. In a general problem of convection, neither magnitude nor direction is known a priori at different positions in the flowing fluid. The problem of ignorance of direction is circumvented by resolving velocity, force, and scalar fluxes in three directions that define the space. The task now is to determine and represent only the magnitudes of these directional flows and fluxes.

Here, the three directions are chosen along Cartesian coordinates. Figure 6.1 shows the considered CV of dimensions Δx_1, Δx_2 and Δx_3 located at (x_1, x_2, x_3) from a fixed origin.

© Springer Nature Singapore Pte Ltd. 2020
A. W. Date, *Analytic Combustion*,
https://doi.org/10.1007/978-981-15-1853-9_6

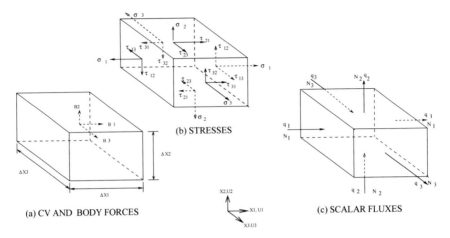

Fig. 6.1 Eulerian specification of the control-volume (CV).

In this chapter, the equation of mass transfer and the energy equation are derived through control-volume analysis because of their importance in combustion analysis. However, the equations of fluid motion, also known as the Navier–Stokes (NS) equations, are presented without derivation[1] in the next section.

6.2   Navier–Stokes Equations

Mass Conservation Equation:

The application of the law of conservation of mass for the bulk fluid mixture gives the *mass conservation equation*. It reads

$$\frac{\partial(\rho_m)}{\partial t} + \frac{\partial(\rho_m\,u_j)}{\partial x_j} = 0 \ , \tag{6.1}$$

where ρ_m is the mixture density.

Momentum Equations $u_i\,(i = 1, 2, 3)$:

The application of Newton's Second law of motion to a CV yields the following equations:

$$\frac{\partial(\rho_m\,u_i)}{\partial t} + \frac{\partial(\rho_m\,u_j\,u_i)}{\partial x_j} = \frac{\partial(\tau_{ij})}{\partial x_j} - \frac{\partial p}{\partial x_i} + \rho_m\,B_i \ , \tag{6.2}$$

---

[1]The N–S equations have been derived in many textbooks. Interested reader may see Ref. [30], for example.

where B_i (N/kg) are the body forces (see Fig. 6.1(a)) and τ_{ij} (N/m^2) are the stresses (see Fig. 6.1(b)). In an incompressible turbulent fluid, both laminar (τ_{ij}^l) and turbulent ($\tau_{ij}^t = -\rho_m \overline{u_i' u_j'}$) stresses are important.[2] Thus,

$$\tau_{ij} = \tau_{ij}^l - \rho_m \overline{u_i' u_j'} \; . \tag{6.3}$$

Equation 6.1 and the three momentum Equations 6.2 contain four (u_1, u_2, u_3 and p) unknowns together with a further six unknown stresses.[3] To close the mathematical problem of predicting fluid motion, therefore, the six stresses must be modeled. For laminar stresses, this modeling is provided by Stokes's stress–strain relation:

$$\tau_{ij}^l = \mu \, S_{ij} = \mu \left[\frac{\partial u_i}{\partial x_j} + \frac{\partial u_j}{\partial x_i} \right] \; , \tag{6.4}$$

where μ is fluid viscosity and S_{ij} is the strain-rate tensor.

Over the past several decades, considerable research has been conducted by specialists for modeling turbulent stresses so they can be as universally represented as the laminar stresses. This subject matter is vast, and several books (see, for example, [22, 70, 90, 93]) describe significant developments in this field. For the purposes of this book, it will suffice to give the simplest model, known as the Boussinesq approximation that mimics the Stokes's relation. It reads

$$-\rho_m \overline{u_i' u_j'} = \mu_t \left[\frac{\partial u_i}{\partial x_j} + \frac{\partial u_j}{\partial x_i} \right] - \frac{2}{3} \delta_{ij} \, \rho_m \, k \; , \tag{6.5}$$

where μ_t is called the *turbulent viscosity* and k is the turbulent kinetic energy. Using Eqs. 6.4 and 6.5 and defining $\mu_{eff} = \mu + \mu_t$, the three momentum Equations 6.2 can be written as

---

[2] Turbulent stresses arise out of Reynolds averaging of instantaneous forms of NS Equations 6.1 and 6.2. The instantaneous velocities and pressure are given the symbols $\hat{u}$ and $\hat{p}$. The same will apply to any other scalar, such as h or ω. Thus, any instantaneous quantity $\hat{\Phi}$, say, is written as $\hat{\Phi} = \Phi + \Phi'$ where Φ' represents fluctuation (positive or negative) over time-mean value Φ. Time-averaging thus implies

$$\overline{\hat{\Phi}} = \frac{1}{T \to \infty} \int_0^T \hat{\Phi} \, dt = \Phi + \frac{1}{T \to \infty} \int_0^T \Phi' \, dt = \Phi \; .$$

Thus, $\overline{\Phi'} = 0$. One important consequence of this definition is that the time-averaged product of two instantaneous quantities are given by

$$\overline{\hat{\Phi}_1 \hat{\Phi}_2} = \Phi_1 \Phi_2 + \overline{\Phi_1' \Phi_2'} \; .$$

Such product terms appear in the convective terms $\partial(\rho_m u_j u_i)/\partial x_j$ of the NS equations. They give rise to what are called turbulent stresses τ_{ij}^t, in analogy with the laminar stresses τ_{ij}^l.

[3] Body forces B_i are to be specified externally when they are present.

$$\frac{\partial(\rho_m u_i)}{\partial t} + \frac{\partial(\rho_m u_j u_i)}{\partial x_j} = \frac{\partial}{\partial x_j}\left[\mu_{eff}\frac{\partial u_i}{\partial x_j}\right] - \frac{\partial p}{\partial x_i} + \rho_m B_i + S_{u_i}, \qquad (6.6)$$

where

$$S_{u_i} = \frac{\partial}{\partial x_j}\left[\mu_{eff}\frac{\partial u_j}{\partial x_i}\right] - \frac{2}{3}\delta_{ij}\frac{\partial(\rho_m k)}{\partial x_j}. \qquad (6.7)$$

Coupled with Eq. 6.1, Eqs. 6.6 are called Reynolds averaged Navier Stokes (RANS) equations. These four equations are used to determine four unknowns u_i (i = 1, 2, 3) and pressure p. In laminar flows, fluid viscosity is taken from property tables. In turbulent flows, however, turbulent viscosity μ_t is a property of the flow itself. Further discussion can be found in Sect. 6.9.

6.3 Equation of Mass Transfer

6.3.1 Species Conservation

In a combustion mixture, several species are present. The number of species k is given by the *postulated* chemical reaction mechanism. The mass conservation principle can also be applied to each species k of the mixture. This principle is stated as

Rate of accumulation of mass $(\dot{M}_{k,ac})$ = Rate of mass in $(\dot{M}_{k,in})$

$-$ Rate of mass out $(\dot{M}_{k,out})$

$+$ Rate of generation within CV .

To apply this principle, let ρ_k be the density of the species k in a fluid mixture of density ρ_m. Similarly, let $N_{i,k}$ be the *mass transfer flux* (kg/m^2-s) of species k in i-direction. Then, with reference to Fig. 6.1(c),

$$\dot{M}_{k,ac} = \frac{\partial(\rho_k \Delta V)}{\partial t},$$

$$\dot{M}_{k,in} = N_{1,k}\,\Delta x_2\,\Delta x_3\,|_{x_1} + N_{2,k}\,\Delta x_3\,\Delta x_1\,|_{x_2}$$
$$+ N_{3,k}\,\Delta x_1\,\Delta x_2\,|_{x_3},$$

$$\dot{M}_{k,out} = N_{1,k}\,\Delta x_2\,\Delta x_3\,|_{x_1+\Delta x_1} + N_{2,k}\,\Delta x_3\,\Delta x_1\,|_{x_2+\Delta x_2}$$
$$+ N_{3,k}\,\Delta x_1\,\Delta x_2\,|_{x_3+\Delta x_3}.$$

Dividing each term by ΔV and letting $\Delta x_1, \Delta x_2, \Delta x_3 \to 0$,

$$\frac{\partial(\rho_k)}{\partial t} + \frac{\partial(N_{1,k})}{\partial x_1} + \frac{\partial(N_{2,k})}{\partial x_2} + \frac{\partial(N_{3,k})}{\partial x_3} = R_k, \qquad (6.8)$$

where R_k is the volumetric generation rate from the chemical reaction. The expressions for this rate were given in Chap. 5.

The total mass transfer flux $N_{i,k}$ is the sum of *convective* flux owing to bulk fluid motion (each species is carried at velocity of the bulk fluid mixture) and *diffusion* flux ($m''_{i,k}$). Thus,

$$N_{i,k} = \rho_k\, u_i + m''_{i,k} \; . \tag{6.9}$$

Under certain restricted circumstances of interest in this book, the diffusion flux is given by Fick's law of mass diffusion

$$m''_{i,k} = -\, D\, \frac{\partial \rho_k}{\partial x_i} \; , \tag{6.10}$$

where D (m²/s) is the mass-diffusivity.[4] Substituting Eqs. 6.9 and 6.10 in Eq. 6.8, it can be shown that

$$\frac{\partial(\rho_k)}{\partial t} + \frac{\partial(\rho_k\, u_1)}{\partial x_1} + \frac{\partial(\rho_k\, u_2)}{\partial x_2} + \frac{\partial(\rho_k\, u_3)}{\partial x_3}$$
$$= \frac{\partial}{\partial x_1}\left(D\, \frac{\partial \rho_k}{\partial x_1}\right) + \frac{\partial}{\partial x_2}\left(D\, \frac{\partial \rho_k}{\partial x_2}\right) + \frac{\partial}{\partial x_3}\left(D\, \frac{\partial \rho_k}{\partial x_3}\right) + R_k \; . \tag{6.11}$$

It is a common practice to refer to species k via its *mass fraction* so $\rho_k = \omega_k\, \rho_m$. Thus, Eq. 6.11 can be compactly written as

$$\frac{\partial(\rho_m\, \omega_k)}{\partial t} + \frac{\partial(\rho_m\, u_j\, \omega_k)}{\partial x_j} = \frac{\partial}{\partial x_j}\left(\rho_m\, D\, \frac{\partial \omega_k}{\partial x_j}\right) + R_k \; . \tag{6.12}$$

When this mass transfer equation is summed over all species of the mixture, the mass conservation equation for the bulk fluid (Eq. 6.1) is retrieved. This is because, from Dalton's law, $\rho_m = \sum_k \rho_k$ (or $\sum \omega_k = 1$) and, $\sum_k N_{i,k} = \rho_m\, u_i$. Thus, on summation over all species k, Eq. 6.12 reduces to

$$\frac{\partial \rho_m}{\partial t} + \frac{\partial(\rho_m\, u_j)}{\partial x_j} = \sum_k R_k \; . \tag{6.13}$$

When compared with Eqs. 6.1, 6.13 shows that $\sum_k R_k = 0$. That is when some species are generated by a chemical reaction, some others are destroyed, so there is no net mass generation in the bulk fluid.

In turbulent flows, the *effective* mass diffusivity is given by $D_{eff} = D + D_t$ where D_t is turbulent mass diffusivity. The latter is given by

---

[4]The mass diffusivity is *defined* only for a binary mixture of two fluids a and b as D_{ab}. Values of D_{ab} are given in Appendix F. In multicomponent gaseous mixtures, however, diffusivities for pairs of species are taken to be nearly equal and a single symbol D suffices for all species. Incidentally, in turbulent flows, this assumption of equal (effective) diffusivities holds even greater validity.

$$D_t = \frac{\nu_t}{Sc_t} \, , \tag{6.14}$$

$\nu_t = \mu_t/\rho_m$ is turbulent kinematic viscosity, and turbulent Schmidt number (Sc_t) is about 0.9 for gaseous mixtures [70].

6.3.2 Element Conservation

In Chap. 3, we noted that the *element conservation principle* applies to a chemical reaction. Thus, let the postulated chemical reaction mechanism be such that species CH_4, O_2, H_2, H_2O, CO_2, CO, N_2, NO, and O are present in the mixture. Then, the mass fraction (denoted by symbol η) of elements C, H, O, and N in the mixture will be related to the mass fractions of species in the following manner:

$$\eta_C = \frac{12}{16} w_{CH_4} + \frac{12}{44} w_{CO_2} + \frac{12}{28} w_{CO} \, , \tag{6.15}$$

$$\eta_H = \frac{4}{16} w_{CH_4} + \frac{2}{2} w_{H_2} + \frac{2}{18} w_{H_2O} \, , \tag{6.16}$$

$$\eta_O = \frac{32}{44} w_{CO_2} + \frac{16}{28} w_{CO} + \frac{32}{32} w_{O_2} + \frac{16}{18} w_{H_2O}$$
$$+ \frac{16}{16} w_O + \frac{16}{30} w_{NO} \, , \tag{6.17}$$

$$\text{and } \eta_N = \frac{14}{30} w_{NO} + \frac{28}{28} w_{N_2} \, . \tag{6.18}$$

In general, therefore, the *mass fraction* η_α of element α in the mixture will be given by

$$\eta_\alpha = \sum_k \eta_{\alpha,k} \, w_k \, , \tag{6.19}$$

where $\eta_{\alpha,k}$ is the mass fraction of element α in the species k. Just as the species are convected, diffused, and generated or destroyed, the elements can also be considered to have been convected and diffused, but *they can never be destroyed or generated* because of the principle of element conservation. Thus, the transport equation for any element α will have no source term. The equation will read as

$$\frac{\partial(\rho_m \, \eta_\alpha)}{\partial t} + \frac{\partial(\rho_m \, u_j \, \eta_\alpha)}{\partial x_j} = \frac{\partial}{\partial x_j} \left(\rho_m \, D \, \frac{\partial \eta_\alpha}{\partial x_j} \right) \, , \tag{6.20}$$

where it is assumed that the diffusion coefficient for the elements is the same as that for the species [62].

6.4 Energy Equation

The first law of thermodynamics, when considered in *rate* form (W/m³), can be written as

$$\dot{E} = \dot{Q}_{conv} + \dot{Q}_{cond} + \dot{Q}_{gen} - \dot{W}_s - \dot{W}_b , \qquad (6.21)$$

where

$$
\begin{aligned}
\dot{E} &= \text{Rate of change of energy of the CV,} \\
\dot{Q}_{conv} &= \text{Net rate of energy transferred by convection,} \\
\dot{Q}_{cond} &= \text{Net rate of energy transferred by conduction,} \\
\dot{Q}_{gen} &= \text{Net volumetric heat generation within CV,} \\
\dot{W}_s &= \text{Net rate of work done by surface forces, and} \\
\dot{W}_b &= \text{Net rate of work done by body forces.}
\end{aligned}
$$

Each term will now be represented by a mathematical expression.

Rate of Change

The rate of change is given by,

$$\dot{E} = \frac{\partial(\rho_m \, e^o)}{\partial t} \qquad e^o = e_m + \frac{V^2}{2} = h_m - \frac{p}{\rho_m} + \frac{V^2}{2} , \qquad (6.22)$$

where e represents specific energy (J/kg), h is specific enthalpy (J/kg) and $V^2 = u_1^2 + u_2^2 + u_3^2$. In the above expression for e^o, contributions from other forms (potential, chemical, electromagnetic, etc.) of energy are neglected.

Convection and Conduction

Following the convention that heat energy flowing *into* the CV is positive (and *vice versa*), it can be shown that

$$\dot{Q}_{conv} = -\frac{\partial \sum (N_{j,k} \, e_k^o)}{\partial x_j} , \qquad (6.23)$$

where $N_{j,k}$ is given by Eq. 6.9. However,

$$
\begin{aligned}
\sum N_{j,k} \, e_k^o &= \sum \left[N_{j,k} \left(h_k - \frac{p}{\rho_m} + \frac{V^2}{2} \right) \right] \\
&= \sum \left[N_{j,k} \, h_k \right] + \left(-\frac{p}{\rho_m} + \frac{V^2}{2} \right) \sum N_{j,k} \\
&= \sum \left[N_{j,k} \, h_k \right] + \left(-\frac{p}{\rho_m} + \frac{V^2}{2} \right) \rho_m \, u_j , \qquad (6.24)
\end{aligned}
$$

where $\sum N_{j,k} = \rho_m u_j$. Using Eq. 6.9,

$$\sum N_{j,k} h_k = \sum \left[\rho_m \omega_k u_j h_k + m''_{j,k} h_k \right]$$
$$= \rho_m u_j \sum (\omega_k h_k) + \sum (m''_{j,k} h_k)$$
$$= \rho_m u_j h_m + \sum (m''_{j,k} h_k) , \qquad (6.25)$$

where h_m is the mixture enthalpy. Thus, using Eqs. 6.24 and 6.25, Eq. 6.23 can be written as

$$\dot{Q}_{conv} = -\frac{\partial}{\partial x_j} \left[\rho_m u_j \left(h_m - \frac{p}{\rho_m} + \frac{V^2}{2} \right) \right] - \frac{\partial (\sum m''_{j,k} h_k)}{\partial x_j}$$
$$= -\frac{\partial (\rho_m u_j e^0)}{\partial x_j} - \frac{\partial (\sum m''_{j,k} h_k)}{\partial x_j} . \qquad (6.26)$$

The conduction contribution is given by Fourier's law of heat conduction, so

$$\dot{Q}_{cond} = -\frac{\partial q_j}{\partial x_j} = \frac{\partial}{\partial x_j} \left[k_{eff} \frac{\partial T}{\partial x_j} \right] , \qquad (6.27)$$

where effective conductivity $k_{eff} = k_m + k_t$ and k_t is the turbulent conductivity. The latter is given by

$$k_t = c p_m \frac{\mu_t}{Pr_t} , \qquad (6.28)$$

where Pr_t is the turbulent Prandtl number. In gaseous mixtures, its value is about 0.9.

Volumetric Generation

In general, volumetric energy generation and dissipation can arise from several sources, such as radiation ($\dot{Q}_{rad}$), body (W_b) and shear (W_s) forces, interaction with the electrical field ($\dot{Q}_{elec}$), or with an interspersed medium ($\dot{Q}_{medium}$). The expressions for such sources are given below.

Radiation: Gas mixtures emit, absorb, and scatter radiation in certain wavelength bands. The $\dot{Q}_{rad}$ represents the net radiation exchange between the control volume and its surroundings. Evaluation of $\dot{Q}_{rad}$, in general, requires solution of integro-differential equations called the *radiation transfer equations* [82]. However, in certain restrictive circumstances, the term may be represented as analogous to $\dot{Q}_{cond}$ with k replaced by *radiation conductivity* k_{rad}, as

$$k_{rad} = \frac{16 \sigma T^3}{a + s} , \qquad (6.29)$$

where σ is the Stefan-Boltzmann constant and a and s are absorption and scattering coefficients, respectively.

Shear and body forces: Following the convention that the work done *on the CV* is negative, it can be shown that

$$-\dot{W}_s = \frac{\partial}{\partial x_1} \left[\sigma_1 u_1 + \tau_{12} u_2 + \tau_{13} u_3\right]$$
$$+ \frac{\partial}{\partial x_2} \left[\tau_{21} u_1 + \sigma_2 u_2 + \tau_{23} u_3\right]$$
$$+ \frac{\partial}{\partial x_3} \left[\tau_{31} u_1 + \tau_{32} u_2 + \sigma_3 u_3\right] \tag{6.30}$$

and

$$-\dot{W}_b = \rho_m \left(B_1 u_1 + B_2 u_2 + B_3 u_3\right) . \tag{6.31}$$

Adding the preceding two equations and making use of Eqs. 6.2 and 6.4, it can be shown that

$$-(\dot{W}_s + \dot{W}_b) = \rho_m \frac{D}{Dt} \left[\frac{V^2}{2}\right] + \mu\,\Phi_v - p\,\nabla\,.\,V , \tag{6.32}$$

where $V^2/2$ is the mean kinetic energy and the *viscous dissipation* function is given by

$$\Phi_v = 2 \left[\left(\frac{\partial u_1}{\partial x_1}\right)^2 + \left(\frac{\partial u_2}{\partial x_2}\right)^2 + \left(\frac{\partial u_3}{\partial x_3}\right)^2\right]$$
$$+ \left(\frac{\partial u_1}{\partial x_2} + \frac{\partial u_2}{\partial x_1}\right)^2 + \left(\frac{\partial u_1}{\partial x_3} + \frac{\partial u_3}{\partial x_1}\right)^2 + \left(\frac{\partial u_3}{\partial x_2} + \frac{\partial u_2}{\partial x_3}\right)^2 . \tag{6.33}$$

Interaction with electrical field: If $\vec{F}_{diff,elec}$ is the diffusion field vector, then

$$\dot{Q}_{elec} = \frac{\vec{F}_{diff,elec} \cdot \vec{F}_{diff,elec}}{\lambda_{elec}} . \tag{6.34}$$

Interaction with interspersed medium: Sometimes the flowing fluid is in thermal contact with an interspersed medium. In such cases

$$\dot{Q}_{medium} = \alpha_{conv,vol} \left(T_{medium} - T\right) , \tag{6.35}$$

where $\alpha_{conv,vol}$ is the volumetric heat transfer coefficient (W/m^3-K) that must be known. The last two sources occur in special cases only and hence will be referred to as $\dot{Q}_{others}$.

Final Form of Energy Equation

Combining Eqs. 6.21–6.35, it can be shown that

$$\frac{\partial \rho_m e^o}{\partial t} + \frac{\partial(\rho_m u_j e^o)}{\partial x_j} = \frac{\partial}{\partial x_j}\left[k_{eff}\frac{\partial T}{\partial x_j}\right] - \frac{\partial(\sum m''_{j,k}h_k)}{\partial x_j}$$

$$+ \rho_m \frac{D}{Dt}\left[\frac{V^2}{2}\right] - p\,\nabla\cdot V + \mu\,\Phi_v$$

$$+ \dot{Q}_{rad} + \dot{Q}_{others}\,. \tag{6.36}$$

Enthalpy and Temperature Forms

In combustion calculations, it is of interest to derive the enthalpy and temperature forms of the energy equation. Thus, using Eq. 6.1, the left-hand side of Eq. 6.36 can be replaced by $\rho_m\,De^o/Dt$. Further, if e^o is replaced by Eq. 6.22, then

$$\rho_m \frac{D e^o}{D t} = \rho_m \frac{D h_m}{D t} + \frac{D}{D t}\left[\frac{V^2}{2}\right] - \frac{D p}{D t} + \frac{p}{\rho_m}\frac{D \rho_m}{D t}$$

$$= \rho_m \frac{D h_m}{D t} + \frac{D}{D t}\left[\frac{V^2}{2}\right] - \frac{D p}{D t} - p\,\nabla\cdot V\,. \tag{6.37}$$

Hence, Eq. 6.36 can also be written as

$$\rho_m \frac{D h_m}{D t} = \frac{\partial}{\partial x_j}\left[k_{eff}\frac{\partial T}{\partial x_j}\right] - \frac{\partial(\sum m''_{j,k}h_k)}{\partial x_j} + \mu\,\Phi_v$$

$$+ \frac{D p}{D t} + \dot{Q}_{rad} + \dot{Q}_{others}\,. \tag{6.38}$$

This is the energy equation in *enthalpy form*. In reacting flows, it will be recalled, the mixture enthalpy (see Eqs. 3.55 and 3.56) is given by

$$h_m(T) = \sum_k \omega_k\,h_k \tag{6.39}$$

$$h_k = h^0_{f,k}(T_{ref}) + \Delta h_{s,k}(T)$$

$$= h^0_{f,k}(T_{ref}) + \int_{T_{ref}}^{T} cp_k\,dT\,. \tag{6.40}$$

Therefore, it follows that

$$\frac{D h_m}{D t} = \sum_k \omega_k \frac{D h_k}{D t} + \sum_k h_k \frac{D \omega_k}{D t}\,.$$

Using Eq. 6.39 and the definition of mixture-specific heat and, further, using the nonconservative form of Eq. 6.12, it can be shown that

$$\frac{D\,h_m}{D\,t} = c_{pm}\,\frac{D\,T}{D\,t} + \sum_k \frac{h_k}{\rho_m}\left(R_k - \frac{\partial m''_{j,k}}{\partial x_j}\right).$$

Therefore, grouping enthalpy-containing terms in Eq. 6.38, we have

$$\rho_m\,\frac{D\,h_m}{D\,t} + \frac{\partial(\sum_k m''_{j,k}\,h_k)}{\partial x_j} = \rho_m\,c_{pm}\,\frac{D\,T}{D\,t} + \sum_k h_k\,R_k - \sum_k m''_{j,k}\,\frac{\partial h_k}{\partial x_j},$$

or, replacing h_k from Eq. 6.40 and using Eqs. 6.10, 6.38 can be written as

$$\rho_m\,c_{pm}\,\frac{D\,T}{D\,t} = \frac{\partial}{\partial x_j}\left[k_{eff}\,\frac{\partial T}{\partial x_j}\right] + \mu\,\Phi_v + \frac{D\,p}{D\,t} + \dot{Q}_{rad} + \dot{Q}_{others}$$
$$- \sum_k (h^0_{f,k} + \Delta h_{s,k})\,R_k - \rho_m\,\sum_k D\,\frac{\partial \omega_k}{\partial x_j}\,\frac{\partial h_k}{\partial x_j}. \qquad (6.41)$$

The following comments are important in this *temperature form* of the energy equation.

1. $-\sum_k h^0_{f,k}\,R_k = \dot{Q}_{chem}$ can be recognized as the net volumetric rate of heat generation from a chemical reaction. For a simple chemical reaction (SCR), $\dot{Q}_{chem} = R_{fu}\,\Delta H_c$.
2. If specific heats of all species were treated equal (that is, $c_{pj} = c_{pm}$) then $\sum_k \Delta h_{s,k}\,R_k = 0$ and the last term is also rendered zero because $\sum_k D\,\partial \omega_k/\partial x_j = 0$.
3. The viscous dissipation $\mu\,\Phi_v$ terms are important only at high Mach numbers (more strictly, Eckert no. $= V^2/c_{pm}\,\Delta T_{ref}$). Likewise, the Dp/Dt term is important in compressible flows particularly in the presence of shocks.

6.5 Two-Dimensional Boundary Layer Flow Model

A wide variety of combustion phenomena, such as flames, ignition, quenching, detonation, gasification, pollutant formation, and fires are strongly influenced by chemical kinetics and fluid mixing effects. In fact, practical equipment is designed or its performance is predicted by accounting for these effects.

To derive practical information, it is necessary to solve the three-dimensional partial differential equations of mass conservation (6.1), mass transfer (6.12), momentum (6.2), and energy (6.38) with appropriate boundary conditions. These equations are called *transport equations* for mass, species, momentum, and energy. Of course, because the equations are strongly coupled and non-linear, *analytical solutions* are impossible. Recourse is, therefore, made to obtain *numerical solutions* in discretized space.

The subject of Computational Fluid Dynamics (CFD) deals mainly with obtaining efficient numerical solutions to the phenomena occurring in practical equipment that always pose a very complex geometry. Fortunately, developments in the past forty years have brought CFD to a considerable level of maturity, and generalized computer codes are now commercially available. Using these codes, phenomena in a wide variety of equipment, such as the combustion chamber of a Gas Turbine, the cylinder of an IC engine, blast furnaces, cement kilns, and cooking stoves can be predicted. The predictions yield, in great detail, distributions of velocities, temperatures, and mass/mole-fractions of species in three-dimensional space and in time. Careful observations of such distributions enable the study of variations in design (usually geometric) parameters, as well as in operating (effects of air-fuel ratio or change of fuel, for example) parameters influencing performance of a practical combustion device.

Description of CFD for the complete set of transport equations is beyond the scope of this book. In fact, until the 1960s, it was also not possible to solve the complete set of equations because of the unavailability of fast computers. Still, much of the value could be derived from simplified forms of the equations by postulating a simplified flow model. One such form, which is particularly relevant to combustion phenomena such as burning adjacent to a solid surface (say, burning of a coal particle) or a free or forced flame, is called the *boundary layer flow model*. Adjacent to a solid (or liquid) surface (see Fig. 6.2(a)), the flow is often unidirectional and parallel to the surface, and major variations of flow properties (velocity, temperature, or mass fraction) occur normal to the surface. The variations mainly occur over a small distance normal to

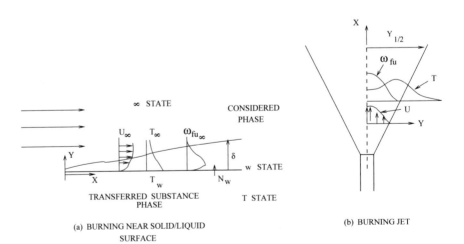

Fig. 6.2 Examples of boundary layer flows

the surface compared with the overall dimensions of the flow. As such, the flow may be regarded as long in the streamwise direction and thin in the direction transverse to the surface. Similarly, a free flame (Fig. 6.2(b)) is long in the direction of the main velocity and thin across the flame. Such long and thin flows are called boundary layer flows.

6.5.1 Governing Equations

Two-dimensional boundary layers have been studied extensively for a wide variety of reacting and nonreacting situations under laminar and turbulent conditions and with and without fluid property variations. For such two-dimensional long and thin flows, $\partial \Phi / \partial z = 0$, $\partial^2 \Phi / \partial x^2 << \partial^2 \Phi / \partial y^2$, and $\partial p / \partial y \simeq 0$ for $\Phi = u$, ω_k and h_m. Here, x is a streamwise coordinate direction and y is a transverse coordinate direction whereas z is perpendicular to x and y. These assumptions simplify the complete transport equations of the previous section. They are reduced to the following forms:

Mass Conservation Equation:

$$\frac{\partial \rho_m}{\partial t} + \frac{\partial (\rho_m u)}{\partial x} + \frac{\partial (\rho_m v)}{\partial y} = 0. \tag{6.42}$$

Momentum Equation:

$$\frac{\partial (\rho_m u)}{\partial t} + \frac{\partial (\rho_m u u)}{\partial x} + \frac{\partial (\rho_m v u)}{\partial y} = -\frac{dp}{dx} + \mu \frac{\partial \tau_{yx}}{\partial y}, \tag{6.43}$$

where shear stress and the streamwise pressure gradient are given by

$$\tau_{yx} = \mu \frac{\partial u}{\partial y} - \rho_m \overline{u' v'}$$

$$= (\mu + \mu_t) \frac{\partial u}{\partial y} = \mu_{eff} \frac{\partial u}{\partial y} \tag{6.44}$$

$$\frac{dp}{dx} = -\rho_m U_\infty \frac{d U_\infty}{dx} \tag{6.45}$$

and U_∞ is the known free-stream velocity. As such, dp/dx is known a priori. For free shear layers, such as jets and wakes, dp/dx = 0. The turbulent stress is modeled in the manner of laminar stress, with μ replaced by μ_t.

Species Equation:

$$\frac{\partial (\rho_m \omega_k)}{\partial t} + \frac{\partial (\rho_m u \omega_k)}{\partial x} + \frac{\partial (\rho_m v \omega_k)}{\partial y} = \frac{\partial}{\partial y} \left[\rho_m D_{eff} \frac{\partial \omega_k}{\partial y} \right] + R_k. \tag{6.46}$$

Table 6.1 Generalised representation of boundary layer equation

Φ	Γ_Φ	S_Φ
1	0	0
u	μ_{eff}	$-dp\,/\,dx + B_x$
ω_k	$\rho_m\,D_{eff}$	R_k
η_α	$\rho_m\,D_{eff}$	0
h_m	k_{eff}/c_{pm}	$Dp/Dt + \dot{Q}_{rad} + \dot{Q}_{others} - \partial(\sum m''_{j,k}\,h_k)/\partial y$
T	k_{eff}/c_{pm}	$(Dp/Dt + \dot{Q}_{rad} + \dot{Q}_{others})/c_{pm}$
		$-\left[\sum_k (h^0_{f,k} + \Delta h_{s,k})\,R_k + \rho_m \sum_k D\,(\partial\omega_k/\partial y)\,(\partial h_k/\partial y) \right]/c_{pm}$

Element Equation:

$$\frac{\partial(\rho_m\,\eta_\alpha)}{\partial t} + \frac{\partial(\rho_m\,u\,\eta_\alpha)}{\partial x} + \frac{\partial(\rho_m\,v\,\eta_\alpha)}{\partial y} = \frac{\partial}{\partial y}\left[\rho_m\,D_{eff}\,\frac{\partial\eta_\alpha}{\partial y} \right]. \qquad (6.47)$$

Energy Equation:

$$\frac{\partial(\rho_m\,h_m)}{\partial t} + \frac{\partial(\rho_m\,u\,h_m)}{\partial x} + \frac{\partial(\rho_m\,v\,h_m)}{\partial y} = \frac{\partial}{\partial y}\left[\frac{k_{eff}}{c_{pm}}\,\frac{\partial h_m}{\partial y} \right] + \mu\left(\frac{\partial u}{\partial y}\right)^2$$

$$+ \frac{Dp}{Dt} + \dot{Q}_{rad} + \dot{Q}_{others}$$

$$- \frac{\partial(\sum m''_{j,k}\,h_k)}{\partial y}, \qquad (6.48)$$

where $Dp/Dt = \partial p/\partial t + u\,dp/dx$. It is a happy coincidence that the above equations can be written in the following general form:

$$\frac{\partial(\rho_m\,\Phi)}{\partial t} + \frac{\partial(\rho_m\,u\,\Phi)}{\partial x} + \frac{\partial(\rho_m\,v\,\Phi)}{\partial y} = \frac{\partial}{\partial y}\left[\Gamma_\Phi\,\frac{\partial\Phi}{\partial y} \right] + S_\Phi, \qquad (6.49)$$

where the values of Γ_Φ and S_Φ are given in Table 6.1 for each Φ. Note that $\Phi = 1$ represents mass conservation Equation 6.42.

6.5.2 Boundary and Initial Conditions

Solution of Eq. 6.49 requires specification of one *initial condition* at x $= x_0$, say and two *boundary conditions* at y $= 0$ and y $\to \infty$ for each Φ. For wall boundary layers and free shear layers, the initial condition must be specified from experimental data or from known exact or approximate analytical solutions. How x_0 is to be selected is explained in Ref. [30].

In free shear layers, $y = 0$ is the symmetry boundary. As such, $\partial \Phi / \partial y \mid_{y=0} = 0$ for all variables. For wall flows, $y = 0$ implies a solid/liquid surface where no-slip condition applies. As such, $u_{y=0} = 0$. For other variables, different types of conditions must be specified, depending on the problem under consideration. Some generalizations of this boundary condition are described in Ref. [30].

At $y \rightarrow \infty$, *free-stream* boundary conditions are to be applied for all variables. The treatment for free and wall flows is the same. Here, the boundary conditions for mass transfer variables are explained in some detail.

With reference to Fig. 6.2(a), the mass transfer takes place from the *transferred substance*[5] phase to the *considered* phase. The two phases are separated by an *interface*. The state of the substance at the interface is in w-state. The T-state defines the transferred substance state. All variations of Φ occur in the considered phase. The ∞-state is defined at $y \rightarrow \infty$. The mass transfer flux at the wall N_w (kg /m²-s) defines the burning rate of the substance in the T-state and is given by [110]:

$$N_w = \rho_{m,w} \, v_w \tag{6.50}$$

$$= g^* \ln (1 + B) \tag{6.51}$$

$$g^* = \frac{- \Gamma_m \, \partial \Psi / \partial y \mid_w}{(\Psi_w - \Psi_\infty)} \tag{6.52}$$

$$B = \frac{(\Psi_\infty - \Psi_w)}{(\Psi_w - \Psi_T)} \, , \tag{6.53}$$

where Ψ is any conserved property.[6] v_w is the surface-normal wall velocity, g^* (kg /m²-s) is *mass transfer coefficient* (analogous to the heat transfer coefficient), and B is the Spalding number.[7] Equation 6.50 serves as a boundary condition for the momentum equation. The rather unusual form of relating wall mass flux to the mass transfer coefficient shown in Eq. 6.51 derives from the Reynolds flow model, which will be described in Sect. 6.7.

As $y \rightarrow \infty$, $\partial \Phi / \partial y$ and $\partial^2 \Phi / \partial y^2$ tend to zero. As such, the Φ profile near the edge of the boundary layer shows asymptotic behavior. The $y \rightarrow \infty$ plane is nonetheless identified[8] with a physical thickness δ_Φ, which varies with x. As such the boundary layer equations are solved with the second set of boundary conditions

---

[5]It is also called the *neighboring* phase.

[6]In the chapters to follow, we shall encounter different conserved properties. A conserved property is one whose transport equation has a *zero source*. In combustion problems, Ψ is formed from the combinations of ω_k and/or h_m. However, η_α is always a conserved property. Similarly, in an inert mixture (air + water vapor, for example), ω_{vap} and ω_{air} will be conserved properties .

[7]In mass transfer literature, B is also called a dimensionless *driving force*.

[8]To a first approximation, thickness δ_Φ may be taken to be the value of y where $(\Phi - \Phi_w)/(\Phi_\infty - \Phi_w) \simeq 0.99$.

$$\frac{\partial \Phi}{\partial y}\Big|_{\delta_\Phi} = \frac{\partial^2 \Phi}{\partial y^2}\Big|_{\delta_\Phi} = 0 \ . \tag{6.54}$$

To identify the essential nature of the solution to the general problem of mass transfer, even further simplifications are often introduced so that simple analytical solutions become possible. These assumptions yield three further flow models: the *Stefan Flow* model (Sect. 6.6) , *Reynolds Flow* model (Sect. 6.7), and *Couette Flow* model (Sect. 6.8).

6.6 One Dimensional Stefan Flow Model

The physical construct of this model is shown in Fig. 6.3. The model represents a Stefan tube, in which the neighboring phase fluid/solid is kept at the bottom. In this model, the fluid in the considered phase is assumed to be *quiescent*, so $u = dp/dx = B_x = 0$. It is also assumed that the neighboring phase fluid is *constantly fed to the tube at the rate equal to the rate of mass transfer*. As such, steady state prevails. Then, the boundary layer Eq. 6.49 reduces to

$$\frac{d}{dy}\left[N_{\Phi,y}\,A\right] = \frac{d}{dy}\left[\rho_m\,v\,A\,\Phi - \Gamma_\Phi\,A\,\frac{d\,\Phi}{dy}\right] = S_\Phi \ , \tag{6.55}$$

where $A(y)$ is the cross-sectional area and $N_{\Phi,y}$ is the convection-diffusion mass flux in direction y (see Eq. 6.9).

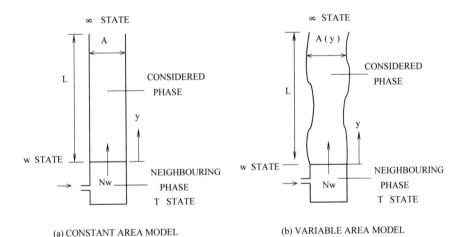

(a) CONSTANT AREA MODEL   (b) VARIABLE AREA MODEL

Fig. 6.3 Stefan flow model

As will be shown in Chap. 10, the solution to this equation takes the form

$$N_w = \frac{\Gamma_m}{L_{eff}} \ln (1 + B)$$

$$B = \frac{(\Phi_\infty - \Phi_w)}{(\Phi_w - \Phi_T)}$$

$$L_{eff} = A_w \int_0^L \frac{dy}{A} \ , \tag{6.56}$$

where, A_w is the interface area at $y = 0$. Note that $L_{eff} = L$ if A is constant. Equation 6.56 is similar to Eq. 6.51 but with $g^* = \Gamma_m/L_{eff}$. In this sense, Eq. 6.56 may be viewed as the conduction/diffusion analog of the convective Eq. 6.51.

6.7 Reynolds Flow Model

This model, developed by Spalding [110], is algebraic, and no differential equations are involved. Even so, the model claims to account for all the effects produced at the w surface in a real boundary layer model. The model construction is shown in Fig. 6.4. The model postulates the existence of two *fictitious* fluxes, g and $(N_w + g)$, at an imaginary surface in the ∞ state. This imaginary surface is assumed to coincide with the boundary layer thickness. The g-flux flows toward the interface carrying with it properties of the ∞ state. In contrast, the $(N_w + g)$ flux flows away from the interface but carries with it the *properties of the w-state*. The two fluxes ensure that there is no net mass creation between the ∞ surface and the w surface, because mass flux N_w enters the considered phase at the w surface. Likewise, no mass is created or destroyed in the transferred substance phase. As such, the same N_w is shown at the w and T surfaces.

With this construct, we invoke the conservation principle for any conserved property Ψ in the region between ∞ surface and the T surface. Then

$$\text{Mass flux in} = g\, \Psi_\infty + N_w\, \Psi_T \quad \text{and} \tag{6.57}$$
$$\text{Mass flux out} = (N_w + g)\, \Psi_w . \tag{6.58}$$

Equating the in- and out-fluxes, we have

$$N_w = g\, \frac{\Psi_\infty - \Psi_w}{\Psi_w - \Psi_T} = g\, B . \tag{6.59}$$

This equation shows that N_w will bear the same sign as B. As such, in combustion, evaporation, or ablation, B will be positive but in condensation, B will be negative. Further, for this equation to be the same as Eq. 6.51, we must establish the relationship between g and g^*. Spalding [110] carried out an extensive analysis of *experimental*

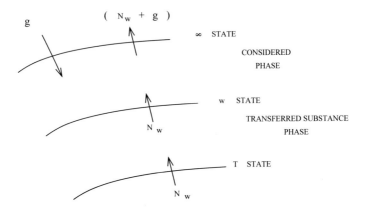

Fig. 6.4 Reynolds flow model

mass transfer data in both external and internal flow boundary layers and showed that

$$\frac{g}{g^*} \simeq \frac{\ln(1+B)}{B} \,. \qquad (6.60)$$

Combining this equation with Eqs. 6.59, 6.51 is readily recovered. It is for this reason that Eqs. 6.51, 6.52, and 6.53 can be trusted as a legitimate boundary condition at y = 0 in a boundary layer flow adjacent to a solid/liquid surface.

Spalding [110] further showed that g^* is nothing but value of g when $B \to 0$, or a case of very low mass transfer rates.[9] As such, Eq. 6.52 can also be written as

$$g^* = \left(\frac{N_w}{B}\right)_{B\to 0} = \frac{-\Gamma_m\, \partial\Psi/\partial y\,|_w}{(\Psi_w - \Psi_\infty)} \,. \qquad (6.61)$$

By invoking the analogy between heat and mass transfer in gaseous flows, Kays [62] has shown that for a given mass transfer situation, g^* can, in fact, be related to heat transfer coefficient α_{conv} for the corresponding heat transfer situation in which $v_w = 0$, or the case of zero mass transfer. This relationship is given by

$$g^* \simeq \left(\frac{\alpha_{conv}}{cp_m}\right) \times PF_c \qquad (6.62)$$

$$\text{and}\ \ PF_c = \left(\frac{Sc}{Pr}\right)^n \times \left(\frac{M_w}{M_\infty}\right)^m \,,$$

where PF_c is the property correction factor, M denotes molecular weight, and Pr and Sc are the Prandtl and Schmidt numbers, evaluated from the mean of the fluid properties in the w and ∞ states. The exponent m $= -2/3$ for the laminar boundary

---

[9]Note that $\ln(1+B) \to B$ as $B \to 0$, irrespective of the sign of B.

layer and m $= -0.4$ for turbulent boundary layer [62]. The value of n depends on the correlation for the particular situation under consideration. As correlations for α_{conv} $(v_w = 0)$ are extensively available (see [51], for example), g^* can be readily evaluated *without solving* any differential equation. Thus, using the Reynolds flow model, N_w can be evaluated by combining Eqs. 6.59, 6.60 and 6.62. The appropriate formula is

$$N_w = \left(\frac{\alpha_{conv}}{cp_m} \right) \times P F_c \times \ln{(1 + B)} . \tag{6.63}$$

6.8   One Dimensional Couette Flow Model

6.8.1   Momentum Equation

This model is the closest approximation to the Boundary Layer flow model. Here it is assumed that streamwise velocity u varies *linearly* with transverse distance y instead of the true non-linear variation seen in Fig. 6.2. Thus, u = c y. Further, in common with the stefan flow model, we set dp/dx $= B_x = 0$ (see Table 6.1 and Fig. 6.5).

Under these assumptions, the steady state form Eq. 6.49 for $\Phi = u$, will read as

$$\frac{d}{dy} \left[N_w\, u - (\mu + \mu_t) \frac{d\,u}{dy} \right] = 0 \quad \text{where} \quad N_w = \rho\, v_w \tag{6.64}$$

Integrating this equation once and noting that $\mu_t = u = 0$ at $y = 0$ and $\mu\,(d\,u/dy)|_{y=0} = \tau_w$, the constant of integration derives to $C = -\tau_w$. Hence,

$$N_w\, u - (\mu + \mu_t) \frac{d\,u}{dy} = -\tau_w \tag{6.65}$$

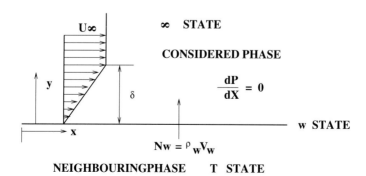

Fig. 6.5   Couette flow model

Integration of this equation across the boundary layer gives

$$\int_0^\infty \frac{du}{N_w\, u + \tau_w} = \int_0^\delta \frac{dy}{\mu + \mu_t} = C_1 \text{ (say)} = \frac{1}{N_w} \ln\left[1 + \frac{N_w\, U_\infty}{\tau_w}\right] \qquad (6.66)$$

where,

$$\frac{N_w\, U_\infty}{\tau_w} = \frac{\rho_w\, V_w\, U_\infty}{\tau_w} = \frac{V_w/U_\infty}{\tau_w/\rho\, U_\infty^2} = \frac{V_w/U_\infty}{C_{f,x}/2} = B_f \qquad (6.67)$$

where C_f is the skin-friction coefficient (or, drag coefficient) and B_f is called the Blowing parameter when v_w is positive. Negative B_f implies boundary layer suction. Thus, Eq. 6.66 can be written as

$$\ln\,(1 + B_f) = C_1\, N_w = C_1\, \rho\, U_\infty\, B_f\,(C_{f,x}/2) \qquad (6.68)$$

Now, as $B_f \to 0$, let $C_{f,x} = C_{f,x,V_w=0}$. Then, assuming C_1 remains independent of whether V_w is finite or zero

$$\frac{C_{f,x,V_w}}{C_{f,x,V_w=0}} \simeq \frac{\ln\,(1 + B_f)}{B_f} \qquad (6.69)$$

The following comments are relevant.

1. This equation is applicable to both laminar and turbulent flow. It is derived for $dp/dx = 0$ but can be taken to be valid for mild pressure gradients as well.
2. Large number of experimental correlations, as well as theoretical developments in the open literature provide information on $C_{f,x,v_w=0}$. Such information can be used to derive guidance for estimating drag in the presence of mass transfer.
3. The equation shows that under positive mass transfer as in evaporation or ablation, the wall shear stress will be smaller than that when there is no mass transfer. This result is often used in estimating the minimum combustion chamber length of a Liquid propellent combustor of Rocket Engine [114].

6.8.2 Scalar Transport Equation

The Coette flow model is now applied to scalar variables $\Phi = \omega_k, \eta_\alpha, h_m$ and T in Table 6.1. Under appropriate assumptions (see Footnote 6) Eq. 6.49 can be reduced to the following conserved property form.

$$\frac{d}{dy}\left[N_w\,(\Psi - \Psi_w) - (\Gamma + \Gamma_t)\,\frac{d\,(\Psi - \Psi_w)}{dy}\right] = 0 \qquad (6.70)$$

where $N_w = N_{\Psi,y}$. Further note that deep inside the transferred substance state $(d\,\Psi/dy)_T = 0$ and in the w-state (that is, $y = 0$), $\Gamma_t = 0$. Thus, integrating equation 6.70 once, the constant of integration C_1 can be expressed as

$$N_w\,(\Psi - \Psi_w) - (\Gamma + \Gamma_t)\frac{d\,\Psi}{dy} = C_1 = N_w\,(\Psi_T - \Psi_w) = -\,\Gamma\,\frac{d\,\Psi}{dy}\Big|_w \quad (6.71)$$

This equation can be manipulated further and upon integration, it can be shown that

$$\frac{1}{N_w}\int_0^\infty \frac{d\,\Psi}{(\Psi - \Psi_T)} = \int_0^\delta \frac{dy}{\Gamma + \Gamma_t} = C_2 \quad \text{(say)} \quad (6.72)$$

Integration of the left-hand side of this equation gives

$$N_w = \frac{1}{C_2}\ln\,(1 + B_\Psi) \quad \rightarrow \quad B_\Psi = \frac{\Psi_\infty - \Psi_w}{\Psi_w - \Psi_T} \quad (6.73)$$

and from Eq. 6.71, we also have

$$N_w = \frac{C_1}{\Psi_T - \Psi_w} = \frac{-\,\Gamma\,(d\,\Psi/dy)_w}{\Psi_T - \Psi_w} \quad (6.74)$$

Now, consistent with the theory of heat transfer, we may write

$$-\,\Gamma\,\frac{d\,\Psi}{dy}\Big|_w = g \times (\Psi_w - \Psi_\infty) \quad (6.75)$$

where g is the mass transfer coefficient (kg/m^2-s). Further, using Eq. 6.74, we have

$$N_w = g \times \left(\frac{\Psi_\infty - \Psi_w}{\Psi_w - \Psi_T}\right) = g \times B_\Psi \quad \text{and} \quad (6.76)$$

$$g = \frac{1}{C_2}\frac{\ln\,(1 + B_\Psi)}{B_\Psi} \quad (6.77)$$

Now, let $g \rightarrow g^*$ as $B_\Psi \rightarrow 0$ and assume that C_2 remains the same for with and without mass transfer. Then

$$\frac{g}{g^*} = \frac{\ln\,(1 + B_\Psi)}{B_\Psi} \quad (6.78)$$

This equation shows that the fictitious g^* flux is given by

$$N_w = g^* \ln{(1 + B_\Psi)} \quad \text{where} \quad \frac{1}{g^*} = C_2 = \int_0^\delta \frac{dy}{\Gamma + \Gamma_t} \qquad (6.79)$$

Thus, g^* may be viewed as the sum of layer-by-layer resistances to mass transfer in the considered phase over the width δ. This interpretation of g^* enables its evaluation from known $\Gamma\,(\Psi, y)$ in a laminar boundary layer and from known $\Gamma_t\,(y)$ from a turbulence model (mixing length (see next section), for example) in a turbulent boundary layer. Thus, the Couette flow model permits the study of property variations as well. In fact, notice that if $\Gamma = \text{const}$ and $\Gamma_t = 0$ then $g^* = \Gamma/\delta$ which is same as the Stefan flow model with $g^* = \Gamma/L_{eff}$.

6.8.3 Special Case of Pure Heat Transfer

If we consider the case of *pure heat transfer in the presence of suction or blowing*, with $\Psi = h_m = c_p\,T$, then from Eq. 6.75, we have

$$- k\,\frac{d\,T}{dy}\Big|_w = g\,c_p\,(T_w - T_\infty) \equiv \alpha_{cof,V_w}\,(T_w - T_\infty) \qquad (6.80)$$

where $\alpha_{cof,V_w} = $ is heat transfer coefficient in the presence of v_w. Therefore, $\alpha_{cof,V_w}/c_p = g$ and, consequently, $\alpha_{cof,V_w=0} = g^*/c_p$. Hence,

$$\frac{g}{g^*} = \frac{\alpha_{cof,V_w}}{\alpha_{cof,V_w=0}} = \frac{St_{x,V_w}}{St_{x,V_w=0}} = \frac{\ln{(1 + B_h)}}{B_h} \quad \rightarrow \quad B_h = \frac{T_\infty - T_w}{T_w - T_T} \qquad (6.81)$$

where $St_x = \alpha_{cof}/(\rho\,c_p\,U_\infty)$ is the local Stanton number. Equation 6.81 is analogous to Eq. 6.69 for the friction coefficient. The equation verifies Eq. 6.78 and, therefore, the definition of Spading number B_h for pure heat transfer can be deduced from Eq. 6.77.

Thus, the Couette flow model justifies both the Stefan flow and Reynolds flow models.

6.9 Turbulence Models

6.9.1 Basis of Modeling

Any turbulent flow is always a three-dimensional and time-dependent phenomenon in which all dependent flow variables, such as velocity, pressure, temperature, and mass fractions, vary randomly in space and time. In a turbulent flow, the fluid continuum holds. As such, the flow is governed by Navier–Stokes Equations 6.1 and 6.2. However, to acknowledge this randomness, the equations are rewritten in *instantaneous* variables[10]

$$\frac{\partial(\rho_m)}{\partial t} + \frac{\partial(\rho_m \, \hat{u}_j)}{\partial x_j} = 0 \, , \tag{6.82}$$

$$\frac{\partial(\rho_m \, \hat{u}_i)}{\partial t} + \frac{\partial(\rho_m \, \hat{u}_j \, \hat{u}_i)}{\partial x_j} = \frac{\partial(\hat{\tau}_{ij})}{\partial x_j} - \frac{\partial \hat{p}}{\partial x_i} \, . \tag{6.83}$$

Time averaging of these equations, coupled with the model for turbulent stress $\tau_{ij}^t = - \rho_m \, \overline{u_i' \, u_j'}$ (Eq. 6.3) results in the time-averaged momentum Equations 6.6.

The central problem of modeling turbulence now is to represent the isotropic turbulent viscosity μ_t in terms of scalar variables characterizing turbulence. In the so called *eddy-viscosity* models, this is done analogous to the manner in which *laminar viscosity* μ is modeled. The latter is represented as

$$\mu = \rho_m \, l_{mfp} \, u_{mol} \, , \tag{6.84}$$

where l_{mfp} is the mean free path length of the molecules and u_{mol} is the average molecular velocity. Likewise, in turbulent flow, the turbulent viscosity is modeled as

$$\mu_t = \rho_m \, l \, |u'| \, , \tag{6.85}$$

where l is the length-scale of turbulence and $|u'|$ is the representative velocity fluctuation. Both l and $|u'|$ will vary from point to point in the flow and, consequently, μ_t will also vary. Thus, whereas μ is a property of the fluid, μ_t is a *a property of the flow*.

6.9.2 *Modeling $|u'|$ and l*

Mixing Length Model

The simplest model for the velocity scales is given by

$$|u'| \propto l \times |\text{mean velocity gradient}| \, , \tag{6.86}$$

so the turbulent viscosity is given by

$$\mu_t = \rho_m \, l_m^2 \left\{ \frac{\partial u_i}{\partial x_j} \left(\frac{\partial u_i}{\partial x_j} + \frac{\partial u_j}{\partial x_i} \right) \right\}^{0.5} \qquad \text{summation,} \tag{6.87}$$

----

[10]Body forces B_i are ignored here.

where l_m is called *Prandtl's mixing length*. For wall bounded flows, l_m is modeled as

$$l_m = \kappa \times n \times (1 - e^{-\xi}) \quad \xi = \frac{1}{26} \frac{n}{\nu} \sqrt{\frac{\tau_w}{\rho_m}} = \frac{n^+}{26} \quad (6.88)$$

$$l_m = 0.2 \times \kappa \times R , \quad \text{if} \quad n^+ > 26 , \quad\quad (6.89)$$

where n is the *normal distance* from the nearest wall and $\kappa \simeq 0.41$, R is a characteristic dimension (radius of a pipe or boundary layer thickness, for example), and wall shear stress τ_w is evaluated from the product of μ and total velocity gradient normal to the wall.

For two-dimensional wall boundary layers, of course, Eq. 6.87 reduces to

$$\mu_t = \rho_m l_m^2 \frac{\partial u}{\partial y} , \quad\quad (6.90)$$

with replacement n = y and R = δ in Eqs. 6.88 and 6.89, where δ is the local boundary layer thickness (see Fig. 6.2(a)). For two-dimensional free shear layers, such as jets or wakes, Eq. 6.90 is used again with

$$l_m = \beta \, y_{1/2} , \quad\quad (6.91)$$

where $y_{1/2}$ is the half-width of the shear layer (see Fig. 6.2(b)) and $\beta = 0.225$ (a plane jet), = 0.1875 (a round jet), = 0.40 (a plane wake).

One Equation Model

In this model, the velocity fluctuation scale is taken as $|u'| \propto \sqrt{k}$, where k represents the *turbulent kinetic energy* $(\overline{u_i' u_i'}/2)$. The distribution of k in a flow is determined from the following transport equation.[11]

$$\frac{\partial(\rho_m k)}{\partial t} + \frac{\partial(\rho_m u_j k)}{\partial x_j} = \frac{\partial}{\partial x_j} \left[(\mu + \mu_t) \frac{\partial k}{\partial x_j} \right] + G - \rho_m \epsilon , \quad\quad (6.92)$$

where G represents the *generation rate* of turbulent kinetic energy and ϵ its *dissipation rate*. These terms are modeled as follows:

$$G = (\mu + \mu_t) \, \Phi_v , \quad\quad (6.93)$$

$$\epsilon = \frac{k^{3/2}}{l} \left[1 - \exp\left(-\frac{Re_n}{5.1} \right) \right] , \quad\quad (6.94)$$

---

[11] The transport equation is derived as follows: First, Eq. 6.83 is multiplied by $\hat{u}_i$. The resulting equation is then *time-averaged* to yield an equation with dependent variable (E + k), where mean kinetic energy E = $V^2/2$. Next, Eq. 6.2 is multiplied by the time-averaged mean velocity u_i to yield an equation with dependent variable E. Subtracting this equation from the first equation yields an equation for k. This equation contains terms that need *modeling*. Equation 6.92 represents this modeled form, with $\rho_m \epsilon$ representing the *turbulent counterpart* of $\mu \, \Phi_v$.

$$Re_n = \frac{n \sqrt{k}}{\nu} \; , \tag{6.95}$$

where Φ_v is given by Eq. 6.33. The turbulent viscosity is now defined as

$$\mu_t = f_\mu \, C_\mu^{0.25} \, \rho_m \, l \, \sqrt{k} \; , \tag{6.96}$$

$$f_\mu = 1 - \exp\left(-\frac{Re_n}{70}\right) \; , \tag{6.97}$$

$$l = \kappa \, C_\mu^{-0.75} \, n \; , \tag{6.98}$$

$$C_\mu = 0.09, \qquad \kappa = 0.41, \tag{6.99}$$

where l is the *integral length scale of turbulence* and n is the normal distance to the wall. The modeling expressions and the associated constants are proposed by Wolfstein [127]. This model successfully mimics the idea that dissipation rate of turbulence kinetic energy dominates at smaller length scales.[12]

Thus, in this one-equation model, Eq. 6.92 is solved (with wall boundary condition $k = 0$) along with the momentum, energy, and species transport equations whereas the length scale of turbulence l is given by Eq. 6.98. In two dimensional wall boundary-layer flows, Eq. 6.92 reduces to

$$\frac{\partial(\rho_m \, k)}{\partial t} + \frac{\partial(\rho_m \, u \, k)}{\partial x} + \frac{\partial(\rho_m \, v \, k)}{\partial y} = \frac{\partial}{\partial y}\left[(\mu + \mu_t)\frac{\partial k}{\partial y}\right]$$
$$+ (\mu + \mu_t)\left(\frac{\partial u}{\partial y}\right)^2$$
$$- \rho_m \, \epsilon \; . \tag{6.100}$$

Two Equation Model

In this model [54, 69], the turbulent viscosity is given by

$$\mu_t = C_\mu \, f_\mu \, \rho_m \, \frac{k^2}{\epsilon^*} \; , \tag{6.101}$$

---

[12]In modeling turbulent flows, the main governing idea is that turbulent kinetic energy generation (G) is maximum at large *integral* length scales denoted by $l = l_{int}$. This energy generation breaks the turbulent eddies to smaller length scales, while the energy is simultaneously convected and diffused. The process continues down to ever smaller scales, until the smallest scale is reached. This smallest scale is called the *Kolmogorov length scale* $l = l_\epsilon$. In this range of smallest scales, through the action of fluid viscosity, all fluctuations are essentially *killed*, resulting in a *homogeneous* and *isotropic* turbulence structure. This annihilation is associated with *turbulent kinetic energy dissipation* to heat. Thus, in turbulent flows, energy generation is directly proportional to μ_t and hence, to l, whereas energy dissipation is associated with fluid viscosity μ and is inversely proportional to l.

where the distribution of k is obtained from Eq. 6.92 but the distribution of dissipation rate ϵ^* is obtained from the following differential equation:

$$\frac{\partial(\rho_m \, \epsilon^*)}{\partial t} + \frac{\partial(\rho_m \, u_j \, \epsilon^*)}{\partial x_j} = \frac{\partial}{\partial x_j} \left[\left(\mu + \frac{\mu_t}{Pr_{\epsilon^*}} \right) \frac{\partial \epsilon^*}{\partial x_j} \right]$$

$$+ \frac{\epsilon^*}{k} \left[C_1 \, G - C_2 \, \rho_m \, \epsilon^* \right] + E_{\epsilon^*} \, , \tag{6.102}$$

$$f_\mu = \exp \left[\frac{-3.4}{(1 + Re_t/50)^2} \right] \, , \qquad Re_t = \frac{\mu_t}{\mu} \, , \tag{6.103}$$

$$\epsilon^* = \epsilon - 2 \, \nu \left(\frac{\partial \sqrt{k}}{\partial x_j} \right)^2 \, , \qquad Pr_{\epsilon^*} = 1.3 \, , \tag{6.104}$$

$$E_{\epsilon^*} = 2 \, \nu \, \mu_t \left(\frac{\partial^2 u_i}{\partial x_j \, \partial x_k} \right)^2 \, , \tag{6.105}$$

$$C_1 = 1.44 \, , \qquad C_2 = 1.92 \left[1 - 0.3 \, \exp \left(-Re_t^2 \right) \right] \, . \tag{6.106}$$

This model permits the wall boundary condition $k = \epsilon^* = 0$. Further refinements of this model are described in Ref. [27].

In *free shear layers*, such as jets and wakes, there is no wall present. As such, the turbulent Reynolds number Re_t is high everywhere in the flow. So, $f_\mu \rightarrow 1$, $\epsilon^* = \epsilon$, $E_{\epsilon^*} = 0$, and $C_2 = 1.92$.

Exercises

1. The law of mass conservation for the bulk fluid is expressed as:

$$\text{Rate of accumulation of mass} \ (\dot{M}_{ac}) = \text{Rate of mass in} \ (\dot{M}_{in})$$
$$- \text{Rate of mass out} \ (\dot{M}_{out}) \, ,$$

where M (kg) refers to bulk mass. Write expressions for $\dot{M}_{in}$, $\dot{M}_{out}$, and $\dot{M}_{ac}$ with respect to a control volume $\Delta V = \Delta x_1 \times \Delta x_2 \times \Delta x_3$, with inflows at (x_1, x_2, x_3) and outflows at $(x_1 + \Delta x_1, x_2 + \Delta x_2, x_3 + \Delta x_3)$. After substitution and letting $\Delta V \rightarrow 0$, demonstrate the correctness of Eq. 6.1.

2. Newton's second law of motion states that for a *given direction*,

$$\text{Rate of accumulation of momentum} \ (Mom_{ac}) = \text{Rate of momentum in} \ (Mom_{in})$$
$$- \text{Rate of momentum out} \ (Mom_{out})$$
$$+ \text{Sum of forces acting on the CV} \ (F_{cv})$$

where Mom represents momentum. Write expressions for each term and demonstrate the correctness of Eq. 6.2 for direction x_1.

3. In the preceding equation, substitute for τ_{11}, τ_{12}, and τ_{13} from Eq. 6.4. Now, assuming *laminar flow of a uniform property fluid*, show that Eq. 6.6 can be written as

$$\rho_m \left[\frac{\partial u_1}{\partial t} + \frac{\partial u_j u_1}{\partial x_j} \right] = \mu \frac{\partial^2 u_1}{\partial x_j^2} - \frac{\partial p}{\partial x_1} + \rho_m B_1 \quad j = 1, 2.$$

Explain why $S_{u_1} = 0$.

4. Explain, through derivations, why $\sum N_{i,k} = \rho_m u_i$.

5. Show that mass concentration of element α in a mixture can be expressed in terms of mole fractions as

$$\eta_\alpha = \sum_k \eta_{\alpha,k} \left(\frac{M_k}{M_{mix}} \right) x_k .$$

How will you evaluate M_{mix}?

6. Consider a simple chemical reaction (SCR) given by

$$1 \text{ kg of fuel} + R_{st} \text{ kg of oxidant} \rightarrow (1 + R_{st}) \text{ kg of product}.$$

Write the relevant Eq. 6.12 for a flow in which the SCR occurs. Show that it is possible to form a conserved scalar equation in which $\Psi = \omega_{fu} - \omega_{ox}/R_{st}$ or $\Psi = \omega_{fu} + \omega_{pr}/(1 + R_{st})$. List the assumptions made.

7. In Eq. 6.41, assuming that the specific heats of all species were equal, show that $\sum_k \Delta h_{s,k} R_k = 0$ and $\sum_k D \partial \omega_k/\partial x_j \, \partial h_k/\partial x_j = 0$.

8. It will be shown in Chap. 9 that in diffusion flames, the fuel stream and the air (oxidizer) streams are physically separated before mixing to cause combustion. In such a flow, it is useful to define the *mixture fraction* f as

$$f = \frac{\text{mass of material originating in the fuel stream}}{\text{mass of the mixture}} .$$

This implies that f = 1 in the fuel stream, f = 0 in the air stream, and 0 < f < 1 in the flow field, where mixing and chemical reaction takes place.
Consider the combustion of hydrocarbon fuel in which the SCR is postulated. Show that

$$f = \left[\omega_{fu} + \frac{\omega_{pr}}{1 + R_{st}} \right]_{mix} = \left[\frac{(\omega_{fu} - \omega_{ox}/R_{st})_{mix} + 1/R_{st}}{1 + 1/R_{st}} \right] .$$

(Hint: There will be no fuel material in the oxidizer stream, and no oxidant in the fuel stream. But, the fuel and oxidant materials may be found in the product stream.)

9. Using the preceding definition, show that the distribution of f is governed by

$$\frac{\partial(\rho_m f)}{\partial t} + \frac{\partial(\rho_m u_j f)}{\partial x_j} = \frac{\partial}{\partial x_j}\left(\rho_m D \frac{\partial f}{\partial x_j}\right) .$$

Is f also a conserved scalar?

10. For a hydrocarbon fuel combustion, show that f can also be written in terms of element mass fractions as

$$f = \eta_C + \eta_H .$$

Show that for the mixture represented by Eqs. 6.15–6.18, f can be evaluated as

$$f = (x_{CH_4} + x_{CO} + x_{CO_2}) \frac{M_C}{M_{mix}} + (4 x_{CH_4} + 2 x_{H_2} + 2 x_{H_2O}) \frac{M_H}{M_{mix}} ,$$

where $M_{mix} = x_{CH_4} M_{CH_4} + x_{CO} M_{CO} + x_{CO_2} M_{CO_2} + x_{H_2} M_{H_2} + x_{H_2O} M_{H_2O} + x_{N_2} M_{N_2} + x_{O_2} M_{O_2} + x_O M_O + x_{NO} M_{NO}$.

11. Determine the mixture fraction f at a point in the combustion flow of methane where $x_{CH_4} = 0.01$, $x_{CO} = 0.001$, $x_{CO_2} = 0.1$, $x_{H_2O} = 0.14$, $x_{O_2} = 0.02$, $x_{H_2} = 295$ ppm, $x_{NO} = 112$ ppm, and $x_O = 65$ ppm.

12. Show that the fuel-air ratio at a point is related to the mixture fraction f as

$$\frac{F}{A} = \frac{f}{1 - f} .$$

Evaluate the equivalence ratio at a point for the data given in the previous problem.

13. Consider the combustion of propane according to the quasi-global reaction mechanism given in Eqs. 5.70–5.73 (with n = 3). Write out the mass transfer equations for the species involved, with appropriate source terms in each equation. How will you evaluate Q_{chem} in the energy equation?

14. Consider mass transfer across a turbulent boundary layer developing on a flat plate. Using the well-known heat transfer correlation, show that the mass transfer coefficient g^* can be estimated from

$$\frac{g^* x}{K/cp_m} = 0.0296 \, Re_x^{0.8} \, Sc^{0.33} \left(\frac{M_w}{M_\infty}\right)^{-0.4} .$$

15. In Eq. 6.92, at a distance n_{eq}, say, the generation and dissipation rates of turbulent kinetic energy become equal (that is, $G = \rho_m \epsilon$) and the turbulence structure there is said to be in local equilibrium with $\mu_t \gg \mu$. In the vicinity of n_{eq}, the velocity (u) parallel to the wall can be modeled as

$$\frac{u}{u_\tau} = \frac{1}{\kappa} \ln\left(\frac{n u_\tau}{\nu}\right) + \text{Const}, \qquad u_\tau = \sqrt{\frac{\tau_w}{\rho_m}} .$$

Show that

$$\frac{u_\tau^2}{k} = \left[\frac{1 - \exp\left(- Re_{n_{eq}}/5.1\right)}{1 - \exp\left(- Re_{n_{eq}}/70\right)}\right] \times C_\mu^{1.25}.$$

In turbulent boundary layers with moderate pressure gradients, it is found that $\tau_w \simeq 0.3\, \rho_m\, k$ in the equilibrium layer. Estimate the value of $Re_{n_{eq}}$.

16. Derive the 2D Boundary layer form of Eq. 6.102 and identify units of each term.
17. Derive one-dimensional differential equations of bulk mass conservation, mass transfer and energy transfer in spherical coordinates assuming only radial transport.

Chapter 7
Thermochemical Reactors

7.1 Introduction

Our interest now is to derive guidance for design or for performance predictions of combustion devices. The word *design* refers to the determination of the geometry (length, diameter, shape) of the device and ensuring safety[1] for the desired product composition. *Performance prediction* implies the reverse—that is, determining product composition for a given geometry under safe operating conditions.

Combustion devices can be long and slender, as in a gas-turbine combustion chamber, a cement kiln, or a pulverized fuel (mainly coal) fired boiler, in which air + fuel enters at one end and products leave at the other. Such devices can be idealized as *plug-flow thermochemical reactors* (PFTCRs). In a PFTCR operating under steady state, properties such as velocity, temperature, pressure, and mass-fractions of species principally vary along the length of the reactor. Therefore, for such reactors, equations of mass, momentum, and energy maybe simplified to their one-dimensional forms.

Devices known as *furnaces* (used in ceramic or steelmaking) are similar but rather *stubby* with their spatial dimensions in the three directions being nearly equal. Such devices are formally called *well-stirred thermochemical reactors* (WSTCRs). Thus, the combustion space in a wood-burning cookstove may also be called a WSTCR. A WSTCR is a stubby PFTCR. Therefore, the governing equations can be reduced to a set of zero-dimensional algebraic equations.

In an IC engine, for example, the greater part of combustion reactions take place when inlet and exhaust valves are closed. Thus, as there is no inflow or outflow, we have a *constant-mass thermochemical reactor* (CMTCR) in which the mixture properties vary with time. The governing equations can, therefore, be simplified to their zero-dimensional, unsteady forms. In general, however, the volume of a CMTCR may or may not vary with time.

---

[1]Nowhere in the device should the temperature exceed a prespecified value, which is often dictated by metallurgical limits.

© Springer Nature Singapore Pte Ltd. 2020
A. W. Date, *Analytic Combustion*,
https://doi.org/10.1007/978-981-15-1853-9_7

Our interest in this chapter is to *model* PFTCRs, WSTCRs, and CMTCRs through simplification of the full three-dimensional time-dependent equations introduced in the previous chapter and by invoking the reaction rate R_k (kg/m^3-s). This will enable calculation of the evolution of the reacting mixture from its initial reactant state to its final product state through time or space. The product state, of course, may or may not be in equilibrium. Simplified models are essentially studied to capture the main ideas. The results obtained from them cannot be relied upon for practical design although they are very illustrative of the main tendencies inherent in the reactor under consideration.

7.2 Plug Flow Reactor

A simple plug-flow reactor is a reactor in which all variations of properties are one-dimensional—that is, significant variations occur only in the flow direction and variations over a cross-section are negligible. Figure 7.1 shows such a reactor.

The reactor is modeled under the following assumptions:

1. The cross-sectional area A and perimeter P are known functions of streamwise distance x.
2. Although all variables (velocity, pressure, temperature, mass fractions, density, etc.) vary in the x-direction, *axial diffusion* of all variables is considered negligible.
3. At the reactor wall, heat transfer (q_w W/m^2), mass transfer (N_w kg/m^2-s), and work transfer ($\dot{W}_{ext}$ W/m^3) rates are assumed known or knowable.

Under these assumptions, and following the methodology of the previous chapter, equations of bulk mass, mass transfer, momentum, and energy are derived.

7.2.1 Governing Equations

Bulk Mass

With reference to Fig. 7.1, for a control volume $\Delta V = A \, \Delta x$, the law of conservation of mass can be written as

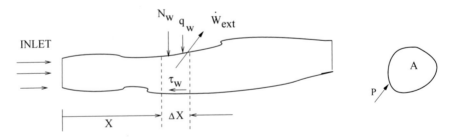

Fig. 7.1 Schematic of a plug flow reactor

$$\dot{m}_x + N_w \, P \, \Delta x - \dot{m}_{x+\Delta x} = \frac{\partial}{\partial t} \left(\rho_m \, A \, \Delta x \right) , \tag{7.1}$$

where $\dot{m} = \rho_m \, A \, u$ is the mixture mass flow rate and u is the cross-sectionally aver-aged axial velocity. Dividing this equation by $A \, \Delta x$,

$$\frac{1}{A} \frac{\dot{m}_x - \dot{m}_{x+\Delta x}}{\Delta x} + N_w \frac{P}{A} = \frac{\partial \rho_m}{\partial t} . \tag{7.2}$$

Letting $\Delta x \to 0$ and rearranging,

$$\frac{\partial \rho_m}{\partial t} + \frac{1}{A} \frac{\partial \dot{m}}{\partial x} = N_w \frac{P}{A} . \tag{7.3}$$

Mass Transfer

Let N_k be the mass flux of species k. Then, writing the law of conservation of mass for the individual species k, we have

$$N_k \, A \mid_x + N_{w,k} \, P \, \Delta x - N_k \, A \mid_{x+\Delta x} + R_k \, A \, \Delta x = \frac{\partial}{\partial t} \left(\rho_k \, A \, \Delta x \right) . \tag{7.4}$$

Dividing by $A \, \Delta x$ and, letting $\Delta x \to 0$, we have

$$-\frac{1}{A} \frac{\partial}{\partial x} [N_k \, A] + N_{w,k} \frac{P}{A} + R_k = \frac{\partial \rho_k}{\partial t} , \tag{7.5}$$

where $\rho_k = \rho_m \, \omega_k$, $N_k = \rho_k \, u$ (because axial diffusion flux is neglected), and $N_k A = \dot{m} \omega_k$. Therefore, on rearrangement, this equation reads as

$$\frac{\partial(\rho_m \, \omega_k)}{\partial t} + \frac{1}{A} \frac{\partial(\dot{m} \, \omega_k)}{\partial x} = R_k + N_{w,k} \frac{P}{A} . \tag{7.6}$$

Momentum

Applying Newton's second law of motion, we have

$$\dot{m} \, u \mid_x + \left(N_w \, P \, \Delta x \right) u - \dot{m} \, u \mid_{x+\Delta x} - \tau_w \, P \, \Delta x + (p \, A)_x - (p \, A)_{x+\Delta x}$$
$$= \frac{\partial}{\partial t} \left(\rho_m \, \Delta V \, u \right) . \tag{7.7}$$

Dividing by $A \, \Delta x$ and, letting $\Delta x \to 0$, we have

$$\frac{\partial(\rho_m \, u)}{\partial t} + \frac{1}{A} \frac{\partial(\dot{m} \, u)}{\partial x} = -\frac{1}{A} \frac{d(pA)}{dx} + (N_w \, u - \tau_w) \frac{P}{A} . \tag{7.8}$$

Here again, the normal-stress contribution is neglected, so axial diffusion is zero.

Energy

Invoking the first law of thermodynamics,

$$\frac{d\,E_{cv}}{dt} = \frac{dQ}{dt}\Big|_{net} - \frac{dW}{dt}\Big|_{net} \;, \tag{7.9}$$

where $E_{cv} = \rho_m\,\Delta V\,e^o_{cv} = \rho_m\,\Delta V\,\left[e + u^2/2\right]$ and e is the *specific internal energy*.

The $dQ/dt\,|_{net}$ will comprise contributions from convection (conduction is neglected), radiation, and wall heat transfer. Thus,

$$\frac{dQ}{dt}\Big|_{net} = \frac{dQ}{dt}\Big|_{conv} + \frac{dQ}{dt}\Big|_{rad}$$
$$+ q_w\,P\,\Delta x + (N_w\,P\,\Delta x)\,h_w \;, \tag{7.10}$$

where h_w is the enthalpy carried by wall mass transfer. Now,

$$\frac{dQ}{dt}\Big|_{conv} = \sum N_k\,e^o_k\,A\,|_x - \sum N_k\,e^o_k\,A\,|_{x+\Delta x}\,. \tag{7.11}$$

But, as $\dot m = A\,\sum N_k$ and, $\dot m\,e = A\,\sum N_k\,e_k$,

$$\sum N_k\,e^o_k\,A = \sum N_k\,\left(e_k + \frac{u^2}{2}\right)\,A = \dot m\,\left(e + \frac{u^2}{2}\right) = \dot m\,e^o\,. \tag{7.12}$$

Therefore, replacing $dQ/dt\,|_{rad} = \dot Q_{rad}\,\Delta V$ for brevity,

$$\frac{dQ}{dt}\Big|_{net} = \dot m\,e^o\,|_x - \dot m\,e^o\,|_{x+\Delta x}$$
$$+ q_w\,P\,\Delta x + (N_w\,P\,\Delta x)\,h_w + \dot Q_{rad}\,\Delta V\,. \tag{7.13}$$

Similarly, $dW/dt\,|_{net}$ can be expressed as

$$\frac{dW}{dt}\Big|_{net} = \dot W_{ext}\,\Delta V + (p\,A\,u\,_{x+\Delta x} - p\,A\,u\,_x) + (\tau_w - N_w\,u)\,P\,\Delta x\,u\,. \tag{7.14}$$

Substituting Eqs. 7.13 and 7.14 in Eq. 7.9 and dividing by $A\,\Delta x$, and letting $\Delta x \to 0$, it follows that

$$\frac{d}{dt}\,(\rho_m\,e^o) + \frac{1}{A}\frac{d}{dx}\,(\dot m\,e^o) = (q_w + N_w\,h_w)\,\frac{P}{A} + (\dot Q_{rad} - \dot W_{ext})$$
$$- p\,\frac{du}{dx}$$
$$- \frac{u}{A}\left[\frac{d(p\,A)}{dx} + (\tau_w - N_w\,u)\,\frac{P}{A}\right]. \tag{7.15}$$

Using the momentum equation 7.8 to replace the terms in the square bracket on the righthand side, replacing e^o by $e + u^2/2$, and, further, making use of mass conservation equation 7.3, we have

$$\frac{d}{dt}(\rho_m\,e) + \frac{1}{A}\frac{d}{dx}(\dot{m}\,e) = (q_w + N_w\,h_w)\frac{P}{A} + (\dot{Q}_{rad} - \dot{W}_{ext})$$

$$- p\frac{du}{dx} + \frac{u^2}{2}N_w\frac{P}{A}. \tag{7.16}$$

However, $e = h_m - p/\rho_m$ (by definition). Therefore, this equation can also be written as

$$\frac{d}{dt}(\rho_m\,h_m) + \frac{1}{A}\frac{d}{dx}(\dot{m}\,h_m) = (q_w + N_w\,h_w)\frac{P}{A} + (\dot{Q}_{rad} - \dot{W}_{ext})$$

$$+ \frac{dp}{dt} + \frac{u}{A}\frac{d(pA)}{dx} + \frac{u^2}{2}N_w\frac{P}{A}. \tag{7.17}$$

Equations 7.3, 7.6, 7.8, and 7.17 thus represent the one-dimensional governing equations for the PFTCR. Their use for design is illustrated through an example.

Problem 7.1 A mixture of Ethane ($C_2\,H_6$) and air (equivalence ratio $\Phi = 0.6$) enters an *adiabatic* plug flow reactor of *constant cross-section* having diameter 3 cm at $T = 1000\,K$, $p = 1$ atm, and $\dot{m} = 0.2$ kg/s. Determine the length required to take the reaction to 80% completion under *steady state*. Assume a single-step global reaction mechanism. Neglect pressure drop and radiation effects, wall mass transfer, and work transfer.

For any hydrocarbon fuel ($C_m H_n$), the single-step global reaction mechanism for degree of reaction ϵ will read as

$$C_m\,H_n + \frac{a_{stoich}}{\Phi}(O_2 + 3.76\,N_2)$$

$$\rightarrow \epsilon\left[m\,CO_2 + \frac{n}{2}H_2O + \left(\frac{a_{stoich}}{\Phi} - m - \frac{n}{2}\right)O_2 + 3.76\frac{a_{stoich}}{\Phi}N_2\right]$$

$$+ (1 - \epsilon)\left[C_m H_n + \frac{a_{stoich}}{\Phi}(O_2 + 3.76\,N_2)\right], \tag{7.18}$$

where stoichiometric ratio is given by $a_{stoich} = m + (n/4)$. In the present example, $m = 2$, $n = 6$, and $\epsilon = 0.8$.

To simplify further, we *define* the total product species Pr and its molecular weight M_{pr} as

$$Pr = \frac{m\,CO_2 + (n/2)\,H_2O}{m + n/2} \tag{7.19}$$

$$M_{Pr} = \frac{m \times M_{CO_2} + (n/2) \times M_{H_2O}}{m + n/2}. \tag{7.20}$$

Then, the reaction mechanism can be written as

$$C_m H_n + \frac{a_{stoich}}{\Phi} (O_2 + 3.76\, N_2)$$

$$\rightarrow \epsilon \left[\left(m + \frac{n}{2} \right) Pr + \left(\frac{a_{stoich}}{\Phi} - m - \frac{n}{2} \right) O_2 + 3.76\, \frac{a_{stoich}}{\Phi} N_2 \right]$$

$$+ (1 - \epsilon) \left[C_m H_n + \frac{a_{stoich}}{\Phi} (O_2 + 3.76\, N_2) \right] . \tag{7.21}$$

Thus, our reaction mechanism has four species: Fuel ($C_m H_n$ or fu), oxygen (O_2), product (Pr), and nitrogen (N_2).

We now invoke the *flow model*. Our rector is adiabatic. Therefore, $q_w = 0$. Also, there is no mass or work transfer at the reactor wall. Therefore, $N_w = \dot{W}_{ext} = 0$. Further, steady state prevails. Therefore, all time derivatives are zero. Finally, pressure variations are considered negligible. Therefore, the momentum equation need not be solved. Also, $Q_{rad} = 0$. Further, for this cylindrical PFTCR, A and P are constants. Thus, the equation of bulk mass Eq. 7.3 simply implies that $\dot{m} = $ constant (given), and therefore, no solution is required. Thus, only the equations for species (Eq. 7.6) and energy (Eq. 7.17) need to be solved.

$$\frac{d\, \omega_{fu}}{d\, x} = - \frac{A}{\dot{m}} \, | R_{fu} | , \tag{7.22}$$

$$\frac{d\, \omega_{O_2}}{d\, x} = - \frac{A}{\dot{m}} \, | R_{O_2} | , \tag{7.23}$$

$$\frac{d\, \omega_{Pr}}{d\, x} = \frac{A}{\dot{m}} \, R_{Pr} , \tag{7.24}$$

$$\frac{d\, \omega_{N_2}}{d\, x} = 0 \quad \text{and} \tag{7.25}$$

$$\frac{\dot{m}}{A} \frac{d\, h_m}{d\, x} = 0 . \tag{7.26}$$

The righthand side of Eq. 7.25 is zero, because nitrogen is inert. However, Eq. 7.26 can be converted to temperature form (see Eq. 6.41) by noting that $h_m = \sum \omega_k \, h_k$, $h_k = h_{fk}^0 + \Delta h_{sk}$ and assuming equal specific heats $cp_k = cp_m$. Thus,

$$\frac{\dot{m}}{A} \frac{d\, h_m}{d\, x} = \frac{\dot{m}}{A} \frac{d}{d\, x} (cp_m T) - \Delta H_c \, | R_{fu} | = 0 , \tag{7.27}$$

where ΔH_c is the heat of combustion. These first-order equations must be solved for known initial conditions at x = 0 until the reaction is completed to the desired degree. To facilitate solution, we make further simplifications that are possible because a single-step global mechanism is used. Such a mechanism is often called the *simple chemical reaction* (SCR) and is written as

$$1 \text{ kg of fuel} + r_{st} \text{ kg of oxygen} \rightarrow (1 + r_{st}) \text{ kg of product,} \tag{7.28}$$

where the stoichiometric ratio r_{st} is defined as

$$r_{st} = \frac{M_{O_2}}{M_{fu}} a_{stoich} = \frac{M_{O_2}}{M_{fu}} \left(m + \frac{n}{4} \right) . \tag{7.29}$$

Then, it follows that

$$R_{O_2} = r_{st} R_{fu} \quad \text{and} \tag{7.30}$$
$$R_{Pr} = -(1 + r_{st}) R_{fu} . \tag{7.31}$$

For ethane, $r_{st} = 3.7333$. Using the preceding derivations, the species and energy equations can be rewritten as

$$\frac{d\omega_{fu}}{dx} = -\frac{A}{\dot{m}} |R_{fu}| , \tag{7.32}$$

$$\frac{d}{dx} \left(\frac{\omega_{O_2}}{r_{st}} \right) = -\frac{A}{\dot{m}} |R_{fu}| , \tag{7.33}$$

$$\frac{d}{dx} \left(\frac{\omega_{Pr}}{1 + r_{st}} \right) = \frac{A}{\dot{m}} |R_{fu}| \quad \text{and} \tag{7.34}$$

$$\frac{d}{dx} (cp_m T) = \frac{A}{\dot{m}} \Delta H_c |R_{fu}| . \tag{7.35}$$

Combining Eqs. 7.33–7.35 with Eq. 7.32, it follows that

$$\frac{d}{dx} \left[\omega_{fu} - \frac{\omega_{O_2}}{r_{st}} \right] = 0 \tag{7.36}$$

$$\frac{d}{dx} \left[\omega_{fu} + \frac{\omega_{Pr}}{(1 + r_{st})} \right] = 0 \tag{7.37}$$

$$\frac{d}{dx} \left[cp_m T + \Delta H_c \omega_{fu} \right] = 0 . \tag{7.38}$$

These equations imply[2] that

$$\left[\omega_{fu} - \frac{\omega_{O_2}}{r_{st}} \right]_x = \left[\omega_{fu} - \frac{\omega_{O_2}}{r_{st}} \right]_i = \left[\omega_{fu} - \frac{\omega_{O_2}}{r_{st}} \right]_L \tag{7.39}$$

---

[2]Equations 7.36–7.38 can be generalized as $d\,\Psi/dx = 0$. In such equations with a zero source terms, Ψ is called a *conserved scalar*, as was mentioned in Chap. 6.

$$\left[\omega_{fu} + \frac{\omega_{Pr}}{(1+r_{st})}\right]_x = \left[\omega_{fu} + \frac{\omega_{Pr}}{(1+r_{st})}\right]_i = \left[\omega_{fu} + \frac{\omega_{Pr}}{(1+r_{st})}\right]_L \qquad (7.40)$$

$$\begin{aligned} \left[c_{pm} T + \Delta H_c \,\omega_{fu}\right]_x &= \left[c_{pm} T + \Delta H_c \,\omega_{fu}\right]_i \\ &= \left[c_{pm} T + \Delta H_c \,\omega_{fu}\right]_L \,, \end{aligned} \qquad (7.41)$$

where subscript i refers $x = 0$ and the subscript L refers to $x = L$. These consequences of the single-step mechanism imply that we need not solve for all variables $\omega_{fu}, \omega_{O_2}$, $\omega_{Pr}, \omega_{N_2}$, and T. Any one variable will suffice and distributions of other variables can be determined from the above relations. *Here, we choose to solve* the temperature Eq. 7.35. To solve this equation, however, R_{fu} must be specified. For hydrocarbons, R_{fu} takes the form (see Eq. 5.80)

$$|R_{fu}| = A^*_{fu}\, \rho_m^{x+y} \exp\left(-\frac{E_{fu}}{T}\right) \omega_{fu}^x \,\omega_{O_2}^y \qquad (7.42)$$

where A^*_{fu}, $E_{fu} = E_a/R_u$, x and y are fuel-specific constants, and density is given by

$$\rho_m = \frac{p\, M_{mix}}{R_u\, T} \quad \text{where} \quad M_{mix} = \left[\sum \frac{\omega_j}{M_j}\right]^{-1}. \qquad (7.43)$$

R_u is a universal gas-constant. For ethane, $E_{fu} = 15098$, $x = 0.1$, $y = 1.65$, and $A^*_{fu} = A_{fu} \times M_{fu}^{1-x} \times M_{O_2}^{-y} = 4.338 \times 10^8$ (see Table 5.8).

To evaluate R_{fu} local mass fractions are required. We determine ω_{fu} from Eq. 7.41 and ω_{O_2} from Eq. 7.39. Thus,

$$\omega_{fu} = \omega_{fu,i} - \frac{c_{pm}}{\Delta H_c}\,(T - T_i)\,, \qquad (7.44)$$

$$\begin{aligned} \omega_{O_2} &= r_{st}\,(\omega_{fu} - \omega_{fu,i}) + \omega_{O_2,i} \\ &= \omega_{O_2,i} - r_{st}\,\frac{c_{pm}}{\Delta H_c}\,(T - T_i)\,. \end{aligned} \qquad (7.45)$$

Substituting in Eq. 7.42, we have

$$\begin{aligned} |R_{fu}| = A^*_{fu} \left[\frac{p\, M_{mix}}{R_u\, T}\right]^{x+y} \exp\left(-\frac{E_{fu}}{T}\right) \\ \times \left[\omega_{fu,i} - \frac{c_{pm}}{\Delta H_c}\,(T - T_i)\right]^x \\ \times \left[\omega_{O_2,i} - r_{st}\,\frac{c_{pm}}{\Delta H_c}\,(T - T_i)\right]^y. \end{aligned} \qquad (7.46)$$

This equation shows that R_{fu} is effectively a function of T only. Therefore, Eq. 7.35 can be solved with initial condition $T = T_i$ at $x = 0$.

Conditions at $x = 0$ and $x = L$

At $x = 0$, only reactants are present. From Eq. 7.21, therefore,

$$m_{mix} = M_{fu} + \frac{a_{stoich}}{\Phi} (M_{O_2} + 3.76\, M_{N_2})$$

$$\omega_{fu,i} = \frac{M_{fu}}{m_{mix}}$$

$$\omega_{O_2,i} = \frac{a_{stoich}}{\Phi} \frac{M_{O_2}}{m_{mix}}$$

$$\omega_{Pr,i} = 0$$

$$\omega_{N_2,i} = 1 - \omega_{fu,i} - \omega_{O_2,i} - \omega_{Pr,i} \ . \tag{7.47}$$

Similarly, at $x = L$, from the righthand side of Eq. 7.21,

$$m_{mix} = (1 - \epsilon)\, M_{fu} + \epsilon \left(m + \frac{n}{2}\right) M_{Pr} + \left[\frac{a_{stoich}}{\Phi} - \epsilon \left(m + \frac{n}{2}\right)\right] M_{O_2}$$

$$+ 3.76 \frac{a_{stoich}}{\Phi} M_{N_2}$$

$$\omega_{fu,L} = (1 - \epsilon) \frac{M_{fu}}{m_{mix}}$$

$$\omega_{O_2,L} = \left[\frac{a_{stoich}}{\Phi} - \epsilon \left(m + \frac{n}{2}\right)\right] \frac{M_{O_2}}{m_{mix}}$$

$$\omega_{Pr,L} = \epsilon \left(m + \frac{n}{2}\right) \frac{M_{Pr}}{m_{mix}}$$

$$\omega_{N_2,L} = \omega_{N_2,i} \ . \tag{7.48}$$

Solution of Eq. 7.35

The temperature equation is now written as

$$\frac{d}{dx}(T) = \frac{A}{\dot{m}} \frac{\Delta H_c}{cp_m} |R_{fu}| = \text{RHS} \ , \tag{7.49}$$

with initial condition $T = T_i$ at $x = 0$. This equation can be solved by Euler's marching integration method as

$$T_{x+\Delta x} = T_x + \Delta x \times \text{RHS}\,(T_x) \ , \tag{7.50}$$

where the righthand side (RHS) is evaluated at T_x.

The choice of the magnitude of Δx is arbitrary. Solutions must be repeated with ever smaller values until the predicted variation of temperature with x is independent of the choice made. The solution algorithm is as follows:

1. Specify fuel constants m, n, A_{fu}, E_{fu}, x , y, and ΔH_c. Specify ϵ, $\dot{m}$ and area A. Evaluate $\omega_{j,i}$ from Eq. 7.47 and $\omega_{j,L}$ from Eq. 7.48. Specify specific heats of all species at $T_{fix} = 1500$ K. Choose Δx and set x = 0 and $T = T_i$.
2. Calculate M_{mix} and $cp_m = \sum \omega_j \, Cp_j(T_{fix})$.
3. Evaluate R_{fu} for T_x and solve Eq. 7.50 to yield $T_{x+\Delta x}$.
4. Evaluate ω_j for j = fu, O_2, and Pr at $T_{x+\Delta x}$ from Eqs. 7.39–7.41. Print values of $T_{x+\Delta x}$, cp_m and ω_j.
5. If ω_{Pr} exceeds $\omega_{Pr,L}$, go to step 7. Otherwise,
6. Set $T_x = T_{x+\Delta x}$ and x = x + Δx and return to step 2.
7. stop.

Tables 7.1 and 7.2 show the results of computations for ethane for the conditions prescribed in the example. The value of heat of reaction at 1000 K is taken as $\Delta H_c = 4 \times 10^7$ J/kg. Although specific heats vary with temperature (and, therefore with x), for all species they are evaluated at $T_{fix} = 1500$ K which is expected to be half of T_i and $T_{blowout}$. However, mixture specific heat cp_m still varies with x (see step 2) because of variation of ω_j with x. Computations are performed with $\Delta x = 0.01$ m or, 1 cm. Computations are continued until *blowoff* occurs and the fuel is exhausted.

For an 80% reaction, $m_{mix} = 830.8$ giving of $\omega_{Pr,L} = 0.1367$ (see Eq. 7.48). From Table 7.2, this value lies between x = 0.78 m and 0.79 m. The estimate of length L for 80% reaction is, therefore, obtained by linear interpolation as

$$\frac{L - 0.78}{0.01} = \frac{0.1367 - 0.124}{0.145 - 0.124},$$

or, L = 0.786 m.

Figures 7.2 and 7.3 show graphical plots of axial variations. It is seen that as the blowoff is approached, R_{fu} rapidly increases with temperature but then rapidly falls again as the fuel is depleted. Thus, in a fuel-lean mixture, the blowoff limit is controlled by the fuel mass fraction.[3] Although the mass-fractions of the species are changing, the values of M_{mix} and $c_{p,mix}$ remain nearly constant.

After a general computer program is written, one can investigate the effects of input parameters on the reactor length required for any fuel with any equivalence ratio. From Eq. 7.49, however, we discover that the rate of change of temperature is controlled by the parameter RHS. Thus, if the reactor diameter $D > 3$ cm is used so A is bigger, the rate of increase of temperature will be faster and, therefore, L for 80% reaction will be shorter. Similarly, if $\dot{m} > 0.2$ kg/s, L will increase. Also because $\dot{m} = \dot{m}_{air} \{1 + \Phi (F/A)_{stoich}\}$, L will increase for $\Phi > 0.6$ for the same air mass flow rate. A richer mixture, therefore, will require a longer length. Finally, if $p < 1$ atm, ρ_{mix} will decrease and hence R_{fu} will also decrease, resulting in longer L. Such deductions are very useful in practical design.

Overall, we learn that for an adiabatic reactor, the blowout length will be given by the following function:

---

[3]For a fuel-rich mixture, the blow-off limit will be controlled by oxygen mass fraction.

Table 7.1 Variation of T, R_{fu}, M_{mix}, and $c_{p,mix}$ with x for ethane combustion (Problem 7.1)

X(m)	T(K)	R_{fu}	M_{mix}	$c_{p,mix}$
0.000E-00	0.100E+04	–	0.289E+02	0.122E+04
0.100E-01	0.100E+04	0.118E+01	0.289E+02	0.122E+04
0.500E-01	0.101E+04	0.127E+01	0.289E+02	0.122E+04
0.100E+00	0.101E+04	0.139E+01	0.289E+02	0.122E+04
0.150E+00	0.102E+04	0.154E+01	0.289E+02	0.122E+04
0.200E+00	0.103E+04	0.171E+01	0.289E+02	0.122E+04
0.250E+00	0.104E+04	0.193E+01	0.289E+02	0.122E+04
0.300E+00	0.106E+04	0.220E+01	0.289E+02	0.122E+04
0.350E+00	0.107E+04	0.254E+01	0.289E+02	0.123E+04
0.400E+00	0.109E+04	0.299E+01	0.288E+02	0.123E+04
0.450E+00	0.110E+04	0.361E+01	0.288E+02	0.123E+04
0.500E+00	0.113E+04	0.448E+01	0.288E+02	0.123E+04
0.550E+00	0.116E+04	0.581E+01	0.288E+02	0.123E+04
0.600E+00	0.120E+04	0.800E+01	0.288E+02	0.124E+04
0.650E+00	0.126E+04	0.121E+02	0.288E+02	0.124E+04
0.700E+00	0.135E+04	0.212E+02	0.287E+02	0.125E+04
0.750E+00	0.155E+04	0.494E+02	0.286E+02	0.127E+04
0.760E+00	0.162E+04	0.614E+02	0.286E+02	0.128E+04
0.770E+00	0.170E+04	0.773E+02	0.286E+02	0.129E+04
0.780E+00	0.181E+04	0.978E+02	0.285E+02	0.130E+04
0.790E+00	0.194E+04	0.122E+03	0.285E+02	0.131E+04
0.800E+00	0.209E+04	0.140E+03	0.284E+02	0.133E+04
0.810E+00	0.221E+04	0.106E+03	0.283E+02	0.134E+04

$$L_{\text{blowout}} = F\left(\dot{m}, D, T_i, p, \Phi, \text{fuel}\right). \tag{7.51}$$

As such, the algorithm developed for solving Problem 7.1 can be run for several values of the parameters.

7.2.2 Nonadiabatic PFTCR

In Problem 7.1, we considered an adiabatic reactor. However, in reality, reactors are nonadiabatic. In fact, the reactor walls are often cooled to protect and to enhance the life of wall linings, as is the case with gas-turbine combustion chambers and rocket motor combustors. To illustrate the analysis of such a reactor, we consider the following problem.

Problem 7.2 Assume that the reactor wall in Problem 7.1 is maintained at $T_w = 500\,^\circ C$ and that heat transfer at the wall is given by $q_w = \alpha\,(T_w - T)$, where

Table 7.2 Variation of ω_{fu}, ω_{O_2} and ω_{Pr} with x for Ethane combustion—Problem 7.1

X(m)	ω_{fu}	ω_{O_2}	ω_{Pr}
0.000E-00	0.360E-01	0.225E+00	0.0000-00
0.100E-01	0.361E-01	0.225E+00	0.198E-03
0.500E-01	0.359E-01	0.224E+00	0.102E-02
0.100E+00	0.357E-01	0.223E+00	0.215E-02
0.150E+00	0.354E-01	0.222E+00	0.338E-02
0.200E+00	0.351E-01	0.221E+00	0.476E-02
0.250E+00	0.348E-01	0.220E+00	0.630E-02
0.300E+00	0.344E-01	0.218E+00	0.805E-02
0.350E+00	0.340E-01	0.217E+00	0.101E-01
0.400E+00	0.335E-01	0.215E+00	0.124E-01
0.450E+00	0.329E-01	0.213E+00	0.152E-01
0.500E+00	0.322E-01	0.210E+00	0.187E-01
0.550E+00	0.312E-01	0.206E+00	0.231E-01
0.600E+00	0.300E-01	0.202E+00	0.291E-01
0.650E+00	0.281E-01	0.195E+00	0.378E-01
0.700E+00	0.251E-01	0.184E+00	0.522E-01
0.750E+00	0.187E-01	0.160E+00	0.826E-01
0.760E+00	0.164E-01	0.151E+00	0.933E-01
0.770E+00	0.136E-01	0.141E+00	0.107E+00
0.780E+00	0.995E-02	0.127E+00	0.124E+00
0.790E+00	0.542E-02	0.110E+00	0.145E+00
0.800E+00	0.124E-03	0.903E-01	0.170E+00
0.810E+00	−0.408E-02	0.747E-01	0.190E+00 BLOW OFF

the heat transfer coefficient $\alpha = 1100\,\text{W/m}^2\text{-K}$. Further, assume that the NO formation is *decoupled* from the combustion reaction. Determine the reactor length required for an 80% complete reaction and the variation of the NO mass fraction along the reactor length. Assume the validity of the Zeldovich mechanism.

In this problem, Eqs. 7.22–7.25 will remain unaltered but the energy equation 7.26 will read as

$$\frac{\dot{m}}{A}\frac{d\,h_m}{d\,x} = \frac{P}{A}\,q_w\,, \tag{7.52}$$

or,

$$\frac{d}{d\,x}(cp_m\,T) = \frac{A}{\dot{m}}\,\Delta H_c\,|\,R_{fu}\,| + \frac{\alpha\,P}{\dot{m}}\,(T_w - T)\,, \tag{7.53}$$

where $P = \pi D$ is the reactor perimeter. This equation now contains an additional source term owing to the wall heat transfer. As such, although conserved scalars with

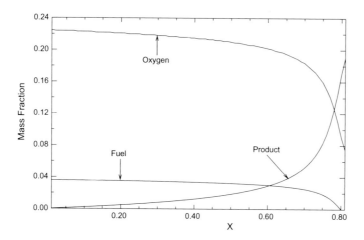

Fig. 7.2 Variation of mass fractions with X(m) for Ethane (Problem 7.1)

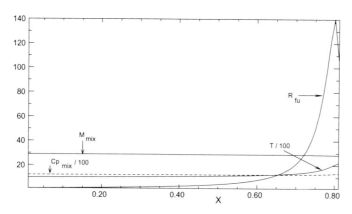

Fig. 7.3 Variation of temperature, R_{fu}, molecular weight, and specific heat with X(m) for Ethane (Problem 7.1)

respect to ω_{fu}, ω_{O_2}, and ω_{pr} can be readily formed (see Eqs. 7.39 and 7.40), a scalar analogous to that in Eq. 7.41 cannot be formed. Because R_{fu} requires knowledge of ω_{fu}, the latter can be obtained only from solution of Eq. 7.22 which is reproduced below:

$$\frac{d\,\omega_{fu}}{d\,x} = -\frac{A}{\dot{m}} \mid R_{fu} \mid . \tag{7.54}$$

As such, for a nonadiabatic reactor, Eqs. 7.53 and 7.54 must be solved simultaneously, using Euler's method. Of course, evaluation of R_{fu} will require ω_{O_2}. The latter is obtained by using Eq. 7.39.

After the fuel and oxygen mass fractions are obtained in this way, NO mass fractions can be obtained using Eqs. 5.67–5.69 because NO does not participate in

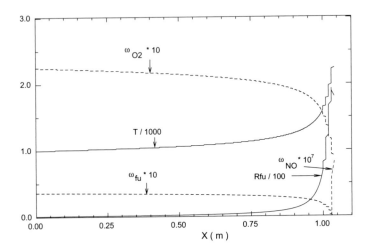

Fig. 7.4 Axial variations in a nonadiabatic PFTCR (Problem 7.2)

the chemical reaction, as given in the problem statement.[4] The applicable equation is

$$\left[\frac{\dot{m}}{\rho_m A}\right] \frac{d}{dx} (\rho_m \, \omega_{NO}) = k_G \, \frac{M_{NO}}{M_{N_2} \, M_{O_2}^{0.5}} \, \rho_m^{1.5} \, \omega_{N_2} \, \omega_{O_2}^{0.5} \, . \qquad (7.55)$$

This equation is derived by replacing the time derivative in Eq. 5.67 as

$$\frac{d}{dt} = u \frac{d}{dx} = \frac{\dot{m}}{\rho_m A} \frac{d}{dx} \, .$$

k_G is obtained as a function of local temperature using Eq. 5.69 and $\omega_{N_2} = \omega_{N_2,i}$ at all times (see Eq. 7.25). The equation is solved with initial condition $\omega_{NO,i} = 0$.

Figure 7.4 shows the results of computations obtained with $\Delta x = 1$ mm. The results are plotted up to blowoff when the fuel mass-fraction reduces to zero. This blowoff now occurs at $x = 1.04$ m (>0.81 m in Problem 7.1) because heat loss from the wall is accounted for. As expected, the fuel and oxygen mass fractions decline with x, but the NO mass fraction (though very small) rises rapidly only in the last 1 mm length. Similarly, T and R_{fu} rise rapidly prior to extinction. In the present case, for 80% reaction, the estimated length is found to be L $\simeq$ 1.02 m (>0.786 m in Problem 7.1) and the NO mass fraction at blowout is $\omega_{NO} \simeq 8.5 \times 10^{-8}$. The efficiency for a nonadiabatic reactor is defined as

---

[4]This assumption of decoupling NO formation from the main combustion reaction is justified on the grounds that NO mass fractions are usually very small and their influence on other mass fractions is negligible.

$$\eta_{comb} = 1 - \frac{\text{heat loss}}{\text{heat input}} = 1 - \frac{\alpha \int_0^L (T - T_w) \, P \, dx}{\Delta H_c \int_0^L | R_{fu} | A \, dx} \, . \tag{7.56}$$

For the present reactor, η_{comb} evaluates to 90.06%.

7.3 Well-Stirred Reactor

7.3.1 Governing Equations

A well-stirred reactor is a *stubby* plug flow reactor, so in all equations of a PFTCR, axial derivatives are replaced as follows:

$$\frac{\partial \Psi}{\partial x} \equiv \frac{(\Psi_2 - \Psi_1)}{\Delta x} \, , \tag{7.57}$$

where states 1 and 2 are shown in Fig. 7.5. Further, it is assumed that as soon as the material enters the reactor, the material instantly attains properties of state 2. Thus, in the entire reactor volume, the material properties are those of state 2. Then, making use of Eq. 7.57, Eqs. 7.3, 7.6, 7.8, and 7.17 applicable to a PFTCR can be modified to read as follows:

Bulk Mass (Eq. 7.3):

$$A \frac{d\rho_m}{dt} + \frac{\dot{m}_2 - \dot{m}_1}{\Delta x} = N_w \, P \, .$$

Dividing by A and noting that $V_{cv} = A \, \Delta x$,

$$\frac{d\rho_m}{dt} = \frac{\dot{m}_1 - \dot{m}_2}{V_{cv}} + N_w \frac{P}{A} \, . \tag{7.58}$$

For a WSTCR of any arbitrary shape, the parameter P/A may essentially be viewed as a *geometry* parameter. As such, the ratio may be replaced by

$$\frac{P}{A} = \frac{P \, \Delta x}{A \, \Delta x} = \frac{\text{surface area}}{\text{volume}} = \frac{A_w}{V_{cv}} \, , \tag{7.59}$$

where A_w is the reactor surface area. Then, Eq. 7.58 may be written as

$$\frac{d\rho_m}{dt} = \frac{\dot{m}_1 - \dot{m}_2 + \dot{m}_w}{V_{cv}} \, , \tag{7.60}$$

where $\dot{m}_w = N_w \, A_w$.

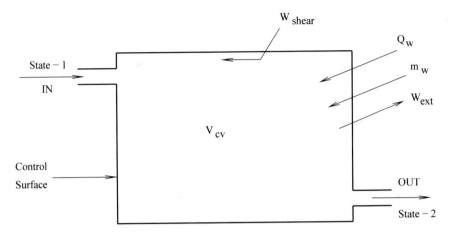

Fig. 7.5 Schematic of a well-stirred reactor

Species (Eq. 7.6):

$$A \frac{d\,(\rho_m\,\omega_k)}{dt} + \frac{(\dot{m}\,\omega_k)_2 - (\dot{m}\,\omega_k)_1}{\Delta x} = N_{w,k}\,P + R_k\,A| :,$$

or, dividing by A and using Eq. 7.59,

$$\frac{d\,(\rho_m\,\omega_k)}{dt} = \frac{(\dot{m}\,\omega_k)_1 - (\dot{m}\,\omega_k)_2 + \dot{m}_{w,k}}{V_{cv}} + R_k\,, \qquad (7.61)$$

where $\dot{m}_{w,k} = N_{w,k}\,A_w$ and $N_{w,k}$ (kg/m^2-s) is the injected mass flux of species k.

Momentum (Eq. 7.8):

$$A \frac{d\,(\rho_m\,u)}{dt} + \frac{(\dot{m}\,u)_2 - (\dot{m}\,u)_1}{\Delta x} = (N_w\,u - \tau_w)\,P - A\left(\frac{p_2 - p_1}{\Delta x}\right),$$

or, dividing by A and using Eq. 7.59 and further noting that $p_1 = p_2$,[5]

$$\frac{d\,(\rho_m\,u)}{dt} = \frac{(\dot{m}\,u)_1 - (\dot{m}\,u)_2 + m_w\,u_2 - \tau_w\,A_w}{V_{cv}}. \qquad (7.62)$$

Energy (Eq. 7.17):

$$A \frac{d}{dt}\,(\rho_m\,h_m) - \frac{(\dot{m}\,h_m)_2 - (\dot{m}\,h_m)_1}{\Delta x} = (q_w + N_w\,h_w)\,P + (Q_{rad} - \dot{W}_{ext})\,A$$

$$+ \frac{Dp}{Dt}\,A + \frac{u^2}{2}\,N_w\,P.$$

---

[5] A WSTCR is assumed to operate at constant pressure.

We note that Dp/Dt = dp/dt because $p_1 = p_2$. Hence, dividing by A and using Eq. 7.59,

$$\frac{d}{dt}(\rho_m h_m) = \frac{(\dot{m}\, h_m)_1 - (\dot{m}\, h_m)_2}{V_{cv}} +$$
$$+ m_w\left(h_w + \frac{u_2^2}{2}\right) + q_w \frac{A_w}{V_{cv}}$$
$$+ Q_{rad} - \dot{W}_{ext} + \frac{dp}{dt}. \tag{7.63}$$

Thus, Eqs. 7.60–7.63 constitute the equations governing operation of a WSTCR.

7.3.2 Steady-State WSTCR

Consider a *spherical* (diameter d) adiabatic WSTCR in which fuel and air enter at temperature T_1 and pressure p_1 (see Fig. 7.6). It is of interest to study the effect of the following parameters on the reactor exit conditions (including *blowoff*): $\dot{m}$, Φ, p_1, T_1, and d. For this purpose, we construct a mathematical model of the reactor under the assumptions that *steady state* prevails. Further, we assume that $N_w = W_{ext} = 0$ and the effect of radiation is negligible. Also, SCR is postulated.

Because our reactor is adiabatic, $q_w = 0$. Also, steady state prevails, so all time derivatives will be zero. Further, radiation effects are neglected. Therefore, $Q_{rad} = 0$. Hence, the reactor performance will be governed by modified forms Eqs. 7.60–7.63, as follows:

$$\dot{m}_1 = \dot{m}_2 = \dot{m}\ (\text{say}). \tag{7.64}$$

$$\dot{m}\,(\omega_{fu,2} - \omega_{fu,1}) = -\,|\,R_{fu}\,|\,V_{cv}, \tag{7.65}$$

$$\dot{m}\,(\omega_{O_2,2} - \omega_{O_2,1}) = -\,|\,R_{O_2}\,|\,V_{cv}, \tag{7.66}$$

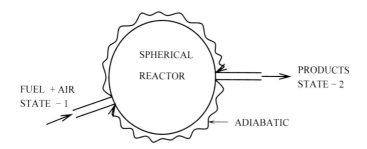

Fig. 7.6 Spherical well-stirred reactor (Problem 7.3)

$$\dot{m} \, (h_{m,2} - h_{m,1}) = 0 \,, \qquad \text{or,}$$

$$\dot{m} \, cp_m \, (T_2 - T_1) = | \, R_{fu} \, | \, \Delta H_c \, V_{cv} \,. \tag{7.67}$$

The momentum equation 7.62 is not relevant because it can only determine u_2 if τ_w and u_1 are known. In our problem, the latter are not given and hence are not of interest.

State-1

The chemical reaction in the reactor may be stated as

$$\text{fu} + \frac{a_{stoich}}{\Phi} \, (O_2 + 3.76 \, N_2) \rightarrow \text{products}$$

Therefore, the mass fractions in state 1 are given by

$$\omega_{fu,1} = \frac{M_{fu}}{M_{fu} + (a_{stoich}/\Phi) \, (32 + 3.76 \times 28)} \tag{7.68}$$

$$\omega_{O_2,1} = \frac{32 \, (a_{stoich}/\Phi)}{M_{fu} + (a_{stoich}/\Phi) \, (32 + 3.76 \times 28)} \tag{7.69}$$

$$\omega_{N_2} = 1 - \omega_{fu,1} - \omega_{O_2,1} \,. \tag{7.70}$$

Further we note that because R_{fu} is defined for a stoichiometric reaction,

$$R_{O_2} = a_{stoich} \left(\frac{32}{M_{fu}} \right) R_{fu} = r_{st} \, R_{fu} \quad \text{and} \quad R_{pr} = -(1 + r_{st}) \, R_{fu} \,.$$

Hence, Eqs. 7.65–7.67 can be written as

$$\omega_{fu,2} - \omega_{fu,1} = - | \, R_{fu} \, | \, \frac{V_{cv}}{\dot{m}} \,, \tag{7.71}$$

$$\omega_{O_2,2} - \omega_{O_2,1} = - | \, R_{fu} \, | \, \frac{V_{cv}}{\dot{m}} \, r_{st} \,, \quad \text{and} \tag{7.72}$$

$$T_2 - T_1 = \frac{\Delta H_c \, V_{cv}}{\dot{m} \, c_{p,mix}} \, | \, R_{fu} \, | = - \frac{\Delta H_c}{cp_m} \, (\omega_{fu,2} - \omega_{fu,1}) \,. \tag{7.73}$$

Combining Eqs. 7.71–7.73, we have

$$\omega_{fu,2} = \omega_{fu,1} - \frac{cp_m}{\Delta H_c} \, (T_2 - T_1) \quad \text{and} \tag{7.74}$$

$$\omega_{O_2,2} = \omega_{O_2,1} - r_{st} \, \frac{cp_m}{\Delta H_c} \, (T_2 - T_1) \,. \tag{7.75}$$

From Eq. 5.80, we have

$$| R_{fu} | = A_{fu} \exp\left(-\frac{E_a}{R_u T_2}\right) \rho_{mix}^{(x+y)} M_{fu}^{(1-x)} M_{O_2}^{-y} \omega_{fu,2}^x \omega_{O_2,2}^y, \qquad (7.76)$$

where A_{fu} is the fuel-specific preexponential constant. Replacing $\omega_{fu,2}$ and $\omega_{O_2,2}$ from Eqs. 7.74 and 7.75, we find that for a given fuel and for given inlet conditions (T_1, p_1, Φ), R_{fu} is a function of T_2 only.

Thus, the reactor performance can now be determined from Eq. 7.73 by solving for T_2 via the following equation

$$\begin{aligned}
T_2 = T_1 + &\left[\frac{\Delta H_c \, V_{cv}}{\dot{m} \, c_{p,mix}}\right] \times A_{fu} \exp\left(-\frac{E_a}{R_u T_2}\right) \\
&\times \left[\frac{p_1 M_{mix}}{R_u T_2}\right]^{(x+y)} M_{fu}^{(1-x)} M_{O_2}^{-y} \\
&\times \left[\omega_{fu,1} - \frac{c_{p,mix}}{\Delta H_c}(T_2 - T_1)\right]^x \\
&\times \left[\omega_{O_2,1} - r_{st} \frac{c_{p,mix}}{\Delta H_c}(T_2 - T_1)\right]^y,
\end{aligned} \qquad (7.77)$$

where V_{cv} for the spherical reactor is $\pi d^3/6$. This is a transcendental equation in T_2. It can be solved by the Newton–Raphson procedure by assigning different values to the model parameters $\dot{m}$, Φ, p_1, T_1, and d. To illustrate the procedure, we consider a problem.

Problem 7.3 Propane (C_3H_8) enters a spherical (d = 8 cm) adiabatic WSTCR (Fig. 7.6) at $T_1 = 298\,K$ and $p_1 = 1\,atm$ with $\Phi = 1$. Determine the exit conditions from this reactor for different mass flow rates $\dot{m} = \dot{m}_{air}(1 + F/A)$, and identify the blowoff limit. Assume $M_{mix} = 29$, and $cp_m = 1230\,J/kg$.

In this problem, the single-step reaction is written as

$$C_3H_8 + \frac{a_{stoich}}{\Phi}(O_2 + 3.76\,N_2) = \text{products}$$

where the stoichiometric coefficient $a_{stoich} = 3 + (8/4) = 5$ and, in our problem $\Phi = 1$. For propane $r_{st} = 5 \times (32/44) = 3.636$, $A_{fu} = 4.836 \times 10^9$, x = 0.1, y = 1.65, and $E_a/R_u = 15098$. Similarly, for our reactor, $V_{cv} = (\pi/6)(0.08)^3 = 2.68 \times 10^{-4}\,m^3$. Also, $p_1 = 1.01325 \times 10^5$ and $T_1 = 298$. Therefore, from Eqs. 7.68–7.70, we have $\omega_{fu,1} = 0.0602$, $\omega_{O_2,1} = 0.219$, and $\omega_{N_2} = 0.7207$. Also, $\Delta H_c = 4.65 \times 10^7\,J/kg$.

With these values, we need to iteratively solve Eq. 7.77 for T_2 for various feasible values of $\dot{m}$. This equation will now read as

Table 7.3 Spherical WSTCR with Propane at $T_1 = 298$ K, $p_1 = 1$ atm, $\Phi = 1$ and d = 8 cm, $T_{ad} = 2394$ K, $\Delta H_c = 4.65 \times 10^7$ J/kg (Problem 7.3)

$\dot{m}$	T_2 (2)	T_2 (3)	$\omega_{fu,2}$	$\omega_{O_2,2}$	R_{fu}	$P_{load} \times 10^5$	Remark
0.100	1299.5	2404	0.005	0.016	20.794	0.06480	
0.150	1386	2350	0.006	0.022	30.348	0.09720	
0.200	1461	2297	0.007	0.027	39.457	0.1296	
0.250	1533	2242	0.009	0.032	47.973	0.1620	
0.300	1606	2183	0.010	0.038	55.725	0.1944	
0.350	1688	2112	0.012	0.045	62.604	0.2268	
0.400	1805	2004	0.015	0.055	67.360	0.2592	
0.410	1849	1962	0.016	0.059	67.354	0.2657	
0.420		1907	0.018	0.064	65.895	0.2721	Blow out

$$T_2 = 298 + \left[\frac{10.132}{\dot{m}}\right] \times 4.836 \times 10^9 \, \exp\left(-\frac{15098}{T_2}\right)$$

$$\times \left[\frac{353.43}{T_2}\right]^{1.75} \times 0.09899$$

$$\times \left[0.0602 - 2.65 \times 10^{-5} \, (T_2 - 298)\right]^{0.1}$$

$$\times \left[0.219 - 9.633 \times 10^{-5} \, (T_2 - 298)\right]^{1.65} \tag{7.78}$$

There are *three* solutions to this equation.[6] The first one, T_2 (1) is a *cold* solution in which the the fuel leaves the reactor almost unreacted, and therefore, is not of interest. The second, T_2 (2), in reality, turns out to be quite *unstable*. The third, T_2 (3) is a *hot* solution, which is of interest.

The computed values T_2 (2) and T_2 (3) are shown in Table 7.3. Figure 7.7, however, shows only the hot solution. It is seen from the figure that as $\dot{m}$ (kg/s) increases, R_{fu} (kg/m³-s), $\omega_{fu,2}$, and $\omega_{O_2,2}$ increase whereas T_2 (K) and ω_{pr} decline. The last entry $\dot{m} = 0.42$ kg/s represents a condition for which no solution is possible; that is, the lefthand and righthand sides of Eq. 7.78 can be balanced only very approximately (say, within less than 10%). This represents *blowout* and no combustion is possible beyond $\dot{m} = 0.42$ kg/s. At this value, $R_{fu} = R_{fu,cr}$ and $T_2 = T_{2,cr}$, where the subscript 'cr' stands for critical or blowout condition.

Such blowout conditions can be discovered for various values of Φ, p_1 (atm), and d (cm) through repeated solutions of Eq. 7.77 for different values of $\dot{m}$ until no solution is possible for each parameter settings.[7] These values are shown in Table 7.4. Thus, blowout can occur under a variety of conditions.

---

[6]The nature of these solutions will be discussed in Sect. 7.3.3. Also see Fig. 7.10.

[7]For each Φ, values of T_{ad} and ΔH_c are evaluated as described in Chap. 3. Figure 7.8 shows these variations. The values can be approximately correlated as given in Appendix B.

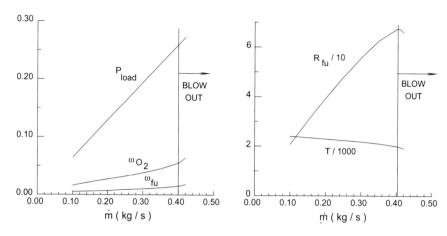

Fig. 7.7 Propane combustion in a spherical WSTCR at $T_1 = 298$ K, $p_1 = 1$ atm, $\Phi = 1$ and $D = 8$ cm (Problem 7.3)

Table 7.4 Blowout conditions for propane with $T_1 = 298$ K (Problem 7.3)

p_1	Φ	d	$\dot{m}$	$T_{2,cr}$	$R_{fu,cr}$	T_{ad}	$\Delta H_c/10^7$	$P_{load}/10^{-5}$	L_{cr}
1.0	1.0	8	0.42	1907	65.89	2394.15	4.65	0.2721	1.318
1.0	0.8	8	0.41	1885	62.75	2066.77	4.65	0.2657	1.142
1.0	0.6	8	0.27	1650	35.84	1706.00	4.65	0.1750	1.047
1.0	1.1	8	0.205	1752	33.05	2302.60	4.08	0.1328	1.398
1.0	1.3	8	0.047	1490	7.773	2131.12	3.23	0.03045	1.575
1.0	1.4	8	0.022	1390	3.658	2050.81	2.90	0.01426	1.67
0.1	1.0	8	0.0075	1900	1.167	2394.15	4.65	0.2733	1.33
0.1	0.8	8	0.0075	1882	1.113	2066.80	4.65	0.2733	1.17
0.1	1.2	8	0.0018	1618	0.291	2215.0	3.62	0.0656	1.503
0.5	1.0	8	0.125	1906	19.223	2394.15	4.65	0.2775	1.345
0.5	1.2	8	0.030	1616	4.767	2215.0	3.62	0.0666	1.529
5.0	1.0	8	7.00	1913	1085.4	2394.15	4.65	0.2763	1.334
1.0	1.0	4	0.0525	1906	65.85	2394.15	4.65	0.2696	1.319
1.0	1.0	16	3.40	1901	65.643	2394.15	4.65	0.2754	1.339

7.3.3 Loading Parameters

Dimensional Loading Parameter

It is of interest to see whether the blowout condition can, in fact, be predicted without the extensive calculations just presented. Guidance for this exercise can be taken from Eq. 7.77, which suggests that the reactor-specific design and operating parameters

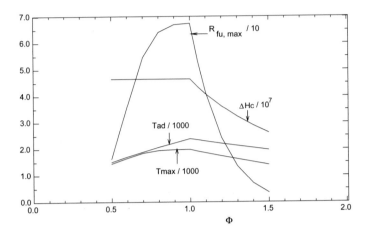

Fig. 7.8 Variation of $R_{fu,max}$, T_{max}, T_{ad}, and ΔH_c with Φ for propane at $T_1 = 298$ K, $p_1 = 1$ atm

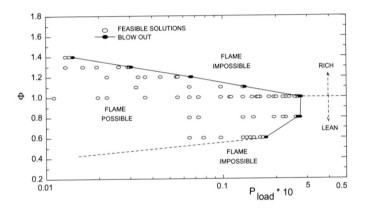

Fig. 7.9 Loading parameter for C_3H_8 in a spherical WSTCR—$T_1 = 298$ K, $p_1 = 1$ atm

governing the balance between the left and righthand sides are:

$$P_{load} = \left[\frac{p_1^{x+y} \, V_{cv}}{\dot{m}} \right]^{-1} \qquad \text{and} \qquad \Phi$$

where Φ is implicit in the values of $\omega_{fu,1}$ and $\omega_{O_2,1}$, and P_{load} has units of kg/m^3-s $\times$ $\{m^2/N\}^{(x+y)}$.

Figure 7.9 shows a plot of Φ versus $P_{load} \times 10^5$. The open circles show the large number (66 for various values of Φ, p_1, and d at different values of $\dot{m}$) of feasible solutions obtained as described earlier (see, for example, the next-to-last column in Table 7.3). These solutions represent *feasible* combustion. The filled circles represent *impossible* combustion, or blowout. In fact, it is observed that the feasible combustion

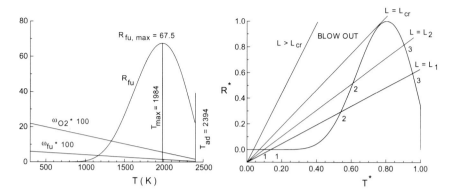

Fig. 7.10 Variation of R_{fu} with T for propane at $T_1 = 298$ K, $p_1 = 1$ atm, $\Phi = 1$

region is surrounded by *an envelope* of the blowout conditions. The envelope appears to take the shape of an asymmetric parabola with its apex at $P_{load} \times 10^5 \simeq 0.27$ at $\Phi = 1$. Such a plot was, in fact, prepared by Longwell and Weiss [75] from experimental data and they found that the plot was indeed universal for a variety of hydrocarbons.[8]

Dimensionless Loading Parameter

The P_{load} parameter is, of course, not dimensionless. Following Spalding [113], however, it will be shown that blowout occurs when a dimensionless *loading parameter* exceeds a critical value. This value will be a function of Φ and will trace the envelope between feasible and impossible combustion.

We have already recognized that for a single-step reaction, R_{fu} is a function of T_2 (see Eq. 7.77). This functional dependence is shown in Fig. 7.10 (figure on the left) for propane for the conditions of Problem 7.3, by way of an example.[9] The left figure shows that at low temperatures, R_{fu} is almost zero but rises rapidly, reaching a maximum, and then again falls off rapidly. This last condition is referred as *extinction* and occurs at $T_{ext} \simeq T_{ad} = 2394$ K for propane. At extinction, both fuel and oxygen mass fractions $\rightarrow 0$ because $\Phi = 1$. The $R_{fu,max} \simeq 67.5$ kg/m^3-s and is indeed very close to $R_{fu,cr} = 65.9$ kg/m^3-s (see the first entry in Table 7.4). Thus, for all practical purposes, $R_{fu,max} \simeq R_{fu,cr}$ (as seen in right figure). The temperature corresponding to $R_{fu,max}$ is $T_{max} = 1984$ K $< T_{ad}$, whereas $T_{cr} = 1908$ K (first entry in Table 7.4) is to the left of T_{max}.

We now consider Eq. 7.73 and rewrite it in the following manner.

---

[8]Longwell and Weiss plotted values of $\dot{m}/(V_{cv} \times p_1^{1.8})$. They found that for $\Phi = 1$, the value of this parameter was $\simeq 0.247 \times 10^{-5}$. Note that the exponent value 1.8 is very close to $x + y = 1.75$ used here.

[9]The variation of R_{fu} with T shown in the left figure is typical for all fuels. Only the magnitudes differ. Note that T_2 is replaced by T on the x-axis.

$$T^* \times L = R^* \, , \tag{7.79}$$

where replacing T_2 by T for generality, we have

$$T^* = \left[\frac{T - T_1}{T_{ad} - T_1} \right] \tag{7.80}$$

$$L = \left[\frac{\dot{m} \, c_{p,mix} \, (T_{ad} - T_1)}{\Delta H_c \, R_{fu,max} \, V_{cv}} \right] \tag{7.81}$$

$$R^* = \left[\frac{R_{fu}}{R_{fu,max}} \right] \tag{7.82}$$

R^* and T^* are fuel-specific parameters, whereas L, now called the *loading parameter*, is a composite parameter that depends on fuel, operating conditions, and the reactor volume. The variation of R^* with T^* is shown on the right hand side of Fig. 7.10. Both R^* and T^* are normalised quantities and thus vary between 0 and 1.

We now draw *load lines* $L = L_1, L_2, \ldots$ on this graph. Typically, the lines make three intersections, which represent three solutions to Eq. 7.79. Intersection 1 maybe called a *cold solution*, in which the fuel leaves at nearly inlet conditions, unreacted. Intersection 3 may be called a *hot solution*, when the fuel has successfully burned in the reactor, and therefore, the products are hot.[10] However, if we increase L, then a state is reached at which the load line is *tangent* to the $R^* \sim T^*$ curve. This is called the *critical load line*, L_{cr}. Clearly, for $L > L_{cr}$, no solution to Eq. 7.79 is possible except the cold solution represented by intersection 1. Thus, the gases in the reactor can burn only when $L \leq L_{cr}$, and $L = L_{cr}$ represents the *blowout* condition. Also, corresponding to $L = L_{cr}$, $T_{cr}^* \simeq 0.78$ at $\Phi = 1.0$. Of course, T_{cr}^* will be a function of Φ (see below).

We have already recorded such blowout solutions in Table 7.4. Therefore, for each of them, the value of L_{cr} is listed in the last column of the table. It is seen that for all solutions at $\Phi = 1$, $L_{cr} \simeq 1.3$. This value was derived by Spalding [113]. For $\Phi < 1$, $L_{cr} < 1.3$, and for $\Phi > 1$, $L_{cr} > 1.3$. In fact, we can approximately correlate L_{cr} as

$$L_{cr} \simeq 0.3 + \Phi \qquad \Phi \geq 1 \, , \text{ and}$$
$$L_{cr} \simeq 0.675 + 0.625 \times \Phi \qquad \Phi < 1 \, . \tag{7.83}$$

Thus, we can regard $L < L_{cr}$ as a *design parameter* for successful combustion. The dependence of L_{cr} on Φ is principally due to $R_{fu,max}$ because the latter is a function of Φ, as shown in Fig. 7.8.

Similarly, from Table 7.4 again, it can be deduced that for $p = 1$ atm, for example, T_{cr}^* can be correlated as

---

[10]The intermediate solution 2, which occurs on the rapidly rising portion of the $R^* \sim T^*$ curve, is essentially unstable, because in a practical reactor, the conditions are never perfectly steady. Thus, for example, if the value of Φ, were to increase slightly above the preset value, then the temperature will also increase, owing to the availability of extra fuel, further increasing R_{fu}. This increase will further increase T and so on. Eventually, solution 3 will be reached in this way. Solution 1 will be reached if the converse were to happen owing to a momentary increase in $\dot{m}$.

$$T_{cr}^* \simeq 1.213 - 0.421 \, \Phi \qquad \text{for all} \quad \Phi \tag{7.84}$$

$$T_{ad}^* = \frac{(T_{ad} - T_1)_\Phi}{(T_{ad} - T_1)_\Phi = 1} \simeq 0.1126 + 0.8874 \times \Phi \qquad \text{for} \quad \Phi < 1$$

$$\simeq 1.4093 - 0.4093 \times \Phi \qquad \text{for} \quad \Phi > 1 \tag{7.85}$$

Thus, T_{cr}^* decreases with increase in Φ, whereas $T_{ad}^* < 1$ for $\Phi \neq 1$.

7.4 Constant-Mass Reactor

In a CMTCR, inflow and outflow are absent. As such, if the reactor volume is V_{cv}, the equations of bulk mass, mass transfer, and energy will take the following forms:

$$\frac{d}{dt}[\rho_m V_{cv}] = 0, \tag{7.86}$$

$$\frac{d}{dt}[\rho_j V_{cv}] = \frac{d}{dt}[\rho_m V_{cv} \omega_j] = R_j V_{cv}, \tag{7.87}$$

$$\frac{d}{dt}[\rho_m V_{cv} e] = \dot{Q}_{rad} - \dot{W}_{exp} + q_w A_w, \tag{7.88}$$

where $q_w A_w$ represents heat transfer at the reactor wall, $\dot{W}_{exp} = p \times dV_{cv}/dt$ represents possible expansion work, and e is the specific energy of the mixture, given by $e = h_m - p/\rho_m$.

It is obvious that if the reactor walls were rigid, V_{cv} will be a constant and $\dot{W}_{exp} = 0$. But, in an IC engine, starting with an instant a few degrees before top dead center (TDC) to the greater part of the expansion stroke, V_{cv} will vary with time; this variation will be known from the RPM of the engine and the dimensions of the crank and the connecting rod connected to the piston. Thus, a CMTCR can be either a constant-volume or a variable-volume reactor. We consider each type in turn.

7.4.1 Constant-Volume CMTCR

To illustrate how a constant-volume CMTCR is analyzed, consider the problem below.

Problem 7.4 A constant volume CMTCR contains a mixture of CO and *moist* air ($x_{H_2O} = 0.02$) with $\Phi = 0.5$. Temperature and pressure are $T_i = 1000$ K and 1 atm respectively. The mixture is ignited so CO burns according to

$$CO + \frac{1}{2} O_2 \underset{k_b}{\overset{k_f}{\rightleftharpoons}} CO_2 .$$

The reaction rates are given by

$$\frac{d}{dt}[CO] = -k_f [CO] [H_2O]^{0.5} [O_2]^{0.25} \qquad \frac{d}{dt}[CO_2] = -k_b [CO_2] ,$$

where

$$k_f = 2.24 \times 10^{12} \exp\left\{-\frac{20134.7}{T}\right\}$$

and

$$k_b = 5 \times 10^8 \exp\left\{-\frac{20134.7}{T}\right\} .$$

Determine time variations of pressure and temperature.

In this problem, H_2O acts as a catalyst. Its presence is required to facilitate oxidation of CO. The H_2O concentration thus remains constant throughout the combustion process. Also, because V_{cv} is constant, Eq. 7.86 gives $d\, \rho_m/dt = 0$ or ρ_m = constant, although p and T will change with time. Hence, Eqs. 7.87 and 7.88 will read as

$$\frac{d\, \omega_{CO}}{dt} = -\frac{|R_{CO}|}{\rho_m} , \tag{7.89}$$

$$\frac{d\, \omega_{O_2}}{dt} = -\frac{|R_{O_2}|}{\rho_m} , \tag{7.90}$$

$$\frac{d\, \omega_{CO_2}}{dt} = \frac{|R_{CO_2}|}{\rho_m} , \tag{7.91}$$

$$\frac{d\, \omega_{H_2O}}{dt} = 0 , \tag{7.92}$$

$$\frac{d\, \omega_{N_2}}{dt} = 0 \tag{7.93}$$

$$\rho_m \frac{d\, e}{dt} = 0 , \qquad \text{or,}$$

$$\frac{d}{dt}(cv_m\, T) = \frac{\Delta H_c}{\rho_m}|R_{CO}| , \tag{7.94}$$

where the form of the energy equation 7.94 is explained as follows:

$$\rho_m \frac{d\, e}{dt} = \rho_m \frac{d}{dt}\left[h_m - \frac{p}{\rho_m}\right] \qquad \text{or}$$

$$= \rho_m \frac{d}{dt}[cp_m\, T + \Delta H_c\, \omega_{CO} - R_{mix}\, T] \qquad \text{or,}$$

$$= \rho_m \frac{d}{dt}[cv_m\, T] + \Delta H_c\, \rho_m \frac{d\, \omega_{CO}}{dt} \qquad \text{where} \quad cv_m = cp_m - R_{mix}$$

$$= \rho_m \frac{d}{dt}[cv_m\, T] - \Delta H_c\, |R_{CO}| .$$

In these equations, the net production rate of CO is given by

$$| R_{CO} | = M_{CO} \times | - k_f [CO] [H_2O]^{0.5} [O_2]^{0.25} + k_b [CO_2] |$$

$$= | - k_f \frac{\rho_m^{1.75}}{M_{H_2O}^{0.5} M_{O_2}^{0.25}} w_{CO} w_{H_2O}^{0.5} w_{O_2}^{0.25}$$

$$+ k_b \frac{\rho_m M_{CO}}{M_{CO_2}} w_{CO_2} | .$$
(7.95)

As usual, from the stoichiometry of CO oxidation, $R_{O_2} = r_{st} R_{CO}$ and $R_{CO_2} = -(1 + r_{st}) R_{CO}$, where $r_{st} = 0.5 (M_{O_2}/M_{CO}) = 0.5714$. Therefore, combining Eqs. 7.89–7.94, we have

$$\left[w_{CO} - \frac{w_{O_2}}{r_{st}} \right] = \left[w_{CO} - \frac{w_{O_2}}{r_{st}} \right]_i ,$$
(7.96)

$$\left[w_{CO} + \frac{w_{CO_2}}{(1 + r_{st})} \right] = \left[w_{CO} + \frac{w_{CO_2}}{(1 + r_{st})} \right]_i ,$$
(7.97)

$$\left[w_{CO} + \frac{c v_m}{\Delta H_c} T \right] = \left[w_{CO} + \frac{c v_m}{\Delta H_c} T \right]_i .$$
(7.98)

From these three relations, w_{CO}, w_{O_2}, and w_{CO_2} can be expressed in terms of temperature T when their values in the initial reactant states are known. The latter is evaluated as follows:

Initial Conditions

The initial reactants are

$$\text{Reactants} = CO + \frac{0.5}{\Phi} (O_2 + 3.76 N_2) + n_{H_2O} H_2O .$$

Thus, the initial number of moles of H_2O can be determined from the given moisture content in air

$$\frac{n_{H_2O}}{1 + 0.5 \times 4.76/\Phi + n_{H_2O}} = x_{H_2O} = 0.02 ,$$
(7.99)

or, $n_{H_2O} = 0.11755$. Therefore, the initial mixture mass is $m_{mix} = 28 + (0.5/0.5) \times (32 + 3.76 \times 28) + 0.11755 \times 18 = 167.39$. Hence, $w_{CO,i} = 28/167.39 = 0.171$ and likewise, $w_{O_2,i} = 0.1958$, $w_{N_2,i} = 0.644$, $w_{H_2O,i} = 0.01294$, and $w_{CO_2,i} = 0$.

With these initial conditions known, we need to solve Eq. 7.94 with $T = T_i = 1000 K$ (given) in which the righthand side is a function of T alone. This equation will read as

$$\frac{d}{dt}(cv_m \, T) = RHS = \frac{\Delta H_c}{\rho_m} \times |\, R_{CO}\,|$$

$$= \Delta H_c \times |\, -k_f \, \frac{\rho_m^{0.75}}{M_{H_2O}^{0.5} \, M_{O_2}^{0.25}} \, \omega_{CO} \, \omega_{H_2O}^{0.5} \, \omega_{O_2}^{0.25}$$

$$+ k_b \, \frac{M_{CO}}{M_{CO_2}} \, \omega_{CO_2} \, |\,, \qquad (7.100)$$

where

$$\omega_{CO} = \omega_{CO,i} - \frac{cv_m}{\Delta H_c}(T - T_i)\,, \qquad (7.101)$$

$$\omega_{O_2} = \omega_{O_2,i} + r_{st}\,(\omega_{CO} - \omega_{CO,i})$$

$$= \omega_{O_2.i} - r_{st} \, \frac{cv_m}{\Delta H_c}(T - T_i)\,, \qquad (7.102)$$

$$\omega_{CO_2} = \omega_{CO_2,i} - (1 + r_{st})\,(\omega_{CO} - \omega_{CO,i})$$

$$= (1 + r_{st})\,\frac{cv_m}{\Delta H_c}(T - T_i)\,, \qquad (7.103)$$

$$\omega_{H_2O,i} = 0.01295\,. \qquad (7.104)$$

To complete evaluation of the righthand side of Eq. 7.100, we must evaluate ρ_m, cv_m and ΔH_c. These are evaluated as follows:

Evaluation of ρ_m, cv_m and ΔH_c

At $T_i = 1000\,\text{K}$, and $\Phi = 1$,

$$\begin{aligned}
\Delta \overline{h}_c &= (\overline{H}_{react} - \overline{H}_{prod})_{1000} \\
&= 1 \times \overline{h}_{CO} + 0.5 \times \overline{h}_{O_2} - \overline{h}_{CO_2} \\
&= 1 \times (-110530 + 21689.5) + \frac{1}{2} \times (0 + 22714.3) \\
&\quad - 1 \times (-393520 + 33396.6) \\
&= 282640 \quad \text{kJ/kmol.}
\end{aligned}$$

Therefore, $\Delta H_c = 282640 \times 1000/28 \simeq 1.017 \times 10^7$ J/kg of CO. Similarly, the AFT for $T_i = 1000$ K and $\Phi = 0.5$ is evaluated at $T_{ad} = 2355.8$ K, and the extinction temperature corresponding to $\omega_{CO} = 0$ is evaluated at $T_{ext} = 2772.85$ K.

Now, $cv_m = \sum_j \omega_j \, cv_j = \sum_j \omega_j \, (cp_j - R_u/M_j)$, where cp_j can be evaluated (see tables in Appendix A) as a function of T at each instant of time during the solution of Eq. 7.100. In the present computations, cp_j values are evaluated at 1500 K for all times, giving $cv_m = 975.7$ J/kg-K. Finally, $\rho_m = p \, M_{mix}/(R_u \, T)$, where $M_{mix} = (\sum_j \omega_j/M_j)^{-1}$. Further, because volume is constant,

$$\frac{p}{R_{mix}\,T} = \left(\frac{p}{R_{mix}\,T}\right)_i$$

Table 7.5 CO-oxidation in a constant volume CMTCR—Problem 7.4

Time (ms)	T^*	T/1000 (K)	w_{CO}	w_{O_2}	$w_{CO_2} \times 10$	p (bars)	M_{mix}
0.0100	0.179E-03	1.000	0.171	0.196	0.369E-03	1.038	27.80
0.1000	0.183E-02	1.002	0.171	0.196	0.377E-02	1.041	27.80
1.0000	0.239E-01	1.032	0.168	0.194	0.493E-01	1.070	27.84
1.5000	0.448E-01	1.061	0.165	0.192	0.923E-01	1.098	27.88
2.0000	0.864E-01	1.117	0.160	0.189	0.178E+00	1.153	27.96
2.0900	0.100E+00	1.136	0.158	0.188	0.206E+00	1.171	27.98
2.2000	0.125E+00	1.169	0.155	0.186	0.257E+00	1.204	28.03
2.4000	0.296E+00	1.402	0.133	0.174	0.610E+00	1.427	28.35
2.4200	0.440E+00	1.597	0.114	0.163	0.906E+00	1.610	28.62
2.4275	0.660E+00	1.894	0.085	0.146	0.136E+01	1.882	29.05
2.4295	0.934E+00	2.266	0.049	0.126	0.192E+01	2.209	29.60
2.4305	0.122E+01	2.655	0.011	0.104	0.251E+01	2.538	30.20
2.4315	0.125E+01	2.689	0.008	0.103	0.257E+01	2.565	30.25
2.4325	0.137E+01	2.857	−0.008	0.093	0.282E+01	2.701	30.52

This relation thus also enables evaluation of p at every instant of time. Equation 7.100 is now solved by Euler's integration as

$$(c v_m T)_{t+\Delta t} = (c v_m T)_t + \Delta t \times RHS (T_t) \qquad (7.105)$$

where the choice of time step Δt is arbitrary. Here, computations are begun with $\Delta t = 0.01$ ms. The time step is gradually reduced as $T^* = (T - T_i)/(T_{ad} - T_i)$ approaches 1.0, to preserve accuracy.[11] Computations are stopped when $w_{CO} < 0$ and extinction is reached.

The results are shown in Table 7.5. The Table shows it takes approximately 2.4297 ms to reach $T^* = 1.0$. Extinction occurs at $t_{ext} \simeq 2.432$ ms when fuel CO is exhausted. Note that in the last 0.005 ms, the pressure rapidly rises from 1.88 bar to nearly 2.64 bar. and the corresponding change in temperature is from 1894 K to 2773 K. This rapid rise of temperature and pressure with corresponding high rate of consumption of oxygen and fuel (CO, in this case) are characteristic of *thermochemical explosion*. In the last column, small changes in Molecular weight are due to the change in the number of moles of the mixture. Finally, the reader may compare the values of T, p, and mass fractions in Table 7.5 with values computed from equilibrium considerations in Table 4.7 for $\Phi = 0.5$.

---

[11] Incidentally, the time required to reach $T^* = 0.1$ is identified with *ignition delay*, as will be discussed in Chap. 3.

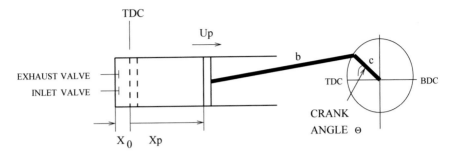

Fig. 7.11   Variable volume CMTCR

7.4.2   *Variable-Volume CMTCR*

In a spark-ignition engine, the fuel + air mixture is ignited following a spark when the piston is at TDC, and combustion may continue well into the expansion stroke. Throughout this process, the mass of the mixture remains constant, but its volume changes with accompanying heat losses to the cooling water through the cylinder walls. In this case, therefore, the equation of bulk mass Eq. 7.86 will read as

$$\frac{d}{dt} [\rho_m V_{cv}] = V_{cv} \frac{d \rho_m}{dt} + \rho_m \frac{d V_{cv}}{dt} = 0 . \tag{7.106}$$

With reference to Fig. 7.11, however, it is obvious that

$$\frac{d V_{cv}}{dt} = A_c \frac{d x_p}{dt} = A_c u_p , \tag{7.107}$$

where u_p is the instantaneous piston speed and A_c is the cross-sectional area of the cylinder.[12] Thus, Eq. 7.106 can be written as

$$\frac{d \rho_m}{dt} = -\rho_m \frac{A_c}{V_{cv}} u_p = -\frac{\rho_m u_p}{x_0 + x_p} \tag{7.111}$$

---

[12]The expressions for V_{cv}, x_p, and u_p can be derived as

$$V_{cv} = A_c (x_0 + x_p) \tag{7.108}$$

$$x_p = c (1 - \cos \theta) + b - \sqrt{b^2 - c^2 \sin^2 \theta} \tag{7.109}$$

$$u_p = \frac{d x_p}{dt} = c \Omega \sin \theta \left[1 + \frac{c \cos \theta}{\sqrt{b^2 - c^2 \sin^2 \theta}} \right] , \tag{7.110}$$

where b and c are lengths of the connecting rod and crank, respectively, x_0 is the clearance at TDC, and $\Omega = 2 \pi N/60$ is the angular velocity of the crank in radians/s, and N is RPM. The crank angle θ is measured clockwise and $\theta = 0$ coincides with TDC.

Similarly, making use of Eq. 7.86, it can be shown that Eqs. 7.87 and 7.88 will reduce to

$$\rho_m \frac{d\omega_j}{dt} = R_j \quad j = \text{fu, ox, pr} \tag{7.112}$$

$$\frac{d}{dt}[\rho_m V_{cv} e] = q_w A_w - \dot{W}_{exp} = q_w A_w - p \frac{dV_{cv}}{dt}, \tag{7.113}$$

where the last term on the righthand side represents the work done. This equation can be further modified as follows:

$$\frac{d}{dt}[\rho_m V_{cv} e] = q_w A_w - p \frac{dV_{cv}}{dt}$$

$$\frac{d}{dt}\left[\rho_m V_{cv}\left(h_m - \frac{p}{\rho_m}\right)\right] = q_w A_w - p \frac{dV_{cv}}{dt}$$

$$\frac{d}{dt}[\rho_m V_{cv} h_m] = q_w A_w - p \frac{dV_{cv}}{dt} + \frac{d}{dt}(p V_{cv})$$

$$\rho_m V_{cv} \frac{dh_m}{dt} = q_w A_w + V_{cv} \frac{dp}{dt}, \tag{7.114}$$

where $h_m = c p_m T + \Delta H_c \omega_{fu}$. Equation 7.114 thus can be further modified to read as

$$\rho_m \frac{d}{dt}[c v_m T] = q_w \frac{A_w}{V_{cv}} + \Delta H_c |R_{fu}| - \frac{p}{V_{cv}} \frac{dV_{cv}}{dt} \tag{7.115}$$

Equations 7.111, 7.112, and 7.114 (or, Eq. 7.115) thus represent the model equations of the variable-volume CMTCR. To close the mathematical problem, however, we need expressions for wall heat transfer q_w and pressure p. In a real engine, the variation of p with time is obtained from measured data. Alternatively, the pressure may be determined from

$$p = \rho_m R_{mix} T. \tag{7.116}$$

Similarly, heat transfer may be determined from

$$q_w = \alpha_{eff}(T_w - T), \tag{7.117}$$

where the heat transfer coefficient α_{eff} is obtained from experimental data. It is a function of time, as well as the location along the cylinder surface. However, it is customary to employ spatially averaged coefficients in a simplified analysis. It is given by [6, 38, 128][13]

---

[13]Borman and Ragland [19] recommend the following empirical formula for engine work:

$$\alpha_{eff}\left(\frac{kW}{m^2\text{-}K}\right) = 0.82\, d^{-0.2}\, U_g^{0.8}\, p^{0.8}\, T^{-0.53}.$$

$$\alpha_{eff} \left(\frac{W}{m^2\text{-}K} \right) = 0.035 \times \frac{k_g}{d} \times \left(\frac{U_g \, d}{\nu_g} \right)^{0.8} \tag{7.118}$$

$$U_g \left(\frac{m}{s} \right) = 2.28 \, \bar{u}_p + 0.00324 \, V_{cv} \, (p - p_{ad}) \, \frac{T_0}{p_0 \, V_{0,cv}}$$

$$\bar{u}_p = 2 \times \text{stroke} \times \frac{RPM}{60}$$

$$p_{ad} = p_0 \times \left(\frac{V_{0,cv}}{V_{cv}} \right)^{\gamma} \quad \rightarrow \quad \gamma \simeq 1.4 \,,$$

where the subscript g refers to gas and the subscript 0 refers to conditions at the start of the compression stroke (or, bottom dead center (BDC) when the intake valve closes). $\bar{u}_p$ is the mean piston speed and d is the cylinder bore diameter. p_{ad} is the pressure corresponding to isentropic compression/expansion $p \, V^{\gamma} = $ Constant.

In deriving the preceding model equations, it is assumed that temperature and mass fractions are uniform over the volume at all times. No variations with axial, radial, and circumferential directions are permitted. Such a model is, therefore, called a *thermodynamic* model. In reality, however, the ignition process creates substantial spatial variations of temperature and mass fractions as the flame front develops and moves, engulfing the unburned mixture. The ignition phenomenon will be analyzed in Chap. 8 employing *flame-front propagation* models. Sometimes, the fuel-air mixture, following pressure and temperature rise during the compression stroke, may *auto-ignite*, causing *engine knock* which may result in damage to the piston-crank assembly. For this reason, accurate prediction of the rate of pressure rise is important. Finally, in Ref. [16], application of a three-dimensional CFD model to Diesel Engine with NO_x prediction is presented.

Exercises

1. Starting with Eq. 7.9, derive Eqs. 7.15–7.17.
2. Derive Eq. 7.55.
3. Derive Eqs. 7.74 and 7.75.
4. A mixture of C_2H_6 and air ($\Phi = 1$) enters a spherical adiabatic well-stirred reactor at 298 K and p = 1 atm at the rate of 0.2 kg/s. The reactor diameter is 8 cm. Assume a global one-step reaction but incorporate the Zeldovich mechanism for NO formation. Assume that the NO kinetics do not influence heat or mass evolutions for the mixture as a whole. Determine the temperature and mass fractions of fuel, O_2, and NO at exit. Take $M_{mix} = 30$, and $cp_{mix} = 1230$ J/kg-K.
5. Develop a well-stirred reactor model (d = 8 cm) to study the effects of Φ and $\dot{m}$ during combustion of CO in *moist* air. Conditions at entry are: T = 1000 K, p = 1 atm, $x_{H_2O} = 0.02$. Evaluate all specific heats at 2000 K and treat them as constant. Take $\Delta H_c = 10^7$ J / kg of CO. Given are:

$$CO + \frac{1}{2} O_2 \underset{k_b}{\overset{k_f}{\rightleftharpoons}} CO_2$$

$$\frac{d \, [CO]}{d \, t} = -k_f \, [CO] \, [H_2O]^{0.5} \, [O_2]^{0.25} \qquad \frac{d \, [CO_2]}{d \, t} = -k_b \, [CO_2]$$

$$k_f = 7.93 \times 10^{14.6} \exp\left(-2.48 \, \Phi\right) \exp\left(-\frac{20134.7}{T}\right)$$

$$k_b = 5.0 \times 10^8 \exp\left(-\frac{20134.7}{T}\right)$$

Identify the blowoff mass flow rate for $\Phi = 0.8$, 1.0, and 1.2 and comment on the result.

6. A mixture of C_3H_8 and air ($\Phi = 0.6$) enters an adiabatic plug-flow reactor (tube diameter 3 cm) at $T = 1000$ K, $p = 1$ atm, and $\dot{m} = 0.2$ kg/s. Determine the length required to take the reaction to 80% completion. Evaluate all specific heats at 1500 K and cp_{fu} from Table B.4. Assume single-step reaction.

7. Repeat the preceding problem so that the Zeldovich mechanism is included with the same assumptions as in preceding Problem 5. Evaluate k_G at 1500 K and determine variation of NO with x till blowout.

8. For the STP conditions at inlet to a WSTCR, plot variation of T_{max} and $R_{fu,max}$ as functions of Φ for $C_{10}H_{22}$ in the manner of Fig. 7.8. Evaluate ΔH_c and T_{ad} from equations given in Appendix B and include their values in the plot with suitable scaling. Assume $M_{mix} = 29$ and $cp_{mix} = 1230$ J/kg-K.

9. For the STP conditions at inlet to a spherical WSTCR (diameter 8 cm), determine *blowout* mass flow rate $\dot{m}$, T_{cr}, P_{load}, and L_{cr} for $C_{10}H_{22}$ in the range $0.5 < \Phi < 1.5$. Assume $M_{mix} = 29$ and $cp_{mix} = 1230$ J/kg-K. Hence, correlate L_{cr} and T_{cr}^* as functions of Φ. (Hint: You will have to carry out complete analysis as described in Problem 7.3 in the text.).

10. A mixture of C_8H_{18} and *moist* air ($\Phi = 1$, $x_{H_2O} = 0.02$) is drawn into an engine at $T = 300$ K and $p = 1$ atm. The compression ratio $CR = 10.5$. Assume one-step reaction at TDC. Write a computer program and calculate variation of T, p, and mass fractions ω_{fu}, ω_{O_2}, ω_{CO_2}, ω_{H_2O} with time until extinction. Assume isentropic compression ($\gamma = 1.4$). Evaluate $\Delta H_c(T_{TDC})$ and T_{ad}, and take $cp_j = cp_m = 1200$ J/kg-K and $M_{mix} = 29$. The objective is to simulate explosion of the mixture assuming a *constant-volume* CMTCR at TDC.

11. Repeat the preceding problem for $CR = 9.0$, 9.5, 10.0, 10.5, and 11.0 and note the times for reaching extinction. Comment on the results.

12. Repeat Problem 10 assuming *variable-volume* CMTCR. The engine RPM is 2200. Bore and stroke are 75 mm each. The crank radius, $c = 3.75$ cm, and connecting rod length, $b = 30$ cm. Assume that at TDC (or $\theta = 0$), the mass fractions and pressure correspond to $T = 1000$ K in Problem 10. Assume single-step reaction and $T_w = 650$ K. Calculate variations of T, p, mass-fractions, α_{eff}, and q_w as a function of the time and the crank angle until the end of the expansion stroke. Also, determine the time (ms) and the corresponding crank angle when fuel or oxygen run out—that is, end of combustion.

Chapter 8
Premixed Flames

8.1 Introduction

Our interest in this chapter is to *look inside* the thermochemical reactors discussed in the previous chapter. Inside the reactors, the formation of visible *flame* is one obvious phenomenon. Flames are of two types; (a) premixed flames and (b) nonpremixed or diffusion flames. A Bunsen burner, shown in Fig. 8.1, is a very good example in which both types of flames are produced. In this burner, air and fuel are mixed in the mixing tube; and this *premixed mixture* burns forming a conical flame of a finite thickness (typically, blue in color). This is called the premixed flame. It is so called because the oxygen required for combustion is obtained mainly from air, which is mixed with the fuel. This premixed combustion releases a variety of species, stable and unstable. Principally, the carbon monoxide resulting from the fuel-rich combustion burns in the outer *diffusion* flame. The oxygen required for combustion in this part of the flame is obtained from the surrounding air by diffusion (or by entrainment). The overall flame shape is determined by the magnitude of the mixture velocity and its profile as it escapes the burner tube, coupled with the extent of heat losses from the tube wall.

Unlike in premixed flames, in pure diffusion flames, the sources of fuel and oxidiser are physically separated. The candle flame shown in Fig. 8.2(a) is an example of such a flame. When the wax is melted, it flows up the wick and is vaporized as fuel. Air flows upward as a result of natural convection and reacts at the outer periphery of the flame, with the vaporized fuel forming the main reaction zone. Air diffuses inward and fuel diffuses outward. In the combustion of hydrocarbon fuels, as in gas turbine combustors, fuel flows through the inner pipe while air flows through the concentric outer pipe. The air again diffuses inward. Figure 8.2(b) shows such a pure diffusion flame. In both these examples, fuel and air are physically separated and not premixed.

In this chapter, premixed flames will be discussed, along with phenomena such as flame stabilization, flame ignition, and flame quenching. These phenomena will be studied through simplified models as was done in the case of thermochemical

© Springer Nature Singapore Pte Ltd. 2020
A. W. Date, *Analytic Combustion*,
https://doi.org/10.1007/978-981-15-1853-9_8

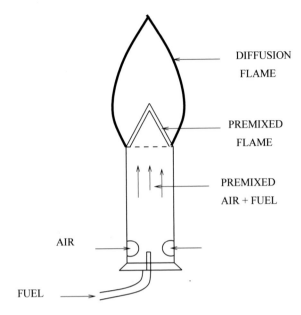

DIFFUSION
FLAME

PREMIXED
FLAME

PREMIXED
AIR + FUEL

AIR

FUEL

Fig. 8.1 The Bunsen burner flame

reactors to capture the main influencing parameters. The models will give information
required for the design of burners used in heating appliances.

The theory of diffusion flames is somewhat involved, because of the prediction
of shape and length of such flames requires solution of multidimensional partial
differential equations. Diffusion flames will be discussed in the next chapter.

8.2 Laminar Premixed Flames

8.2.1 Laminar Flame Speed

In the discussion of premixed flames, the notion of *laminar flame speed* denoted by
the symbol S_l proves useful. Its definition can be understood from Fig. 8.3(a). Let V_u
be the *average* velocity of the unburned mixture escaping the vertical tube. Let the
conical premixed flame be formed with semiangle α.

S_l **is defined as the apparent average velocity that the unburned mixture will
acquire if it were to pass through the conical flame surface.**

Clearly, therefore, the mass conservation principle will require that
$\rho_u \times V_u \times A_{pipe} = \rho_u \times S_l \times A_{cone}$, or

$$S_l = \frac{A_{pipe}}{A_{cone}} \, V_u = V_u \sin \alpha \ . \tag{8.1}$$

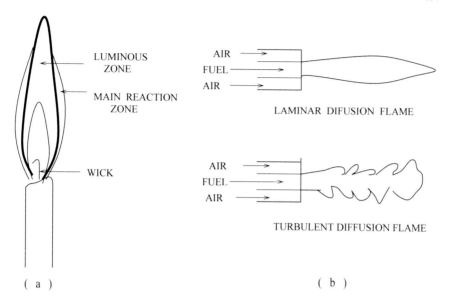

Fig. 8.2 Examples of diffusion flames

Thus, S_l is essentially the velocity of the unburned mixture across the premixed flame. Of course, it cannot physically exist, because the mixture will have already burned in the premixed region. It is thus a notional or apparent quantity, having units of velocity. Later we shall discover that S_l is a property of the fuel that denotes its *reactivity*. The greater the value of S_l, more readily that fuel will burn.

When $S_l < V_{sound}$, it is called a *deflagration* wave. When $S_l > V_{sound}$, detonation occurs. In the practical design of burners, it is ensured that $S_l \ll V_u$, to avoid flame *flash-back*. Flashback is a state in which the flame propagates in the direction opposite

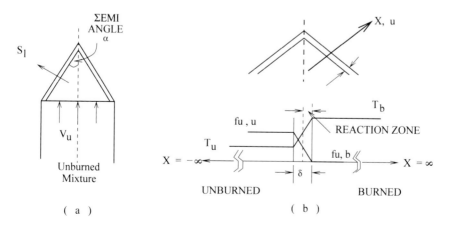

Fig. 8.3 (a) Definition of laminar flame speed, (b) local coordinate system

to that of V_u. In effect, from Eq. 8.1, it is clear that a burner must be designed with as small α as possible or with V_u as large as possible. Similarly, air-fuel mixtures are often carried through a pipe over long distances. If a source of ignition occurs downstream, the velocity V_u of the mixture in the vicinity of the source must be well above S_l to avoid flashback.[1]

8.2.2 Approximate Prediction of S_l and δ

To develop the theory for determining S_l, consider Fig. 8.3(b), in which the x-coordinate is normal to the plane of the premixed flame. It is further assumed that the flow across the flame of thickness δ is one dimensional, so the equation of bulk mass will read as

$$\frac{d\,(\rho_m\,u)}{dx} = 0 \quad \text{or} \quad \rho_m\,u = \rho_u\,S_l = G \;(\text{constant.}) \tag{8.2}$$

S_l is multiplied by the density of the unburned mixture ρ_u. This is because S_l is a fictitious velocity associated with the unburned condition. Assuming a simple chemical reaction (SCR)

$$1 \text{ kg of fuel} + R_{st} \text{ kg of oxidant air} \rightarrow (1 + R_{st}) \text{ kg of products}$$

where R_{st} is the stoichiometric air-fuel ratio, we have[2]

$$\frac{d}{dx}\left[\rho_m\,u\,\omega_{fu} - \rho_m\,D\,\frac{d\,\omega_{fu}}{dx}\right] = -\mid R_{fu}\mid, \tag{8.3}$$

$$\frac{d}{dx}\left[\rho_m\,u\,\omega_{ox} - \rho_m\,D\,\frac{d\,\omega_{ox}}{dx}\right] = -\mid R_{ox}\mid, \tag{8.4}$$

$$\frac{d}{dx}\left[\rho_m\,u\,\omega_{pr} - \rho_m\,D\,\frac{d\,\omega_{pr}}{dx}\right] = R_{pr}, \quad \text{and} \tag{8.5}$$

$$\frac{d}{dx}\left[\rho_m\,cp_m\,u\,T - k_m\,\frac{d\,T}{dx}\right] = \Delta H_c\mid R_{fu}\mid. \tag{8.6}$$

Axial diffusion is now included because significant variations of mass fraction and temperature are to be expected in the thin premixed flame of thickness δ. It is also assumed that the mass diffusivity D of fuel, oxidant and product is the same and, further, the mass diffusivity D equals thermal diffusivity: $\alpha_m = k_m/(\rho_m\,cp_m)$.

---

[1] Of course, here the reference is being made to *average* V_u. In reality, near the tube/duct walls, the local velocity tends to be zero, and therefore, S_l may exceed the unburned velocity there. Methods for avoiding flashback are not discussed here.

[2] The subscript ox in Eq. 8.4 now refers to oxidant or air. In Eq. 8.6, all specific heats are assume equal ($cp_j = cp_m$).

This is called the *unity Lewis number* (Le = Pr/Sc = 1), an assumption that holds reasonably well for gases. With these assumptions, it is easy to show that Eqs. 8.3 to 8.6 can be represented by a single equation

$$\frac{d}{dx}\left[\rho_m u \Phi - \Gamma \frac{d\Phi}{dx}\right] = 0 , \tag{8.7}$$

where

$$\Gamma = \frac{k_m}{cp_m} = \rho_m D \quad \left(\frac{kg}{m\text{-}s}\right) \tag{8.8}$$

and

$$\begin{aligned}
\Phi &\equiv w_{fu} - \frac{w_{ox}}{R_{st}} \\
&= w_{fu} + \frac{w_{pr}}{(1 + R_{st})} \\
&= cp_m T + \Delta H_c w_{fu} \\
&= cp_m T (1 + R_{st}) - \Delta H_c w_{pr} \\
&= \text{constant.}
\end{aligned} \tag{8.9}$$

Thus, T, w_{fu}, w_{ox}, and w_{pr} bear a proportionate relationship at every x. We can, therefore, solve any of Eqs. 8.3 to 8.6. We choose Eq. 8.6. Because G is constant and cp_m is assumed constant, this equation can be written as

$$G \frac{dT}{dx} - \frac{d}{dx}\left[\Gamma \frac{dT}{dx}\right] = \frac{\Delta H_c}{cp_m} |R_{fu}| . \tag{8.10}$$

Integrating this equation from $-\infty$ to $+\infty$,

$$G (T_b - T_u) - \Gamma \left[\frac{dT}{dx}\Big|_{\infty} - \frac{dT}{dx}\Big|_{-\infty}\right] = \frac{\Delta H_c}{cp_m} \int_{-\infty}^{\infty} |R_{fu}| dx . \tag{8.11}$$

However, notice that $dT/dx |_{\pm\infty} = 0$. This is because, at $-\infty$, unburned conditions are assumed to prevail with $T = T_u$, whereas at $+\infty$, only burned conditions prevail with $T = T_b = T_{ad}$. As such,

$$G (T_b - T_u) = \frac{\Delta H_c}{cp_m} \int_{-\infty}^{\infty} |R_{fu}| dx , \tag{8.12}$$

where R_{fu} is function of T and, hence, of x. To make further progress, we simplistically assume that the temperature variation over length δ is *linear*. That is,

$$\frac{T - T_u}{T_b - T_u} = \frac{x}{\delta} . \tag{8.13}$$

Hence, Eq. 8.12 can be written as

$$G \, (T_b - T_u) = \frac{\Delta H_c}{cp_m} \frac{\delta}{(T_b - T_u)} \int_{T_u}^{T_b} | R_{fu} | \, dT \, . \tag{8.14}$$

We now define the integrated average R_{fu} as

$$\overline{R}_{fu} \equiv \left[\int_{T_u}^{T_b} | R_{fu} | \, dT \right] / (T_b - T_u) \, . \tag{8.15}$$

Thus, Eq. 8.14 can be written as

$$G \, (T_b - T_u) = \frac{\Delta H_c}{cp_m} \, \delta \, \overline{R}_{fu} \, . \tag{8.16}$$

Our next task is to obtain an estimate of δ. To do this, Eq. 8.6 is integrated from x $= -\infty$ to the flame midplane at x $= \delta/2$. Thus, noting that $dT/dx \, |_{-\infty} = 0$, we have

$$G \, (T_{\delta/2} - T_u) - \Gamma \left. \frac{d \, T}{dx} \right|_{\delta/2} = \frac{\Delta H_c}{cp_m} \int_{-\infty}^{\delta/2} | R_{fu} | \, dx \, . \tag{8.17}$$

We now make another drastic assumption: we assume that over the distance $0 < x < \delta/2$, $R_{fu} \simeq 0$. This assumption derives from our knowledge that at low temperatures, R_{fu} is indeed very small (see Fig. 7.10 in Chap. 7). Thus, the right hand side of Eq. 8.17 is zero. Further, from the assumed temperature profile Eq. 8.13,

$$T_{\delta/2} = 0.5 \, (T_b + T_u) \qquad \text{and} \qquad \frac{d \, T}{dx} |_{\delta/2} = \frac{(T_b - T_u)}{\delta} \, .$$

With these assumptions, Eq. 8.17 reduces to

$$\delta = 2 \, \frac{\Gamma}{G} \, . \tag{8.18}$$

Substituting this equation in Eq. 8.16, it is easy to derive that

$$S_l = \frac{G}{\rho_u} = \sqrt{\frac{2}{\rho_u^2} \frac{\Delta H_c}{cp_m} \frac{\Gamma}{(T_b - T_u)} \overline{R}_{fu}} \tag{8.19}$$

and

$$\delta = \frac{2 \, \Gamma}{\rho_u \, S_l} \, , \tag{8.20}$$

where Γ is given by Eq. 8.8.

Further Simplification

From Eq. 8.9, we note that

$$cp_m T (1 + R_{st}) - \Delta H_c \omega_{pr} = \text{constant.} \tag{8.21}$$

Writing this equation for the burned and the unburned states and noting that $\omega_{pr,b} = 1$ and $\omega_{pr,u} = 0$, it follows that

$$\Delta H_c = cp_m (1 + R_{st}) (T_b - T_u) . \tag{8.22}$$

Substituting this equation in Eq. 8.19, we have

$$S_l = \sqrt{2 (1 + R_{st}) \frac{\Gamma}{\rho_u^2} \overline{R}_{fu}} . \tag{8.23}$$

Thus, this equation confirms our assertion that S_l is a quantity having units of velocity and that it is essentially a property of the fuel, as $\overline{R}_{fu}$ and R_{st} are fuel-specific properties.

Although the foregoing development is very approximate, with several drastic assumptions, Eq. 8.23 nonetheless enables prediction of some interesting effects. For example, we know that $R_{fu} \propto p^{x+y}$ where $x + y = 1.75$ for most hydrocarbon fuels (except methane, for which $x + y = 1.0$; see Table 5.8). However, $\rho_u \propto p$. Therefore, $S_l \propto \sqrt{p^{(x+y)-2}}$ or $S_l \propto p^{-0.125}$ for general hydrocarbons and $S_l \propto p^{-0.5}$ for methane. The experimental data for methane obtained by Andrews and Bradley [5] in fact show that $S_l = 43\ p^{-0.5}$ cm/s.

Besides pressure, S_l is also influenced by Φ and T_u. In Fig. 7.8 of Chap. 7, we noted that $R_{fu,max}$ is a function of Φ. However, $\overline{R}_{fu}$ is related to $R_{fu,max}$. This is shown in Fig. 8.4 for propane and with $T_u = 298$ K. It is seen that the ratio $\overline{R}_{fu}/R_{fu,max}$ varies between 0.2 and 0.34 for $0.6 < \Phi < 1.5$. Similar variations are observed for other fuels as well. Thus, in general, we can conclude that

$$S_l = F (p, T_u, \Phi, \text{fuel}) . \tag{8.24}$$

Experimental correlations for S_l, to be discussed shortly, indeed show this functional dependence.

8.2.3 Refined Prediction of S_l and δ

In the analysis of the previous subsection, we were able to obtain a closed form solution for S_l because drastic assumptions were made. For example, the temperature profile over length δ was taken to be linear and R_{fu} was considered to make no contribution over $0 < x < \delta/2$. The linear temperature profile also implies linear

Fig. 8.4 Variation of
$R_{fu,max}$, $\overline{R}_{fu}$ and
$(\overline{R}_{fu}/R_{fu,max}) \times 10$ with Φ
for propane at p = 1 atm and
$T_u = 298$ K

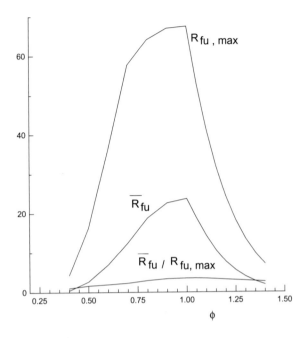

mass fraction variations of fuel, oxygen, and product by virtue of relation Eq. 8.9. In order to bring greater realism, however, any of Eqs. 8.3 to 8.6 must be solved by a numerical method on a computer. Here, we choose Eq. 8.6 by way of an example.

In this method, a domain $0 < x < x_0$ is selected such that $x_0 \to \infty$ or is very large. On this domain, a large number of points (called nodes) are now chosen, as shown in Fig. 8.5(a). Midway between consecutive nodes, a vertical dotted line is drawn to demarcate a *control volume* of dimension $\Delta x = \Delta x \times 1 \times 1$, where unity dimensions are in the y and z directions. Figure 8.5(b) shows one such typical node P, along with neighbouring nodes to the west (W) and to the east (E). Location w is midway between P and W and, likewise, location e is midway between P and E.

Finite Difference Equation

Equation 8.6 is integrated over length of the control volume surrounding P—that is, form w to e. Then, noting that $\rho_m u = G = $ constant (see Eq. 8.2), we have

$$\int_w^e \frac{d}{dx}\left(G\,T - \frac{k_m}{cp_m}\frac{dT}{dx}\right) dx = \frac{\Delta H_c}{cp_m}\int_w^e |R_{fu}|\,dx\,,$$

or, with $\Gamma_m = k_m/cp_m$,

$$G\,(T_e - T_w) - \Gamma_m\left(\frac{dT}{dx}\bigg|_e - \frac{dT}{dx}\bigg|_w\right) = \frac{\Delta H_c}{cp_m}\int_w^e |R_{fu}|\,dx\,.$$

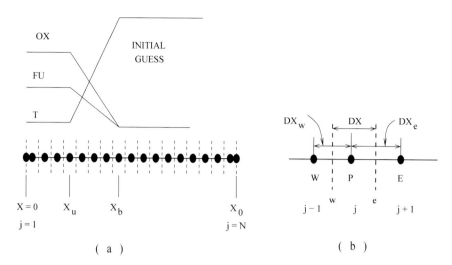

Fig. 8.5 Numerical solution of Eq. 8.5

This equation is now discretized using using the upwind difference scheme (UDS) for the convective term [30, 88].

$$G\,(T_P - T_W) - \Gamma_m \left(\frac{T_E - T_P}{DXe} - \frac{T_P - T_W}{DXw} \right)$$
$$= \frac{\Delta H_c}{cp_m}\,|\,R_{fu,P}\,|\,DX\,, \tag{8.25}$$

where $|\,R_{fu}\,|$ is taken to be constant over length DX. This equation can be rearranged to read as

$$AP\,T_P = AE\,T_E + AW\,T_W + S\,. \tag{8.26}$$

Such an equation is derived for each node P. In general, therefore, designating P ($\equiv j$), E ($\equiv j+1$), and W ($\equiv j-1$) where j is a running node index $j = 2, 3, \ldots, N - 1$, we have N − 2 equations with $j = 1$ at $x = 0$ and $j = N$ at $x = x_0$. Thus,

$$AP_j\,T_j = AE_j\,T_{j+1} + AW_j\,T_{j-1} + S_j\,, \tag{8.27}$$
$$AE_j = \frac{\Gamma_{m,j+1/2}}{x_{j+1} - x_j}\,,$$
$$AW_j = G + \frac{\Gamma_{m,j-1/2}}{x_j - x_{j-1}}\,,$$
$$AP_j = AE_j + AW_j\,, \quad \text{and}$$
$$S_j = \frac{\Delta H_c}{cp_m}\,|\,R_{fu,j}\,|\,DX_j\,,$$

where $DX_j = x_{j+1/2} - x_{j-1/2}$. This equation set is solved for j = 2 to j = N − 1 by an iterative solver called the *Tri-Diagonal Matrix Algorithm* [30, 88] with the boundary conditions $T_1 = T_u$ (known) at x = 0 and $dT/dx|_{x_0} = 0$, or $T_N = T_{N-1}$. At each iteration, the value of Γ_m is evaluated using the equation

$$\Gamma(T) = 2.58 \times 10^{-5} \left(\frac{T}{298}\right)^{0.7}, \tag{8.28}$$

where T is evaluated from the most recently predicted T_j. Knowing T_j, the mass fractions are evaluated as

$$\omega_{fu,j} = \omega_{fu,u} - \frac{cp_m}{\Delta H_c}(T_j - T_u), \tag{8.29}$$

$$\omega_{ox,j} = \omega_{ox,u} - R_{st}\frac{cp_m}{\Delta H_c}(T_j - T_u), \tag{8.30}$$

$$\omega_{O_2,j} = \omega_{O_2,u} - r_{st}\frac{cp_m}{\Delta H_c}(T_j - T_u), \tag{8.31}$$

where r_{st} is stoichiometric coefficient[3] based on O_2. Once the mass fractions are evaluated, it is possible to evaluate $R_{fu,j}$. Further, ρ_m is evaluated from $\rho_{m,j} = p\,M_{mix,j}/(R_u\,T_j)$.

Solutions are obtained for an *assumed* value of G. The converged solution enables evaluation of S_l by integration of Eq. 8.6 from 0 to x_0. The result is

$$G = \frac{\Delta H_c}{cp_m}\frac{\int_0^{x_0}|R_{fu}|\,dx}{(T_b - T_u)} \quad \text{and}$$

$$S_l = \frac{\Delta H_c}{cp_m}\frac{\sum_{j=2}^{j=N-2}|R_{fu,j}|\,DX_j}{\rho_u\,(T_N - T_1)}. \tag{8.32}$$

When the domain length is chosen properly, of course, $\omega_N = \omega_{fu,b} \to 0$ and $T_N \to T_b$.

Initial Guess

Computations are begun with an initial guess for ω_{fu}, ω_{ox}, and T as shown in Table 8.1.

The domain length is chosen to be 3 mm, where $x_u = 0.1$ mm and $x_b = x_u + \delta$ mm and $\delta = 1$ mm are chosen.[4] The values of T_{ad} and of ΔH_c are taken from Appendix B. Computations are performed with equally spaced nodes (N = 502) so DX = 3/500 = 0.006 mm.

Solutions

Solutions are repeated by assigning different values to G (or, in effect, S_l) until S_l predicted from Eq. 8.32 matches with the one specified. Figure 8.6 shows computed

---

[3] Stoichiometric coefficient based on oxidant air is R_{st} = air/fuel ratio.
[4] For premixed flames, $\delta \simeq 1$ mm as will be found shortly.

Table 8.1 Initial guesses of variables

Domain	ω_{fu}	ω_{ox}	T
$x < x_u$	$\omega_{fu,u}$	$\omega_{ox,u}$	T_u
$x_u < x < x_b$	$\omega_{fu,u}\left[\frac{x-x_b}{x_u-x_b}\right]$	$\omega_{ox,u}\left[\frac{x-x_b}{x_u-x_b}\right]$	$T_u + (T_{ad} - T_u)\left[\frac{x-x_u}{x_b-x_u}\right]$
$x > x_b$	0	0	T_{ad}

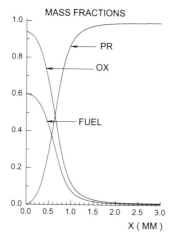 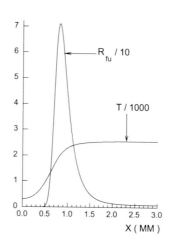

Fig. 8.6 Predicted Profiles for propane $T_u = 298$ K, $p = 1$ atm, $\Phi = 1$. Predicted $S_l = 0.438$ m/s

profiles for a propane (C_3H_8) flame. For $T_u = 298$ K, $p = 1$ atm, and $\Phi = 1$, the value of S_l predicted in this way is 0.438 m/s or 43.8 cm/s. It is also clear from the figure that both the fuel and the oxidant mass fractions decline, whereas the product mass fraction and T increase rapidly over the reaction zone that is demarcated by region of rapid variation of R_{fu}. This region is identified with flame thickness δ. It is difficult to exactly estimate δ. However, if it is assumed that the region between $0.99 \times \omega_{fu,u}$ and $0.01 \times \omega_{fu,u}$ represents flame thickness, then Fig. 8.6 shows that the predicted $\delta \simeq 1.3$ mm. Similar calculations have been performed for different values of Φ, T_u, and p. These will be considered in the next subsection.

Usually, predictions of S_l obtained with the one-step reaction model fail to match the experimental data for different fuels under all conditions of T_u, p_u, and Φ. In such cases, more detailed chemistry models, such as the quasi-global or multi-step reaction models, must be considered. These models, however, require solution of the mass transfer equation for several species prescribed in the reaction mechanism. The solution of such an equation set is typically obtained by Newton's method [106, 107].

Table 8.2 Values of constants in Correlation 8.33

Fuel	B_1 (cm/s)	B_2 (cm/s)	Φ^*
Methanol	36.92	−140.51	1.11
Propane	34.22	−138.65	1.08
Isooctane	26.32	−84.72	1.13
Gasolene	27.58	−78.34	1.13

8.2.4 Correlations for S_l and δ

The experimentally derived correlations for S_l do, in fact, reflect the functional relationship derived from our simplistic and refined theories (see function Eq. 8.24) as will be shown later. For example, for fuel used in IC engines and gas-turbine applications, Meghalachi and Keck [81] give the following correlation valid for $0.8 < \Phi < 1.5$, $0.4 < p_u(\text{atm}) < 50$ and $298 T_u(\text{K}) < 750$:

$$S_l \left(\frac{\text{cm}}{\text{s}} \right) = S_{l,ref} \left[\frac{T_u}{T_{ref}} \right]^a \left[\frac{p}{p_{ref}} \right]^b (1 - 2.1\, \omega_{dil}) , \tag{8.33}$$

$$S_{l,ref} \equiv B_1 + B_2 \, (\Phi - \Phi^*)^2 ,$$

$$a \equiv 2.18 - 0.8 \, (\Phi - 1) , \quad \text{and}$$

$$b \equiv -0.16 + 0.22 \, (\Phi - 1) , \tag{8.34}$$

where $T_{ref} = 298\,\text{K}$, $p_{ref} = 1\,\text{atm}$, and the values of fuel-dependent constants are given in Table 8.2 fore some fuels. Sometimes, combustion products are mixed with the reactant mixture to act as dilutant. ω_{dil} in Eq. 8.34 is the mass fraction of the dilutant present in the reacting mixture.[5] The value of δ can now be obtained from Eq. 8.20.

Gottgens et al. [41] have provided following correlation:

$$S_l \left(\frac{\text{cm}}{\text{s}} \right) = A \left[\frac{\Phi}{R_{st} + \Phi} \right]^x \times \exp \left(\frac{B}{T_0} \right) \times \left(\frac{T_u}{T_0} \right) \left[\frac{T_{ad} - T_0}{T_{ad} - T_u} \right]^y , \tag{8.35}$$

$$T_0 = -\frac{C}{\ln (p/D)} , \quad \text{and} \tag{8.36}$$

$$\delta = z \times \left[\frac{\Gamma\,(T_0)}{\rho_u\, S_l} \right] \times \left[\frac{T_{ad} - T_u}{T_0 - T_u} \right] . \tag{8.37}$$

The values of A, B, C, and D are given in Table 8.3, and $\Gamma\,(T_0)$ is obtained from Eq. 8.28.

---

[5]In industrial boilers, such flue gas recirculation is known to reduce NO_x formation by 50% to 80% [122].

Table 8.3 Values of Constants in Correlation 8.35

Fuel	A (cm/s)	D (bars)	C (K)	B (K)	x	y	z
Methane	22.176	3.1557×10^8	23873.0	6444.27	0.565175	2.5158	0.994
Propane	1274.89	2.2501×10^6	17223.5	1324.78	0.582214	2.3970	0.951
Ethane	1900.41	4.3203×10^6	18859.0	506.973	0.431345	2.18036	1.051

Comparison of Predictions

It will be useful to compare the predictive capabilities of the preceding correlations with our theoretical analysis (Eq. 8.19 and numerical method with a single-step reaction). This comparison is presented in Table 8.4 for propane (C_3H_8) for various combinations of T_u, p, and Φ. The first three row-blocks in the table show variations of S_l with Φ at three values of pressure p = 1, 2, and 5 atm and T_u = 298 K.

From Table 8.4, it is seen that all results predict the main trend correctly; that is, S_l is maximum when Φ is somewhat greater than 1.0. Also, S_l decreases with an increase in pressure, as was already noted. Overall, however, results from neither the correlations nor the simplified numerical analyses agree with each other. For this reason, multistep detailed chemistry models are required for prediction purposes.

The fourth and fifth row-blocks show the effect of T_u and Φ at p = 1 atm. It is seen that S_l increases with an increase in T_u, whereas the variation with Φ is similar to that observed earlier. The last row-block shows the effect of T_u at p = 5 atm. Again, S_l decreases with increase in p.

Correlations for Methane $\Phi = 1$

It will be recalled (see Table 5.8) that exponents x and y for methane (x + y = 1) are different from those for other hydrocarbons (x + y = 1.75). As such, special correlations are needed for methane. All correlations are for $\Phi = 1$ and express S_l in cm/s. Further, the effect of dilution is not accounted.

Andrews and Bradly [5] provide the following correlations.

$$S_l \left(\frac{cm}{s}\right) = 10 + 3.71 \times 10^{-4} \, T_u^2 \quad \text{for} \quad p = 1 \quad \text{atm}$$

$$= 43 \times p^{-0.5} \quad \text{for} \quad T_u = 298 \text{ K} \quad p > 5 \quad \text{atm} \quad (8.38)$$

Hill and Huang [47] suggest the following correlation.

$$S_l \left(\frac{cm}{s}\right) = 32.9 \left(\frac{T_u}{T_{ref}}\right)^{1.8} \left(\frac{p}{p_{ref}}\right)^{-0.299} \quad (8.39)$$

Correlations for Hydrogen

Coffee et al. [26] have provided following correlations for 40% H_2 and 60% Air (that is, $\Phi \simeq 1.57$) to reflect effect of T_u (K) and p (atm).

Table 8.4 Comparison of S_l predictions for propane

Φ	T_u (K)	p (atm)	S_l (Eq. 8.33) (m/s)	S_l (Eq. 8.35) (m/s)	S_l (Eq. 8.19) (m/s)	S_l numerical (m/s)
0.6	298.0	1.0	0.0228	0.112	0.155	0.2501
0.7	298.0	1.0	0.141	0.209	0.191	0.3372
0.8	298.0	1.0	0.233	0.318	0.211	0.4023
0.9	298.0	1.0	0.297	0.411	0.222	0.4275
1.0	298.0	1.0	0.333	0.469	0.225	0.4384
1.1	298.0	1.0	0.342	0.487	0.148	0.3204
1.2	298.0	1.0	0.322	0.474	0.095	0.1850
1.3	298.0	1.0	0.275	0.448	0.059	0.1030
0.6	298.0	2.0	0.0192	0.079	0.142	0.2180
0.7	298.0	2.0	0.121	0.151	0.175	0.3073
0.8	298.0	2.0	0.203	0.318	0.194	0.3675
0.9	298.0	2.0	0.262	0.322	0.203	0.3910
1.0	298.0	2.0	0.298	0.371	0.206	0.4020
1.1	298.0	2.0	0.310	0.385	0.136	0.2980
1.2	298.0	2.0	0.297	0.371	0.0876	0.2005
1.3	298.0	2.0	0.258	0.346	0.054	0.1210
0.6	298.0	5.0	0.0151	0.0395	0.126	0.2140
0.7	298.0	5.0	0.0987	0.0908	0.156	0.2710
0.8	298.0	5.0	0.168	0.161	0.173	0.3170
0.9	298.0	5.0	0.222	0.224	0.181	0.3500
1.0	298.0	5.0	0.258	0.261	0.184	0.3600
1.1	298.0	5.0	0.274	0.270	0.121	0.2680
1.2	298.0	5.0	0.267	0.256	0.078	0.1830
0.6	400.0	1.0	0.0475	0.193	0.32	0.4520
1.0	400.0	1.0	0.633	0.714	0.415	0.7350
1.2	400.0	1.0	0.584	0.727	0.196	0.3800
0.6	500.0	1.0	0.083	0.296	0.578	0.7500
1.0	500.0	1.0	1.03	1.019	0.686	1.1140
1.2	500.0	1.0	0.917	1.044	0.352	0.6750
1.0	400.0	5.0	0.489	0.398	0.339	0.5930
1.0	500.0	5.0	0.796	0.568	0.561	0.9050

$$S_l \left(\frac{cm}{s} \right) = 44.2 + 0.01906 \times T_u^{1.655} \quad \text{for} \quad 200 < T_u < 600 \quad \text{and} \quad (8.40)$$

$$= 261.3 \times p^{0.0976} \quad \text{for} \quad 0.1 < p < 10 . \quad (8.41)$$

The exponent of p is positive. Based on detailed chemistry model, analytical results with $T_u = 298$ K and p = 1 atm obtained by Coffee et al. [25] can be correlated as

$$S_l \left(\frac{cm}{s} \right) = 298.73 - 139.22 \, (\Phi - 1.8)^2 \quad \text{for} \quad 0.4 < \Phi < 2.4 \quad (8.42)$$

8.3 Turbulent Premixed Flames

The unburned air-fuel mixture velocity issuing from burners used in furnaces and engines is typically very high, so turbulence is inevitably present. It is well known that turbulence causes enhanced rates of heat, mass, and momentum transfer (that is, $\mu_t \gg \mu$, and, hence, $\Gamma_t \gg \Gamma$) resulting in an enhanced rate of heat release in flames. Recall that the flame speed is $\propto \sqrt{\Gamma}$ (see Eq. 8.19). As such, turbulent flame speeds (denoted by S_t) are higher than laminar flame speeds for a given fuel.[6] Another geometric explanation can be had from Fig. 8.7. The figure shows that the actual flame area A_{flame} is greater than the *apparent* area A_{app}. Of course, the flame area itself fluctuates with time. As such, A_{flame} is a time-averaged area. Thus, if the flame speed across the flame area is taken to be essentially laminar speed S_l, then it follows that

$$S_t = \frac{\dot{m}}{\rho_u \, A_{app}} \simeq S_l \frac{A_{flame}}{A_{app}} \quad \text{with} \quad A_{flame} > A_{app} , \quad (8.43)$$

where $\dot{m} = \rho_u \, A_{duct} \, V_u$.

Although these simple explanations suffice to illustrate why S_t should be greater than S_l, prediction of the exact magnitude of S_t under a variety of circumstances has been found to be quite difficult because the turbulent premixed flame demonstrates substantially different structures for different unburned conditions. One empirical way in which these structures can be characterized is through the use of the Damkohler number (Da), which is defied as

$$Da = \frac{\text{Characteristic flow mixing time}}{\text{Characteristic Chemical time}} = \frac{l_{int}/u'}{\delta_l/S_l} , \quad (8.44)$$

where l_{int} and u' represent estimates of turbulence length scale and velocity fluctuation (in rms sense) respectively in the flame region. The flow mixing time scale is inversely proportional to the fluid mixing rate and, likewise, the chemical time scale

---

[6]Unlike S_l, which is a characteristic property of the fuel, S_t very much depends on the local conditions—that is, burner geometry, unburned conditions, mixture velocity, and the levels of turbulence.

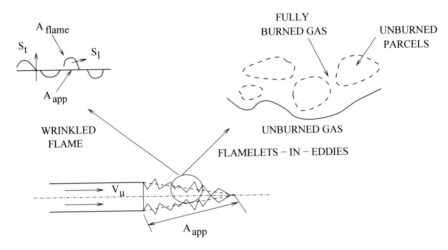

Fig. 8.7 Turbulent premixed flame

is inversely proportional to the chemical reaction rate. As such, a large value of Da implies that chemical reaction rates dominate over fluid mixing rates and vice versa.

In Chap. 6, it was shown that turbulence length scale varies between the smallest *dissipative* Kolmogorov scales l_ϵ to the largest integral scales l_{int}, at which production of turbulent kinetic energy from the mean kinetic energy is maximum. Thus, the *wrinkled flame* structure shown in Fig. 8.7, for example, is associated with $\delta_l \leq l_\epsilon$, whereas the *flamelets-in-eddies* structure is associated with $l_\epsilon < \delta_l < l_{int}$.

Andrews et al. [4] have reported a variety of empirical relations for S_t in the wrinkled-flame region. In this region, $l_{int} \gg \delta_l$ and hence, Da > 1. The larger the turbulence Reynolds number $Re_t = u' l_{int}/\nu$, the larger is the value of l_{int}. In IC engines, typically, Da $\simeq 500$ and $Re_{int} \simeq 100$. Damkohler interpreted $A_{flame} = A_{app} + A_{wrinkle}$. As such,

$$\frac{S_t}{S_l} = \frac{A_{app} + A_{wrinkle}}{A_{app}} = 1 + \frac{A_{wrinkle}}{A_{app}} \simeq 1 + C \left[\frac{u'}{S_l}\right]^n . \tag{8.45}$$

From experimental evaluations,[7] values of C and n can be determined. For small values of u'/S_l, $1 < n < 2$ and C $= 1$. For large values of u'/S_l when Da $\gg 1$, n $= 0.7$ and C $= 3.5$.

Recently, Filayev et al. [39] have studied turbulent premixed flames in a slotted (length L $= 50.8$ mm and width W $= 25.4$ mm) Bunsen burner in which turbulence was generated by means of a grid placed in the supply duct. Experiments were conducted with a stoichiometric methane-air mixture with $3 < V_u < 12$ m/s and $0.28 < u'/S_l < 8.55$. They reported following correlation:

---

[7]Experimental evaluations require time averaging of the flame area and this procedure introduces several uncertainties in the reported data.

$$\frac{S_t}{S_l} = 1 + B_1 \left[\left(\frac{u'}{S_l} \right) - B_2 \left(\frac{u'}{S_l} \right)^2 \right]^{0.5} \left[\frac{V_u}{S_l} \right] \left[\frac{l_{int}}{\delta_l} \right]^{0.5} \left[\frac{W}{S_l} \right]^{0.5}, \qquad (8.46)$$

with $B_1 = 0.002$ and $B_2 = 0.16$ and estimated laminar flame thickness $\delta_l = 0.35$ mm. This correlation clearly shows that S_t depends on a large number of flow- and geometry-dependent parameters; hence, generalized, easy-to-use correlations cannot be obtained. Also, correlations for spherical flames (of interest in IC engines) are likely to be different from those for plane flames. For weakly wrinkled flames of methane and ethylene, experimentally determined values of S_t / S_l have been reported in Ref. [97]. The values range from 2.2 to 6.0.

In the flamelets-in-eddies regime ($Da \rightarrow 1$), the burning rate R_{fu} is determined no longer by chemical kinetics, but by turbulence. In this regime, laminar unburned flamelets are engulfed in a turbulent envelope. The flamelets break down into smaller ones by the action of turbulence. As the flamelets break down, their surface area increases, permitting chemical reaction. This description led Spalding [111] to propose the Eddy-Break-Up (EBU) model for R_{fu}. (This model will be introduced in footnote 12, Chap. 9.) Using this model, equations presented in Sect. 8.2.3 can be solved with an appropriate evaluation of Γ_{eff} by solving one-dimensional forms of equations for the turbulent kinetic energy k and its dissipation rate ϵ. The solutions will yield S_t. Using this method, Karpov et al. [60] have predicted S_t. Typical predicted values varied between 0.8 and 2.4 m/s (or, $2 < S_t / S_l < 6$).

8.4 Flame Stabilization

In Chap. 7, we observed that the value of the loading parameter L must be maintained below the critical value L_{cr} for stable combustion. With Bunsen and other industrial burners, if $V_u \gg S_l$, however, flame liftoff occurs, resulting in blowoff. In gas turbine combustion chambers, reliable combustion must be ensured under wide variations of load and air-fuel ratios without endangering blowoff. The blowoff limits can be substantially enhanced by *flame stabilization*. Such stabilization can be achieved mechanically and/or fluid-dynamically. Figure 8.8 shows two typical examples.

In mechanical stabilization, typically a bluff body (a V-gutter, a sphere, or an ellipsoidal strut) is used. Figure 8.8(a) shows a V-gutter of width B inside a duct of dimension D. Fluid dynamic stabilization (Fig. 8.8(b)) involves injecting air to oppose the oncoming fuel-air mixture. Essentially, both the mechanical and fluid dynamic stabilization involves fluid recirculation, so hot burned products mix with unburned gas and sustain combustion. In the recirculation zone, the temperature is essentially uniform and equals the burned gas temperature, $T_b \simeq T_{ad}$. When such stabilization is used, the blowoff velocity is much greater than that without stabilization.

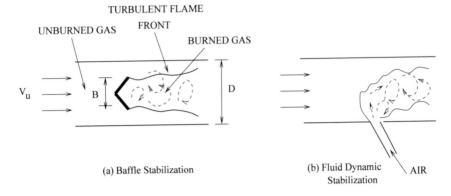

Fig. 8.8 Examples of flame stabilization

To predict blowoff velocity V_{bo} under bluff-body stabilization, we carry out a dimensional analysis. Thus, using phenomenology

$$V_{bo} = F\ (B, D, T_b - T_u, cp_m, \rho_u, \mu_u, \Delta H_c\ R_{fu,max})\ . \tag{8.47}$$

Here $R_{fu,max}$ is included because of its association with the blowoff phenomenon (see Chap. 7) and $R_{fu,max} \simeq R_{fu,cr}$. If we take five dimensions—mass (M), length (L), time (t), heat (H), and temperature (T), then we expect three dimensionless groups because there are eight variables in Eq. 8.47. Thus, we have

$$\left[\frac{V_{bo}\ \rho_u\ cp_m\ (T_b - T_u)}{B\ \Delta H_c\ R_{fu,max}} \right] = F\ \left\{ \left[\frac{B}{D} \right]\ ,\ \left[\frac{\rho_u\ V_{bo}\ B}{\mu_u} \right] \right\}\ \text{and}$$

$$L_B \qquad = F\ \{\ AR\ ,\qquad Re\}\ . \tag{8.48}$$

The parameter L_B is clearly the *loading parameter* for a stirred-reactor comprising the recirculation zone. AR is the aspect ratio and Re is the familiar Reynolds number. Usually, Re is large, and turbulent conditions prevail. As such, the effect of laminar viscosity μ_u is negligible and must be removed from Eq. 8.47. Then, Eq. 8.48 will reduce to $L_B = F\ (AR)$ only.

Case of $AR \to 0$

Baffles are typically plane or conical. We consider the latter in Fig. 8.9. In this figure, the burned gas region of length l is *idealized* to be cylindrical.[8] This region exchanges mass, momentum, and heat with the large unburned gas region. Then, if C_D is the drag coefficient of the baffle, the force at the imaginary cylinder surface will be given by

---

[8]Length l is purely fictitious and is introduced to make progress with the analysis in Ref. [113]. Its estimate will be provided shortly.

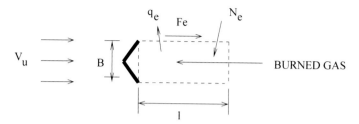

Fig. 8.9 Bluff body flame stabilization, $B \ll D$ $(AR \to 0)$

$$F_e = C_D \times \frac{1}{2} \rho_u V_u^2 A_c = \tau_e A_e \ , \tag{8.49}$$

where the cross-sectional area $A_c = \pi B^2 / 4$, surface area $A_e = \pi B l$, and τ_e is the shear stress. Upon rearrangement, the momentum flux across the cylindrical interface may be written as

$$\frac{\tau_e}{V_u} = \frac{C_D}{8} \left[\frac{B}{l} \right] \rho_u V_u \quad \left(\frac{\text{kg}}{\text{m}^2\text{-s}} \right). \tag{8.50}$$

We now invoke the *Reynolds analogy* between heat and momentum transfer. The analogy states[9] that

$$St \simeq \frac{f}{2} \quad \text{or} \quad \frac{\alpha_e}{\rho \, cp_m \, V_u} = \frac{\tau_e}{\rho \, V_u^2}$$

where St is the Stanton number, f is friction factor, and α_e is the notional convection heat transfer coefficient at the interface. Thus, upon rearrangement again,

$$\frac{\alpha_e}{cp_m} = \frac{\tau_e}{V_u} = \frac{C_D}{8} \left[\frac{B}{l} \right] \rho_u V_u \quad \frac{\text{kg}}{\text{m}^2\text{-s}}. \tag{8.51}$$

We now assume that the burned gas region is characterized by *uniform* temperature T_r and fuel mass fraction $\omega_{fu,r}$. Then, the outward heat flux q_e (W/m^2) and the inward fuel mass flux N_e (kg/m^2−s) are given by

$$q_e = \alpha_e (T_r - T_u) \quad \text{and} \quad N_e = g_e (\omega_{fu,u} - \omega_{fu,r}) , \tag{8.52}$$

where g_e is the notional mass transfer coefficient introduced in Chap. 6. If we now, as usual, make the assumption that Lewis number $= 1$ (that is, Pr $=$ Sc, $g_e = \alpha_e / cp_m$, and ω_{fu} and T are linearly related), then we need to consider only the heat transfer.

---

[9] Strictly, $St \, Pr^m = f/2$. However, for gases, $Pr^m \simeq 1$.

Hence,

$$Q_e = q_e \, A_e = \alpha_e \, A_e \, (T_r - T_u) = R_{fu} \, \Delta H_c \, V_{recirc} \,, \qquad (8.53)$$

where volume of the recirculation region $V_{recirc} = (\pi/4) \, B^2 \, l$.

Combining Eqs. 8.51 and 8.53 and eliminating α_e, it can be shown that

$$\left[\frac{C_D}{2} \right] \left[\frac{B}{l} \right] \left[\frac{\rho_u \, cp_m \, V_u \, (T_r - T_u)}{B \, \Delta H_c} \right] = R_{fu} \,. \qquad (8.54)$$

If we now divide both sides of this equation by $R_{fu,max}$ and define $T^* = (T_r - T_u) \, / \, (T_b - T_u)$, then it can be shown that

$$L_{eff} = \left[\frac{C_D}{2} \right] \left[\frac{B}{l} \right] \left[\frac{\rho_u \, cp_m \, V_u \, (T_b - T_u)}{B \, \Delta H_c \, R_{fu,max}} \right] = \frac{R^*}{T^*} \,. \qquad (8.55)$$

From Eq. 7.79, clearly the lefthand side of this equation can be interpreted as the *effective* loading parameter (L_{eff}, say) with the quantity in the third bracket representing the loading parameter L (see Eq. 7.81). Now, from Eq. 7.83, at extinction or blowout, $L_{eff,cr} \simeq 1.3$ for $\Phi = 1$. Therefore, with $V_u = V_{bo}$,

$$L_{B,cr} = \left[\frac{\rho_u \, cp_m \, V_{bo} \, (T_b - T_u)}{B \, \Delta H_c \, R_{fu,max}} \right] \simeq 1.3 \left[\frac{2}{C_D} \right] \left[\frac{l}{B} \right] \,. \qquad (8.56)$$

From fluid dynamic studies [51, 126], it is known that the drag coefficient C_D and the recirculation ratio (l/B) depend on the baffle shape and the Reynolds number. But, at high Reynolds numbers, typical values are $0.6 < C_D < 2$ and $3 < (l/B) < 6$. As such, $L_{B,cr} \gg 1.3$. Obviously, the presence of a baffle enhances the blowout limit, and one can obtain stable combustion at unburned velocities much in excess of those limited by the conditions expressed in Eq. 7.83. A good baffle is one that has a low C_D and a high recirculation ratio (l/B).

Further, it is also possible to relate V_{bo} to the laminar flame speed S_l. Thus, using Eq. 8.19, it can be shown that the extinction velocity is given by

$$V_{bo} = 2 \left(\frac{L_{B,cr}}{C_D} \right) \left(\frac{l}{B} \right) \left(\frac{\rho_u \, B}{\Gamma_m} \right) \left(\frac{R_{fu,max}}{\overline{R}_{fu}} \right) S_l^2 \,. \qquad (8.57)$$

This equation shows that because $\rho_u \propto p$, $V_{bo} \propto B \, p$. Experimentally, it is found that $V_{bo} \propto B^{0.85} \, p^{0.95}$ [113], indicating that large baffles are required at low pressures. This result is important because performance of high-altitude systems can be predicted by experimental tests on a small ground-level system. We have also observed that $L_{B,cr}$ and $R_{fu,max} \, / \, \overline{R}_{fu}$ are functions of Φ. Thus, V_{bo} also depends on Φ.

Case of $AR \to 1$

For this case, the recirculation zone will extend up to the duct wall. We may regard it as a well-stirred reactor for which L_B is given by Eq. 7.81. Defining $\dot{m} = \rho_u V_u A_c$ and $V_{cv} = A_c \times l$, we have

$$L = \left[\frac{\dot{m}\, cp_m\, (T_{ad} - T_1)}{\Delta H_c\, R_{fu,max}\, V_{cv}} \right] = \left[\frac{\rho_u\, V_u\, A_c\, cp_m\, (T_b - T_u)}{\Delta H_c\, R_{fu,max}\, A_c\, l} \right] . \tag{8.58}$$

Experimental data for the case $B \to D$ suggest that $l \simeq 3 \times D = 3\, B$. Therefore, $L_{B,cr}$ can be given by

$$L_{B,cr} = \left[\frac{\rho_u\, V_{bo}\, cp_m\, (T_b - T_u)}{\Delta H_c\, R_{fu,max}\, B} \right] \simeq 3 . \tag{8.59}$$

Again, this value of $L_{B,cr}$ is in excess of 1.3.

8.5 Externally Aided Ignition

It is very easy to ignite a premixed mixture of fuel and air by means of an electric spark, as done in gasolene (or spark ignition) engines. The basic idea is that when spark energy is fed into a small volume of the mixture, its temperature will attain a high enough value to sustain combustion in the rest of the mixture even after the spark is removed. This occurs as a result of *flame propagation*. The propagating flame releases energy in excess of the heat losses to the surroundings and thus sustains combustion of the entire unburned mixture through progressive engulfment.

Besides an electrical spark, a mixture can be ignited in a variety of ways, including a glow-plug, a pilot flame, or a laser beam. Even an iron-shove striking a stone in a mine is known to cause underground fires. Accidental fires due to doffing of a nylon shirt or by a poor electrical contact are also known to occur in paint shops or oil refineries. These examples show that mixture ignition requires extremely small amounts of energy (in milliJoules). It is, therefore, of interest to determine the *minimum* external energy required for ignition.

Ignition is a transient phenomenon in which, by convention, the minimum ignition energy E_{ign} is defined as the energy required to raise the temperature of a *critical* mass (m_{cr}) of the mixture from its unburned state (having temperature T_u) to its burned state (having temperature T_b). Minimum energy will be required under adiabatic conditions. Hence, $T_b = T_{ad}$. Thus,

$$E_{ign} \equiv m_{cr}\, cp_m\, (T_b - T_u) = m_{cr}\, cp_m\, (T_{ad} - T_u) . \tag{8.60}$$

In a large volume of the mixture, if the energy was fed through a *point source* (an electric spark), the flame propagation will be spherical. If a *plane source* (a pilot flame) is used, the propagation will be linear, forming a plane front. Figure 8.10 shows the two types of propagation.

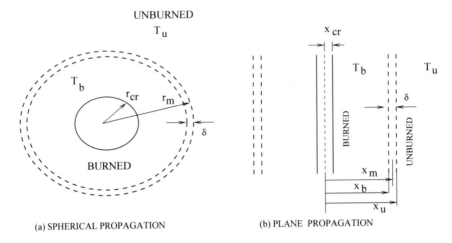

(a) SPHERICAL PROPAGATION   (b) PLANE PROPAGATION

Fig. 8.10 Spherical and plane ignition

8.5.1 Spherical Propagation

In the spherical propagation, the critical mass is defined as

$$m_{cr} \equiv \rho_b \left(\frac{4}{3} \pi r_{cr}^3 \right) , \qquad (8.61)$$

where, ρ_b is the density of the burned mixture and r_{cr} is the critical radius of the sphere. To determine E_{ign}, therefore, the task is to estimate r_{cr} because $T_b = T_{ad}$ (and, hence ρ_b) is known a priori for the fuel.

We assume that the unburned mixture is stagnant and that the propagation is only radial. Thus, the ignition phenomenon will be governed by the following one dimensional equations of mass transfer and energy:

$$A \frac{\partial}{\partial t} (\rho_m \, \omega_{fu}) = \frac{\partial}{\partial r} \left[A \rho_m \, D \, \frac{\partial \omega_{fu}}{\partial r} \right] + A \, R_{fu} , \qquad (8.62)$$

$$A \frac{\partial}{\partial t} (\rho_m \, \omega_{O_2}) = \frac{\partial}{\partial r} \left[A \rho_m \, D \, \frac{\partial \omega_{O_2}}{\partial r} \right] + A \, R_{O_2} , \qquad (8.63)$$

$$A \frac{\partial}{\partial t} (\rho_m \, \omega_{pr}) = \frac{\partial}{\partial r} \left[A \rho_m \, D \, \frac{\partial \omega_{pr}}{\partial r} \right] + A \, R_{pr} , \qquad (8.64)$$

$$A \frac{\partial}{\partial t} (\rho_m \, h_m) = \frac{\partial}{\partial r} \left[A \, k_m \, \frac{\partial T}{\partial r} \right] + A \, R_{fu} \, \Delta H_c , \qquad (8.65)$$

where $A = 4 \pi r^2$ and $h_m = cp_m \, (T - T_u)$. Further, if an SCR is assumed with Lewis number $= 1$, then the above equations will reduce to the following conserved property equation:

$$\frac{\partial}{\partial t} \left(\rho_m \, \Phi \right) = \frac{1}{r^2} \frac{\partial}{\partial r} \left[\Gamma_m \, r^2 \, \frac{\partial \Phi}{\partial r} \right] ,$$ (8.66)

where

$$\Phi = \omega_{fu} - \frac{\omega_{O_2}}{r_{st}}$$ (8.67)

$$= \omega_{fu} + \frac{\omega_{pr}}{1 + r_{st}}$$ (8.68)

$$= cp_m \left(T - T_u \right) + \Delta H_c \, \omega_{fu} .$$ (8.69)

These linear relationships suggest that we may solve any of Eqs. 8.62, 8.63, 8.64, or 8.65. We choose the energy equation as usual. To solve this equation, we need an *initial condition* at t = 0. In our problem, though, the initial location r_{cr} of the burned mixture is unknown. In fact, this is the objective of ignition analysis. Although such an analysis can certainly be carried out, the resulting algebra becomes tedious for a spherical system. Hence, in the next subsection, this analysis is carried out for a *plane* system for which A = 1. The results from the analysis will then be used to explain the results for spherical propagation.

8.5.2 Plane Propagation

For the plane system (see Fig. 8.10(b)), using unit dimensions in y and z directions, the critical mass is given by

$$m_{cr} = \rho_b \left(2 \, x_{cr} \right) ,$$ (8.70)

and the energy Eq. 8.65 will read as

$$\rho_m \frac{\partial T^*}{\partial t} = \frac{\partial}{\partial x} \left[\Gamma_m \frac{\partial T^*}{\partial x} \right] + \frac{R_{fu} \, \Delta H_c}{cp_m \left(T_b - T_u \right)} ,$$ (8.71)

where

$$T^* = \frac{T - T_u}{T_b - T_u} .$$ (8.72)

Following Spalding [113], we multiply Eq. 8.71 by Γ_m. Then,

$$\rho_m \, \Gamma_m \frac{\partial T^*}{\partial t} = \Gamma_m \frac{\partial}{\partial x} \left[\Gamma_m \frac{\partial T^*}{\partial x} \right] + \frac{\Gamma_m \, R_{fu} \, \Delta H_c}{cp_m \left(T_b - T_u \right)}$$ (8.73)

Further, we define a dimensionless burn rate as

$$\Psi \equiv \frac{\Gamma_m \, R_{fu}}{\int_0^1 \Gamma_m \, R_{fu} \, dT^*} \; . \tag{8.74}$$

Using this definition, Eq. 8.73 will read as

$$\rho_m \, \Gamma_m \, \frac{\partial T^*}{\partial t} = \Gamma_m \, \frac{\partial}{\partial x} \left[\Gamma_m \, \frac{\partial T^*}{\partial x} \right] + \frac{\Delta H_c \, \Psi}{cp_m \, (T_b - T_u)} \int_0^1 \Gamma_m \, R_{fu} \, dT^* . \tag{8.75}$$

To develop analytical solution, Spalding [113] made a simplification based on a remarkable observation of the properties of combustion mixtures in the range $(1000 < T(K) < 2500)$. He showed that, in this range, the value of $\rho_m \, \Gamma_m$ remains nearly constant.[10] Thus,

$$\rho_m \, \Gamma_m \simeq \kappa \, \rho_u \, \Gamma_u \; , \tag{8.76}$$

where assuming $T_u = 400\,\text{K}$, for example, $\kappa \simeq 0.64$ for pure air and, for typical combustion products $\kappa \simeq 0.80$. Thus, Eq. 8.75 can be written as

$$\kappa \, \rho_u \, \Gamma_u \, \frac{\partial T^*}{\partial t} = \Gamma_m \, \frac{\partial}{\partial x} \left[\Gamma_m \, \frac{\partial T^*}{\partial x} \right] + \frac{\Delta H_c \, \Psi}{cp_m \, (T_b - T_u)} \int_0^1 \Gamma_m \, R_{fu} \, dT^* , \tag{8.77}$$

or, upon rearrangement,

$$(\kappa \, \rho_u \, \Gamma_u \, Z^2) \, \frac{\partial T^*}{\partial t} = (\Gamma_m \, Z) \, \frac{\partial}{\partial x} \left[\Gamma_m \, Z \, \frac{\partial T^*}{\partial x} \right] + \Psi , \tag{8.78}$$

where the dimensional quantity Z is defined as

$$Z \equiv \left\{ \frac{cp_m \, (T_b - T_u)}{\Delta H_c \int_0^1 \Gamma_m \, R_{fu} \, dT^*} \right\}^{0.5} \quad (\frac{\text{m}^2\text{-s}}{\text{kg}}). \tag{8.79}$$

It is obvious that for a given fuel, $\Gamma_m Z$ (m) is a dimensional constant. As such, let

$$X^* = \frac{x}{\Gamma_m \, Z} \qquad \tau = \frac{t}{\kappa \, \rho_u \, \Gamma_u \, Z^2} \; .$$

Then, Eq. 8.78 can be written as

$$\frac{\partial T^*}{\partial \tau} = \frac{\partial^2 T^*}{\partial X^{*2}} + \Psi \; . \tag{8.80}$$

---

[10]The reader can verify this from properties of air.

We now seek the solution of this equation. With reference to Fig. 8.10(b), it is obvious that starting with initial thickness $\pm x_{cr}$, the flame front will move to the right, as well as to the left of $x = 0$, and the flame front thickness $\delta (t) = x_u (t) - x_b (t)$ will be a function of time. To facilitate analysis, we assume that temperature variation over the small length δ is linear at every instant, so that T^* for positive x is given by

$$T^* = 1 \quad x < x_b$$
$$= \frac{x_u - x}{x_u - x_b} = \frac{X_u^* - X^*}{\delta^*} \quad x_b < x < x_u$$
$$= 0 \quad x > x_u .$$

$$(8.81)$$

Thus, integrating Eq. 8.80 from $x = 0$ to $x = \infty$ and noting that $\partial T^*/\partial X^* |_{X^*=0}$ and $\infty = 0$, we have

$$\frac{\partial}{\partial \tau} \left[\int_0^\infty T^* dX^* \right] = \frac{\partial T^*}{\partial X^*} \Big|_\infty - \frac{\partial T^*}{\partial X^*} \Big|_0 + \int_0^\infty \Psi \, dX^*$$
$$= \int_0^\infty \Psi \, dX^* .$$

$$(8.82)$$

The integrand on the lefthand side of this equation can be evaluated by parts as

$$\int_0^\infty T^* dX^* = \int_0^{X_b^*} 1 \, dX^* - \int_{X_b^*}^{X_u^*} \frac{X^* - X_u^*}{\delta^*} dX^* + \int_{X_u^*}^\infty 0 \, dX^*$$
$$= X_b^* - \left\{ \frac{1}{\delta^*} \left(\frac{1}{2} X^{*2} - X^* X_u^* \right) \right\}_{X_b^*}^{X_u^*}$$
$$= X_b^* + \frac{\delta^*}{2}$$

$$(8.83)$$

$$= X_m^* ,$$

$$(8.84)$$

where x_m is the distance of the midplane of the flame front as shown in Fig. 8.10(b). Similarly, using Eq. 8.74 and noting that $dX^* = -\delta^* dT^*$, the righthand side of Eq. 8.82 will evaluate to

$$\int_0^\infty \Psi \, dX^* = \frac{- \int_1^0 \Gamma_m R_{fu} \delta^* \, dT^*}{\int_0^1 \Gamma_m R_{fu} \, dT^*}$$
$$= \delta^* \left[\frac{\int_0^1 \Gamma_m R_{fu} \, dT^*}{\int_0^1 \Gamma_m R_{fu} \, dT^*} \right]$$
$$= \delta^* .$$

$$(8.85)$$

Substituting Eqs. 8.84 and 8.85 in Eq. 8.82, we have

$$\frac{d\,X_m^*}{d\,\tau} = \delta^* , \qquad (8.86)$$

where at $\tau = 0$, $X_m^* = X_u^* = X_b^* = X_{cr}^*$. This is because, initially, the flame thickness δ is zero. Equation 8.86 has two unknowns, x_m and δ, whose variations with time are not known. Therefore, we need a second equation connecting the two quantities. This second equation is derived by integrating Eq. 8.80 from $x = x_m$ to $x = \infty$. Thus,

$$\int_{X_m^*}^{\infty} \frac{\partial T^*}{\partial \tau}\, dX^* = \frac{\partial T^*}{\partial X^*}\Big|_\infty - \frac{\partial T^*}{\partial X^*}\Big|_{x_m^*} + \int_{X_m^*}^{\infty} \Psi\, dX^* . \qquad (8.87)$$

As in the case of determination of S_l, however, we may assume that for $x_m < x < x_u$, $R_{fu} \simeq 0$ because temperatures beyond $x = x_m$ are very low. As such, the last term in Eq. 8.87 is set to zero. Also, using Eq. 8.81,

$$\frac{\partial T^*}{\partial X^*}\Big|_\infty = 0 \quad \text{and} \quad -\frac{@T^*}{@X^*}\Big|_{X_m^*} = \frac{1}{*} . \qquad (8.88)$$

Therefore, Eq. 8.87 reduces to

$$\int_{X_m^*}^{\infty} \frac{\partial T^*}{\partial \tau}\, dX^* = \frac{1}{\delta^*} . \qquad (8.89)$$

The lefthand side of this equation is now evaluated using the Leibniz rule because x_m is a function of time. Thus,

$$\begin{aligned}
\int_{X_m^*}^{\infty} \frac{\partial T^*}{\partial \tau}\, dX^* &= \frac{\partial}{\partial \tau}\left[\int_{X_m^*}^{\infty} T^*\, dX^*\right] + T_{X_m^*}^* \frac{d\,X_m^*}{d\,\tau} \\
&= -\frac{\partial}{\partial \tau}\left[\int_{1/2}^{0} \delta^*\, T^*\, dT^*\right] + T_{X_m^*}^* \frac{d\,X_m^*}{d\,\tau} \\
&= -\frac{\partial}{\partial \tau}\left[\delta^* \left\{\frac{T^{*2}}{2}\right\}_{1/2}^{0}\right] + T_{X_m^*}^* \frac{d\,X_m^*}{d\,\tau} \\
&= \frac{\partial}{\partial \tau}\left[\frac{\delta^*}{8}\right] + \frac{1}{2}\frac{d\,X_m^*}{d\,\tau} ,
\end{aligned} \qquad (8.90)$$

where $T_{X_m^*}^* = 1/2$. Thus, Eq. 8.89 can be written as

$$\frac{1}{8}\frac{d\,\delta^*}{d\,\tau} + \frac{1}{2}\frac{d\,X_m^*}{d\,\tau} = \frac{1}{\delta^*} . \qquad (8.91)$$

Substituting Eq. 8.86 in this equation, we have

$$\frac{1}{8} \frac{d \delta^*}{d \tau} + \frac{\delta^*}{2} = \frac{1}{\delta^*} \ . \tag{8.92}$$

This first order equation with initial condition $\delta^* = 0$ at $\tau = 0$ has the following solution:

$$\delta^* = \sqrt{2} \left[1 - \exp(-8\,\tau) \right]^{0.5} \ . \tag{8.93}$$

This solution shows that initially δ^* will increase with time, but at large times it will reach a steady state value, given by

$$\delta_{ss}^* = \sqrt{2} \ . \tag{8.94}$$

Similarly, substituting solution Eq. 8.93 in Eq. 8.86, we have

$$\frac{d\,X_m^*}{d\,\tau} = \sqrt{2} \left[1 - \exp(-8\,\tau) \right]^{0.5} \ . \tag{8.95}$$

The solution to this equation will read as

$$X_m^* = X_{m,cr}^* + \int_0^\tau \left[\sqrt{2} \, \{ 1 - \exp(-8\,\tau) \} \right]^{0.5} d\tau \ , \tag{8.96}$$

where $X_{m,cr}^*$ is the constant of integration and is *unknown*. Evaluation of X_m^* requires numerical integration. From this evaluation, $X_b^* = X_m^* - \delta^*/2$ and $X_u^* = X_m^* + \delta^*/2$ can be evaluated at every instant, as variation of δ^* with time is known from Eq. 8.93.

Solutions for X_m^* can be generated in this manner by assuming different values of $X_{m,cr}^*$. However, it is obvious that for the correct $X_{m,cr}^*$, X_b^* must not be negative at any instant, because $\delta^* = 0$ at $t = 0$. Figure 8.11 shows the typical behavior of the solutions for δ^*, X_m^*, X_u^*, and X_b^* for $X_{m,cr}^* \simeq 0.4605$. For this value, X_b^* is zero at $\tau \simeq 0.16$. Thus, using the definition of Z, the physical magnitude of $x_{m,cr}$ is given by

$$x_{m,cr} = X_{m,cr}^* \, \Gamma_m \, Z \simeq 0.4605 \, \Gamma_m \left\{ \frac{cp_m \, (T_b - T_u)}{\Delta H_c \int_0^1 \Gamma_m \, R_{fu} \, dT^*} \right\}^{0.5} \ . \tag{8.97}$$

However, because $\delta^* = 0$ at $t = 0$, this $x_{m,cr}$ also equals the required x_{cr}. Further, making use of Eqs. 8.15 and 8.19, $x_{m,cr}$ can be related to S_l in the following manner:

$$x_{cr} = x_{m,cr} \simeq 0.4605 \sqrt{2} \, \frac{\Gamma_m}{\rho_u \, S_l} = 0.651 \, \frac{\Gamma_m}{\rho_u \, S_l} \ . \tag{8.98}$$

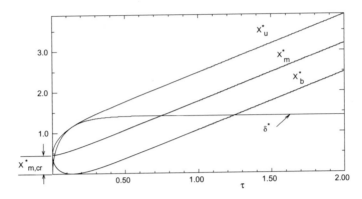

Fig. 8.11 Variation of X_m^*, X_m^*, X_m^*, δ^* with time—plane ignition

Table 8.5 Values of minimum ignition energies at p = 1 atm

Fuel	T_u (K)	T_b (K)	E_{ign} (milli-Joules)
C_7H_{16}	444	2274	3.2
	298	2274	14.5
C_3H_6	477	2334	1.4
	298	2334	5.5
C_5H_{12}	444	2272	2.3
	298	2272	7.8

Thus, from Eq. 8.70, the *minimum size* of the slab for ignition must be $2 \times x_{cr}$. Hence, the minimum ignition energy per unit area is estimated as

$$E_{ign} = \rho_b \times (2 \times x_{cr})\, cp_m\, (T_b - T_u)$$

$$\simeq 1.3 \times \left(\frac{\rho_b\, \Gamma_b}{\rho_u\, S_l} \right) cp_m\, (T_b - T_u)\,, \qquad (8.99)$$

where $\Gamma_b \simeq \Gamma_m$.

Result for Spherical Propagation

The results for spherical propagation have the same form as those for plane propagation. Only the constants differ. Thus, under steady state, $\delta_{ss}^* = \sqrt{2}$ (as before) but the dimensionless critical radius is estimated to be $r_{cr}^* = r_{m,cr}^* \simeq \sqrt{3}$. As such, the critical radius of the mixture required for ignition is

$$r_{cr} = r_{m,cr} \simeq \sqrt{3} \times \sqrt{2}\, \frac{\Gamma_b}{\rho_u\, S_l} = 2.45\, \frac{\Gamma_b}{\rho_u\, S_l}\,, \qquad (8.100)$$

and the minimum ignition energy is

$$E_{ign} = \rho_b \times \left(\frac{4\,\pi}{3} \times r_{cr}^3 \right) cp_m \, (T_b - T_u)$$

$$\simeq 61.56 \times \left(\frac{\rho_b \, \Gamma_b^3}{\rho_u^3 \, S_l^3} \right) cp_m \, (T_b - T_u) \,. \qquad (8.101)$$

This equation shows that for a given fuel, $E_{ign} \propto \rho^{-2} \propto p^{-2}$. Thus, at high altitudes, the spark energy required will be greater than that required at the ground level. Similarly, at a given pressure, fuels with a smaller S_l will be harder to ignite, because $E_{ign} \propto S_l^{-3}$. Table 8.5 shows estimates of E_{ign} for a few hydrocarbons at p = 1 atm.

8.6 Self- or Auto-ignition

8.6.1 Ignition Delay and Fuel Rating

In contrast to externally aided ignition, a reacting mixture can also ignite and burst into a flame using it's *internal energy*, as is found in a Diesel engine. Such ignition is called *self- or spontaneous- or auto-ignition*.

In a Diesel engine, liquid fuel is injected into air near the end of the compression stroke. Depending on the compression ratio, the pressure and temperature attain a high value. The liquid fuel, in the form of very fine droplets, heats up and vaporizes and, after some *time delay*, almost instantly ignites without any external means. Thereafter, the entire fuel-air mixture is engulfed by flame propagation, whose rate is controlled by turbulence and mixing. The time delay[11] to ignition is called the *ignition delay* and is of consequence in determining engine performance. For example, when the ignition delay is long, there is an accumulation of unburned fuel arriving from the injector; when ignition finally occurs, it results in a *bang* or *knock*. To avoid knock, therefore, the fuel-air mixture must be highly reactive, so the delay is short.

Ignition delay of Diesel fuels[12] is characterized by the Cetane number (CN). Cetane, or n-Hexadecane, is assigned CN = 100 and isocetane is arbitrarily assigned CN = 15. The cetane number is calculated as CN = (% cetane) + 0.15 (% isocetane). For Diesel fuel, $35 < CN < 56$. Both n-hexadecane and iso-cetane have the same chemical formula, $C_{16}H_{34}$. The higher the CN value, smaller is the ignition delay. A high CN effectively reduces the value of the activation energy, E_a. Hence, $E_{a,eff}$ is inversely proportional to CN.

Knock (or detonation) can also occur in spark-ignition engines when a stagnant gas mixture in the corner regions of the engine cylinder is compressed, owing to

---

[11]The total time delay is a combination of physical time delay required for atomization, heating up and vaporization, and the chemical time delay, which is dominated by chemical kinetics.

[12]Chemical formula: $C_{10}H_{20}$ to $C_{15}H_{28}$; average formula: $C_{12}H_{23}$.

a propagating flame, as well as by piston movement. To avoid knock, therefore, the fuel-air mixture *must be resistant* to self-ignition. This requirement, which is the opposite of that in Diesel engines, is met by iso-octane (C_8H_{16}) which has a long delay period. n-heptane (C_7H_{16}), on the other hand, has a short delay period. Thus, gasolene (n- octane, with average formula C_8H_{18}) with octane number (ON) 91 possesses the same anti-knock rating of a mixture of 91% (by volume) isoctane (ON = 100) and 9% (by volume) n-heptane (ON = 0). The percentages here do not imply the actual proportions of the two species in the mixture. They simply refer to knock- or detonation-resistance properties.

8.6.2 Estimation of Ignition Delay

It is of interest to estimate the ignition delay period for different hydrocarbon fuels. The delay period can be experimentally measured in an adiabatic constant-mass (constant-volume or constant-pressure) reactor in which a fuel-air mixture at raised pressure (p_i) and temperature (T_i) is left alone to ignite on its own.[13] After a slow rate of rise in temperature, there will be an exponential (almost sudden) increase in the rate of rise in temperature, with sudden fall in mass fractions of both fuel and oxygen. The mixture will extinguish when it runs out of either fuel or oxygen.

An approximate theory of ignition delay can be developed by assuming an SCR for a $C_m H_n$ fuel in a CMTCR (see Chap. 7).[14] Then, the only equation that needs to be solved is

$$\frac{dT}{dt} = \frac{\Delta H_c}{\rho_m\, C_m}\, |R_{fu}|, \tag{8.102}$$

where C_m equals cv_m or cp_m depending on whether a constant-volume or a constant-pressure CMTCR is considered, and R_{fu} is given by (see Eq. 5.80):

$$R_{fu} = A^* \exp\left(-\frac{E_a}{R_u\, T}\right) \rho_{mix}^{(x+y)}\, w_{fu}^x\, w_{O_2}^y \tag{8.103}$$

$$w_{fu} = w_{fu,i} - \frac{C_m}{\Delta H_c}\, (T - T_i) \tag{8.104}$$

$$w_{O_2} = w_{O_2.i} - r_{st}\, \frac{C_m}{\Delta H_c}\, (T - T_i) \tag{8.105}$$

$$\rho_{mix} = \frac{p\, M_{mix}}{R_u\, T} \qquad M_{mix} = \sum \left(\frac{w_j}{M_j}\right)^{-1} \tag{8.106}$$

---

[13]In practice, ignition delay studies are carried out in a *continuous flow device* for $300 < T_i < 1300\,K$ [11, 84] and in a *shock tube* for $1200 < T_i < 2500\,K$ [109, 121].

[14]More elaborate models of ignition delay consider multistep mechanisms with several species (see, for example, [42]).

$$p = p_i \quad \text{for Constant-pressure CMCTR}$$

$$= p_i \left(\frac{R_{mix} T}{R_{mix,i} T_i} \right) \quad \text{for constant-volume CMCTR.} \qquad (8.107)$$

In these expressions, values of A^*, E_a/R_u, x, and y are read from Table 5.8 and $r_{st} = 32 \, a_{stoich}/(12m + n)$. In Eq. 8.106, the subscript j refers to fu, O_2, N_2, CO_2, and H_2O, where N_2 is of course inert and mass fractions of H_2O and CO_2 are determined from stoichiometry. The righthand side of Eq. 8.102 is thus a function of T only. After evaluating T_{ad} and ΔH_c for given T_i and Φ, the equation is to be solved with initial conditions

$$T = T_i$$

$$\omega_{fu,i} = \frac{M_{fu}}{M_{fu} + 137.28 \, a_{stoich}/\Phi} \quad \text{and}$$

$$\omega_{O_2,i} = \frac{32 \, a_{stoich}/\Phi}{M_{fu} + 137.28 \, a_{stoich}/\Phi} \ .$$

Table 8.6 shows a typical transient for C_7H_{16} at $T_i = 700$ C, p = 1 atm and $\Phi = 0.5$ in a constant-pressure CMCTR. Computations were begun with time step $\Delta t = 0.01$ ms. The step size was progressively reduced to 0.001 ms. Solutions are tabulated for a few intermediate time steps. The second column of the Table

Table 8.6 C_7H_{16} constant-pressure combustion: $T_i = 700$ C, p = 1 atm, $\Phi = 0.5$, $T_{ad} = 2081.17$ K, $T_{ext} = 2143.8$ K, $\Delta H_c = 44.76$ MJ/kg, $cp_m = 1225.25$ J/kg-K

Time (ms)	T^* (−)	T/1000 (C)	ω_{fu} (−)	ω_{O_2} (−)	R_{fu} (kg/m^3-s)
0.010	0.648E−03	0.701E+00	0.320E−01	0.226E+00	0.726E+00
0.020	0.130E−02	0.701E+00	0.320E−01	0.225E+00	0.733E+00
0.050	0.331E−02	0.704E+00	0.319E−01	0.225E+00	0.755E+00
0.100	0.679E−02	0.708E+00	0.318E−01	0.225E+00	0.794E+00
0.300	0.230E−01	0.725E+00	0.314E−01	0.223E+00	0.996E+00
0.500	0.443E−01	0.749E+00	0.307E−01	0.221E+00	0.132E+01
0.750	0.854E−01	0.795E+00	0.295E−01	0.217E+00	0.222E+01
0.815	0.101E+00	0.812E+00	0.290E−01	0.215E+00	0.266E+01
1.000	0.175E+00	0.894E+00	0.267E−01	0.207E+00	0.576E+01
1.100	0.263E+00	0.992E+00	0.241E−01	0.198E+00	0.122E+02
1.150	0.372E+00	0.111E+01	0.208E−01	0.186E+00	0.261E+02
1.160	0.410E+00	0.115E+01	0.196E−01	0.182E+00	0.324E+02
1.170	0.460E+00	0.121E+01	0.181E−01	0.177E+00	0.419E+02
1.180	0.529E+00	0.129E+01	0.160E−01	0.169E+00	0.572E+02
1.190	0.633E+00	0.140E+01	0.129E−01	0.158E+00	0.836E+02
1.200	0.804E+00	0.159E+01	0.766E−02	0.140E+00	0.127E+03
1.205	0.927E+00	0.173E+01	0.392E−02	0.127E+00	0.147E+03
1.208	0.100E+01	0.182E+01	0.146E−02	0.118E+00	0.142E+03

Table 8.7 Effect of T_i, p and Φ on ignition delay for C_7H_{16} constant-pressure Combustion

T_i (C)	p (Atm)	Φ_i	t_{ign} (ms (Num))	t_{ign} (ms (Eq. 8.109))
700	1	0.5	0.815	0.7243
800	1	0.5	0.235	0.206
900	1	0.5	0.080	0.0724
1000	1	0.5	0.0333	0.030
700	5	0.5	0.2415	0.2166
1000	5	0.5	0.0102	0.00898
700	10	0.5	0.144	0.1289
1000	10	0.5	0.00649	0.00534
700	1	0.8	0.875	0.744
1000	1	0.8	0.039	0.0338
700	1	1.0	0.889	0.741
1000	1	1.0	0.0414	0.0351

shows values of $T^* = (T - T_i)/(T_{ad} - T_i)$. It is seen that it takes 1.208 ms to reach $T^* = 1.0$ when the fuel, in this lean mixture combustion, almost runs out while the oxygen mass fraction is reduced to 0.118. In the context of the study of ignition delay, however, the time (0.815 ms) required to reach $T^* = 0.1$ may be considered important. This is because T, ω_{fu}, and ω_{O_2} have changed very little in this period. After t = 0.815 ms and for $T^* > 0.1$ significant changes are observed. As such, time t_{ign} to reach $T^* = T^*_{ign} = 0.1$ may be regarded as the ignition delay [113].

The preceding computations are repeated for different values of T_i, p, and Φ and the results for t_{ign} corresponding to $T^* = 0.1$ are shown in the fourth column of Table 8.7. It is seen that at constant p and Φ, t_{ign} decreases with an increase in T_i, as would be expected. The ignition delay also decreases with an increase in pressure at constant T_i and Φ. However, at constant p and T_i, t_{ign} increases with Φ. To explain these tendencies, Spalding [113] developed a simple theory.

As noted in Table 8.6, for $T^* < 0.1$, as the fuel and oxygen mass fractions and T change very little, the values of dT/dt will be influenced mainly by the exponential term in Eq. 8.103. As such, to avoid numerical integration, we may write

$$
\frac{dT}{dt} \simeq A^{**} \exp\left(-\frac{E_a}{R_u\,T}\right) \text{ with } A^{**} = \left(\frac{\Delta H_c}{\rho_m\,C_m}\right) A^* \, (\rho_{mix}^{(x+y)}\,\omega_{fu}^x\,\omega_{O_2}^y)_i
$$

$$
= A^{**} \exp\left\{-\frac{E_a}{R_u}\left(\frac{1}{T_i} + \frac{1}{T} - \frac{1}{T_i}\right)\right\}
$$

$$
\simeq A^{**} \exp\left(-\frac{E_a}{R_u\,T_i}\right) \times \exp\left\{\frac{E_a\,(T - T_i)}{R_u\,T_i^2}\right\}. \tag{8.108}
$$

It is now a simple matter to integrate this equation. The result is

$$t_{ign} = \frac{1}{A^{**}} \left(\frac{R_u T_i^2}{E} \right) \exp \left(-\frac{E_a}{R_u T_i} \right) \times \left[1 - \exp \left\{ -\frac{E_a (T_{ign} - T_i)}{R_u T_i^2} \right\} \right]$$

$$\simeq \frac{1}{A^{**}} \left(\frac{R_u T_i^2}{E} \right) \exp \left(-\frac{E_a}{R_u T_i} \right). \tag{8.109}$$

Notice that the term involving T_{ign} is dropped and the quantity in the square bracket $\rightarrow 1$. This is because the term can be written as

$$\frac{E_a (T_{ign} - T_i)}{R_u T_i^2} = \left(\frac{E_a}{R_u T_i^2} \right) \times \left(\frac{T_{ign} - T_i}{T_{ad} - T_i} \right) \times (T_{ad} - T_i)$$

$$= \left(\frac{E_a}{R_u T_i^2} \right) \times T_{ign}^* \times (T_{ad} - T_i) > 1.0 \tag{8.110}$$

For most hydrocarbons, this term will exceed unity for $T_{ign}^* \simeq 0.1$ and, thus, the exponential of the term can, therefore, be dropped. The values of t_{ign} estimated from Eq. 8.109 are shown in the last column of Table 8.7. The values are in remarkable agreement with the numerically computed values.

Finally, similar computations are carried out for n-octane and the results are presented in Table 8.8. The table confirms that, under the same conditions of p, T_i, and Φ, the ignition delay for C_8H_{18} is longer than that for n-heptane.

Table 8.8 Effect of T_i, p and Φ on ignition delay for C_8H_{18} constant-pressure combustion

T_i (C)	p (Atm)	Φ_i	t_{ign} (ms (Num))	t_{ign} (ms (Eq. 8.109))
700	1	0.5	7.43	6.69
800	1	0.5	1.326	1.201
900	1	0.5	0.324	0.290
1000	1	0.5	0.100	0.876
700	5	0.5	0.2235	2.00
1000	5	0.5	0.0325	0.0262
700	10	0.5	1.33	1.190
1000	10	0.5	0.0348	0.0261
700	1	0.8	7.54	6.56
1000	1	0.8	0.114	0.094
700	1	1.0	7.50	6.417
1000	1	1.0	0.119	0.0953

8.7 Flammability Limits

In Table 8.4, it was observed that S_l is a function of Φ. S_l is at a maximum for equivalence ratio (Φ) slightly greater than 1, but for very lean and very rich mixtures, S_l is very small. In fact, mixtures with very small S_l are extremely hard to ignite and to propagate a flame. As such, below a minimum and above a maximum value of Φ, flames cannot be sustained. Values of Φ_{min} and Φ_{max} thus define the *flammability limits* of a fuel-air mixture. Experimentally, these values are determined for fuel by initiating a flame in a vertical glass tube of about 5 cm diameter and 1 m long (essentially a plug flow reactor in which heat losses from the tube wall are present) and observing whether the flame propagates along the length or not. Table 8.9 gives flammability limits for few fuels.

It is obvious that for $\Phi < \Phi_{min}$, the fuel will run out much earlier than the oxygen and thus the flame will not propagate. Conversely, for $\Phi > \Phi_{max}$, oxygen will run out earlier than fuel. This is because, under these conditions, the loading parameter L exceeds L_{cr}. We recall that for hydrocarbon fuel, $R_{fu} \propto w_{fu}^x \times w_{O_2}^y$, where under adiabatic conditions without heat loss from the tube,

$$\omega_{fu,cr} = \omega_{fu,u} - \frac{cp_m}{\Delta H_c}(T_{cr} - T_u) \text{ and} \tag{8.111}$$

$$\omega_{O_2,cr} = \omega_{O_2,u} - r_{st}\frac{cp_m}{\Delta H_c}(T_{cr} - T_u), \tag{8.112}$$

and, at the inlet to the tube, $\omega_{fu,u}$ and $\omega_{O_2,u}$ will be functions of Φ. Also, at the critical point, we note (see Eq. 7.84) that

$$\begin{aligned}(T_{cr} - T_u)_\Phi &= \frac{(T_{cr} - T_u)_\Phi}{(T_{ad} - T_u)_\Phi} \times \frac{(T_{ad} - T_u)_\Phi}{(T_{ad} - T_u)_{\Phi=1}} \times (T_{ad} - T_u)_{\Phi=1}, \\ &= T_{cr}^* \times T_{ad}^* \times (T_{ad} - T_u)_{\Phi=1}.\end{aligned} \tag{8.113}$$

Table 8.9 Flammability limits at $p = 1$ atm and $T_u = 298$ K

Fuel	Φ_{min}	Φ_{max}
CH_4	0.46	1.64
C_2H_6	0.50	2.72
C_3H_8	0.51	2.83
C_8H_{18}	0.51	4.25
$C_{10}H_{22}$	0.36	3.92
CO	0.34	2.52
H_2	0.14	2.54

Therefore, upon substitution, Eqs. 8.111 and 8.112 will read as,

$$\omega_{fu,cr} = \frac{M_{fu}}{M_{fu} + (a_{stoich}/\Phi)\,(M_{O_2} + 3.76\,M_{N_2})} \\ - T_{cr}^* \, T_{ad}^* \, \frac{cp_m}{\Delta H_c}\,(T_{ad} - T_u)_{\Phi=1} \qquad (8.114)$$

$$\omega_{O_2,cr} = \frac{(a_{stoich}/\Phi)\,M_{O_2}}{M_{fu} + (a_{stoich}/\Phi)\,(M_{O_2} + 3.76\,M_{N_2})} \\ - T_{cr}^* \, T_{ad}^* \, r_{st}\, \frac{cp_m}{\Delta H_c}\,(T_{ad} - T_u)_{\Phi=1}. \qquad (8.115)$$

Thus, Φ_{min}, for example, can be determined by setting $\omega_{fu,cr} = 0$, so using the simplification 8.22 for ΔH_c, we have

$$\Phi_{min} \simeq R_{st}\left[\frac{1 + R_{st}}{T_{cr}^* \, T_{ad}^*} - 1\right]^{-1}, \qquad (8.116)$$

where R_{st} is the stoichiometric A/F ratio for the fuel. Likewise, to determine Φ_{max}, we set $\omega_{O_2,cr} = 0$ giving the relation

$$\Phi_{max} \simeq \frac{1 + R_{st}}{T_{cr}^* \, T_{ad}^*} - R_{st}. \qquad (8.117)$$

Φ_{min} and Φ_{max} estimated by Eqs. 8.116 and 8.117 will of course be approximate because adiabatic conditions are used in their derivations. For propane (C_3H_8), for example, $R_{st} = 15.6$, and using Eqs. 7.84 and 7.85 for T_{cr}^* and T_{ad}^*, respectively, iterative evaluations give $\Phi_{min} \simeq 0.611$ (with $T_{cr}^* \, T_{ad}^* = 0.626$) and $\Phi_{max} \simeq 5.3$ (with $T_{cr}^* \, T_{ad}^* = 0.793$). Both values are greater than values given in Table 8.9. Nonetheless, the simplified analysis predicts correct tendencies.

8.8 Flame Quenching

To determine flammability limits, a 5 cm diameter tube is used, as mentioned in the previous section. If the tube diameter is gradually reduced, a stage will be reached at which flame propagation is no longer possible, irrespective of the value of Φ. This phenomenon is called *flame quenching*, and the tube diameter at which it occurs is called the *quenching distance*. Flame quenching will also occur when the flame front in a larger-diameter tube is made to pass through an orifice.[15] In this case, the orifice diameter is again called a quenching distance. If the mixture flowed in a duct of the rectangular cross-section of high aspect ratio and if quenching or

---

[15]The miner's safety lamp known as Davy's lamp uses this principle for arresting flame propagation.

Table 8.10 Quenching distance at p = 1 atm, $T_u = 293$ K and $\Phi = 1$ [19]

Fuel	d (mm)
H_2	0.6
CH_4	1.9
C_3H_8	2.1
C_8H_{16}	2.0
CH_3OH	1.8
C_2H_2	2.3

extinction[16] occurred, then the length of the shorter side (or, the slot-width) is called the quenching distance.

If the flame is to continue to propagate in a tube or duct, the energy released from combustion must keep the temperature in the reaction zone high enough to sustain the reaction. When the tube diameter becomes small, the heat transfer to the tube wall becomes large, and hence, the temperature drops, which further reduces the reaction rate. This, in turn, reduces the temperature further. Eventually, the flame quenches.

Experimentally, the quenching distance is determined by what is known as the *flashback* method. In Sect. 8.2, it was mentioned that in a Bunsen burner, flashback will occur when $V_u \ll S_l$. Thus, consider a burner tube fitted with an orifice of diameter d at its exit. For a given mixture with mass flow rate $\dot{m}$ and given equivalence ratio Φ, let a stable flame be formed. Now, switch off the flow and observe whether flashback occurs. Repeat the experiment with different values of $\dot{m}$ and note the value of $\dot{m}_{fb}$ at which flashback does occur. Repeat this experiment for different values of Φ. A plot of Φ versus $\dot{m}_{fb}$ can now be made. Extrapolation of this curve to $\dot{m}_{fb} = 0$ gives the value of Φ for a given d. Repeating the experiment for different values of d gives the required $\Phi \sim d$ relationship for a given fuel, and d is the quenching distance. Usually, values of d are tabulated for $\Phi = 1$. These are shown in Table 8.10. The table shows that for a highly reactive fuel (very high S_l) such as H_2, the quenching distance is very small. For hydrocarbons, however, it is large.

To develop a simplified theory, consider a rectangular slot burner of width d, as shown in Fig. 8.12. The breadth of the slot is much greater than d. A premixed propagating flame of width δ is considered, with average volumetric fuel burn rate $\overline{R}_{fu}$. Then, to estimate the quenching distance, the one-dimensional energy balance is considered for unit length along with the breadth. Thus, the energy released in the volume $\delta \times d \times 1$ is equated to the heat loss by conduction to the walls from this volume. Thus,

$$\overline{R}_{fu} \, \Delta H_c \, (\delta \times d) \simeq \kappa \, (k_m \times \delta) \, \frac{(T_b - T_w)}{d/2} \,, \qquad (8.118)$$

---

[16]Flame extinction can be brought about in several ways. One method is to blow a flame away from its reactants, as is done by hard blowing. Another way is to dilute or cool the flame temperature by adding water. It is also possible to reduce the reaction kinetic rates by adding suppressants, such as halogens. Yet another method, used specifically in oil well fires, is to blow out a flame by explosive charges so the flame is starved of oxygen. The addition of carbon dioxide extinguishes a flame through the same effect. Our interest here is to study flame extinction due to passage-size reduction.

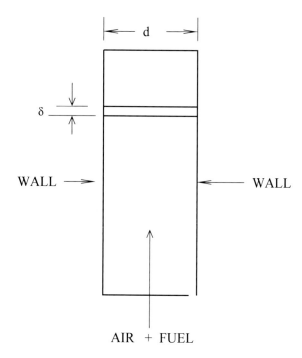

Fig. 8.12 Flame front in a rectangular duct of width d

where T_w is the wall temperature and T_b is the average temperature in the flame zone. In reality, the temperature in this zone will vary in the direction normal to the walls when conduction to the walls is accounted for. To account for this departure from reality, however, the factor κ is introduced. If it is now assumed that $T_w \simeq T_u$,

$$d^2 = 2\,\kappa \left[\frac{k_m\,(T_b - T_u)}{\overline{R}_{fu}\,\Delta H_c} \right]$$

$$= 2\,\kappa \left[\Gamma_m\,\frac{cp_m\,(T_b - T_u)}{\overline{R}_{fu}\,\Delta H_c} \right]$$

$$= 2\,\kappa \left[\Gamma_m\,\frac{2\,\Gamma_m}{\rho_u^2\,S_l^2} \right] \qquad \text{from Equation 8.19}$$

$$= 4\,\kappa \left[\frac{\Gamma_m^2}{\rho_u^2\,S_l^2} \right]$$

$$d = 2\,\sqrt{\kappa} \left[\frac{\Gamma_m}{\rho_u\,S_l} \right], \qquad\qquad (8.119)$$

where the value of κ for a given fuel can be evaluated from Table 8.10. Equation 8.119 shows that the quenching distance depends on all factors such as pressure, T_u, Φ and so forth on which S_l depends, and confirms that d is inversely proportional to

S_l. Andrews and Bradley [5] have provided experimental data for some fuels and showed their dependence on Φ.

Exercises

1. Prove Eq. 8.1 for a conical premixed flame. Also, show that the equation will apply to a flame issuing from a slot of width W and length L.
2. Starting from Eq. 8.13, derive Eqs. 8.19 and 8.20.
3. By carrying out dimensional analysis, prove functional dependence shown in Eq. 8.48.
4. For the case of AR $\rightarrow$ 1, derive Eq. 8.54 for a *Plane Baffle* of width W and height B. List all assumptions made.
5. Verify the assumption made in Eq. 8.76, for a mixture of burnt products with $w_{N_2} = 0.73$, $w_{CO_2} = 0.15$, $w_{H_2O} = 0.12$. Hence, estimate the value of κ for $T_u = 400$ K and 800 K. (Hint: Use $\rho_m = \sum \rho_j$, $cp_m = \sum cp_j\, w_j$, $k_m = \sum k_j\, x_j$).
6. Using Eqs. 8.19 and 8.20, calculate laminar flame speed and flame thickness of CH_4 for $\Phi = 0.8$, 1.0 and 1.2 when $T_u = 300$ K and p $= 1$ atm. Compare your results with Eq. 8.35. Assume $M_{mix} = 29$, $cp_{mix} = 1230$ J/kg-K and $k_{mix} = 0.1$ W/m-K
7. Repeat the preceding problem for isooctane (C_8H_{18}) with $T_u = 298$ K and p $= 0.2$, 1.0, 5.0 atm and compare your results with Eq. 8.33.
8. In an IC engine (fuel C_8H_{18}, $\Phi = 1.05$), calculate laminar flame speed when p $= 13.4$ atm and $T_u = 560$ K. Use Eq. 8.33 for S_l. Assuming spherical flame propagation, estimate minimum ignition energy, critical radius, and steady state flame thickness. Evaluate Γ_b using Eq. 8.28.
9. Compare the minimum ignition energies of gasolene C_8H_{16} at ground level ($T_u = 298$ K), and at a height of 6000 m, using p and T_u conditions given in Appendix E (Table E.1). Use Eq. 8.33 for S_l.
10. Fuel is injected at the axis of a 25 mm diameter duct through an orifice of 2 mm diameter. The mass flow rate through the duct is 0.03 kg/s. The unburned mixture is at p $= 1$ atm, T $= 300$ K, and $M_{mix} = 29.6$. Estimate the length of the conical flame zone until the edge of the flame touches the duct wall. Assume $S_t = 5$ m/s.
11. Consider combustion of propane-air mixture at p $= 2$ atm, $T_u = 350$ K, $\sqrt{(u')^2} = 4$ m/s, and $l_m = 5$ mm. Determine flow and chemical time scales for $\Phi = 0.6, 0.8$, and 1.0. In each case, evaluate S_l and flame thickness δ using correlations Eqs. 8.33 and 8.35. Hence, comment whether the reaction is governed by *fast* or *slow* chemistry in each case.
12. A turbulent methane-air flame ($\Phi = 1$) is stabilised in the wake of a flame holder (D $= 3.2$ mm) in a duct with $AR \rightarrow 0$. The unburned conditions are, p $= 1$ atm and $T_u = 298$ K. Take $C_D = 1.1$ and recirculation ratio l/D $= 6$. Estimate extinction velocity (m/s) using Eqs. 8.56 and 8.57. Evaluate all mixture properties at 0.5 ($T_b + T_u$) using properties of dry air. Evaluate S_l using Eq. 8.23.

13. In Fig. 8.11, variations of $X^*_{m,u,b}$ are seen to be linear with τ at large times. Hence, show that $d\,(x_{m,u,b})/dt = (\rho_u/\rho_{mix})\,S_l$.

14. Using Eq. 8.110 and taking $T^*_{ign} = 0.1$, verify whether $E_a\,(T_{ign} - T_i)/(R_u\,T_i^2) > 1$ for methane, propane, heptane, and iso-octane assuming $T_i = 750\,K$ and $\Phi = 1$.

15. Evaluate flammability limits from Eqs. 8.116 and 8.117 for C_2H_6 combustion in air, and compare their values with those given in Table 8.9. (Hint: You will need to establish *blowout* solutions and $T^*_{cr}\, T^*_{ad} = F(\Phi)$ in the manner of exercise problem 9 in Chap. 7.).

16. Using the data given in Table 8.10, and Eq. 8.35 for S_l, evaluate value of κ for methane. Hence, determine quenching distance for methane at 5 atm and 30 °C for $0.6 < \Phi < 1.5$. Plot variation of δ, S_l, and d with Φ.

Chapter 9
Diffusion Flames

9.1 Introduction

Diffusion flames were introduced in the previous chapter (see Fig. 8.2) as flames in which fuel and air are physically separated. Diffusion flames occur in the presence of flowing or stagnant air such that the fuel moves outward and air moves inward to create a reaction zone at the flame periphery. Figure 9.1 shows a typical *laminar* diffusion flame configuration[1] in stagnant surroundings.

Fuel issues from a round nozzle (diameter D) with axial velocity U_0 into stagnant surroundings. As there are no effects present to impart three dimensionality to the flow, the flow is two dimensional and axisymmetric. The fuel burns by drawing air from the surroundings forming a flame that is visible to the eye. The flame develops to length L_f and assumes a shape characterized by flame radius $r_f(x)$, which is a function of axial distance x.

The energy release is thus principally governed by the *mixing process* between air and fuel. As such, the chemical kinetics are somewhat less important, being very fast. For this reason, diffusion flames are often termed *physically controlled* flames.[2] Flames formed around burning volatile liquid droplets or solid particles (whose volatiles burn in the gaseous phase) are thus physically controlled diffusion flames.

The two main objectives of developing a theory of diffusion flames are to predict

1. The flame length L_f and
2. The shape of the flame, that is, the flame radius $r_f(x)$.

Prediction of L_f is important because the engineer would like to ensure that the flame is contained well within the combustion space, for reasons of safety. Prediction

---

[1]It is assumed that the fuel jet momentum is sufficiently high so flame remains horizontal and is not lifted upward owing to natural convection. Also, it is assumed that there is no vertical crosswind. Flames so formed are also called Burning Jets.

[2]Diffusion flames are also called non-premixed flames.

© Springer Nature Singapore Pte Ltd. 2020
A. W. Date, *Analytic Combustion*,
https://doi.org/10.1007/978-981-15-1853-9_9

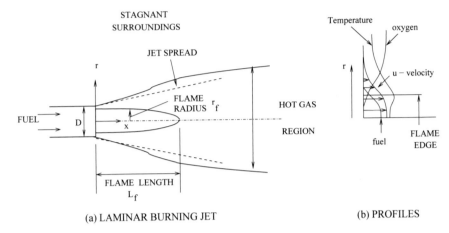

(a) LAMINAR BURNING JET   (b) PROFILES

Fig. 9.1 Laminar diffusion flame configuration

of r_f and its maximum value are important for the same reason, but also to understand the nature of the mixing patterns.

Because the burning jet situation represents a *steady* long and thin flow, the boundary layer equations introduced in Chap. 6 must be invoked.[3] The fast chemical kinetics permit assumption of a simple chemical reaction (SCR) given by

$$1 \text{ kg of fuel} + R_{st} \text{ kg of oxidant air} = (1 + R_{st}) \text{ kg of product.}$$

Under these assumptions, the boundary layer equations for a round jet flame must be written in cylindrical polar coordinates x and r. These equations will read as follows:

$$\frac{\partial}{\partial x} (\rho_m\, u\, r) + \frac{\partial}{\partial r} (\rho_m\, v\, r) = 0 \,, \tag{9.1}$$

$$\frac{\partial}{\partial x} (\rho_m\, u\, r\, u) + \frac{\partial}{\partial r} (\rho_m\, v\, r\, u) = \frac{\partial}{\partial r} \left[\mu_{eff}\, r\, \frac{\partial u}{\partial r} \right] \,, \tag{9.2}$$

$$\frac{\partial}{\partial x} (\rho_m\, u\, r\, \omega_{fu}) + \frac{\partial}{\partial r} (\rho_m\, v\, r\, \omega_{fu}) = \frac{\partial}{\partial r} \left[\rho_m\, D_{eff}\, r\, \frac{\partial \omega_{fu}}{\partial r} \right] - r\, |\, R_{fu}\, | \,, \tag{9.3}$$

$$\frac{\partial}{\partial x} (\rho_m\, u\, r\, \omega_{ox}) + \frac{\partial}{\partial r} (\rho_m\, v\, r\, \omega_{ox}) = \frac{\partial}{\partial r} \left[\rho_m\, D_{eff}\, r\, \frac{\partial \omega_{ox}}{\partial r} \right] - r\, |\, R_{ox}\, | \,, \tag{9.4}$$

---

[3]The boundary layer assumption is strictly valid a small distance away from the nozzle exit. At small axial distances, the flow width is comparable and hence upstream flow properties are influenced by downstream ones. Thus, at small axial distances, the flow is essentially elliptic.

$$\frac{\partial}{\partial x}\left(\rho_m\, u\, r\, h_m\right) + \frac{\partial}{\partial r}\left(\rho_m\, v\, r\, h_m\right) = \frac{\partial}{\partial r}\left[\frac{k_{eff}}{c_{pm}}\, r\, \frac{\partial h_m}{\partial r}\right] + r\mid R_{fu}\mid \Delta H_c\,, \quad (9.5)$$

where sensible enthalpy $h_m = c_{pm}\,(T - T_{ref})$ and it is assumed that the specific heats of all species are equal. These equations are applicable to both laminar and turbulent burning jets—but in the latter, it will be recalled, vector variables u and v and, scalar variables h_m, ω_{ox}, and ω_{fu} represent their time-averaged values. As usual, diffusivities for fuel and air are assumed equal. In the energy equation 9.5, the viscous dissipation, Q_{rad} and Dp/Dt terms are ignored. Dp/Dt = 0 because steady flow prevails and because dp/dx = 0 for the jet. Hence, the pressure gradient also does not appear in the momentum equation 9.2.

In a real flame, fluid properties will be functions of local temperature and composition and, in turbulent flames, they will further be functions of scalar quantities characterizing turbulence. Our equations are thus strongly coupled and nonlinear. Exact analytical solutions are therefore not possible, and numerical solutions must be sought. Computer codes for doing this are available [28, 112]. However, in the sections to follow, simple analytical solutions will be developed, making simplifying assumptions to capture the main tendencies exhibited by the solutions. The solutions will enable prediction of L_f and $r_f(x)$ and identification of parameters on which their values depend.

9.2 Laminar Diffusion Flames

9.2.1 Velocity Prediction

For laminar flames, $\mu_{eff} = \mu_m$. In a real flame, the fluid properties ρ_m and μ_m will be functions of temperature and composition. To simplify the analysis, however, it will be assumed that they are *uniform*; that is, not functions of x and r. Then, Eqs. 9.1 and 9.2 can be written as

$$\frac{\partial u}{\partial x} + \frac{\partial v}{\partial r} + \frac{v}{r} = 0 \quad \text{and} \qquad (9.6)$$

$$u\,\frac{\partial u}{\partial x} + v\,\frac{\partial u}{\partial r} = \frac{\nu_m}{r}\,\frac{\partial}{\partial r}\left[r\,\frac{\partial u}{\partial r}\right]. \qquad (9.7)$$

The boundary conditions are $v = \partial u/\partial r = 0$ at $r = 0$ (that is, jet centerline) and, for stagnant surroundings, $u = 0$ as $r \to \infty$. These equations can be solved using the *similarity* method [98] which converts the partial differential equations into ordinary ones. Thus, we introduce the *stream function* ψ and a similarity variable η with the following definitions:

$$\psi\,(x, \eta) \equiv \nu_m\, x\, f\,(\eta) \quad \text{and} \quad \eta \equiv C \times \frac{r}{x}\,, \qquad (9.8)$$

where C is an arbitrary constant whose value will be determined in the course of the analysis. If we now define u and v in terms of ψ as

$$u \equiv \frac{1}{r}\frac{\partial \psi}{\partial r} \quad \text{and} \quad v \equiv -\frac{1}{r}\frac{\partial \psi}{\partial x}, \tag{9.9}$$

then it can be verified that these definitions satisfy Eq. 9.6. Using Eqs. 9.8 and 9.9, it can further be shown[4] that

$$u = \frac{C^2 \nu_m}{x}\left[\frac{f'}{\eta}\right], \tag{9.10}$$

$$v = -\frac{C \nu_m}{x}\left[\frac{f}{\eta} - f'\right], \tag{9.11}$$

$$\frac{\partial u}{\partial r} = \frac{C^3 \nu_m}{x^2}\left[\frac{f''}{\eta} - \frac{f'}{\eta^2}\right], \tag{9.12}$$

$$\frac{\partial u}{\partial x} = -\frac{C^2 \nu_m}{x^2}\left[f''\right], \quad \text{and} \tag{9.13}$$

$$\frac{\nu_m}{r}\frac{\partial}{\partial r}\left[r\frac{\partial u}{\partial r}\right] = \frac{C^4 \nu_m^2}{\eta x^3}\frac{d}{d\eta}\left(f'' - \frac{f'}{\eta}\right), \tag{9.14}$$

where $f' = df/d\eta$, $f'' = d^2 f/d\eta^2$, and so on. Making the preceding substitutions in the momentum equation 9.7, it can be shown that

$$\frac{f f'}{\eta^2} - \frac{f f''}{\eta} - \frac{f'^2}{\eta} = \frac{d}{d\eta}\left[f'' - \frac{f}{\eta}\right]. \tag{9.15}$$

This equation can also be written as

$$\frac{d}{d\eta}\left[f'' - \frac{f}{\eta} + \frac{f f'}{\eta}\right] = 0. \tag{9.16}$$

Therefore, integrating this equation once from $\eta = 0$ to η and noting that the boundary conditions at r = 0 (or at $\eta = 0$) can be interpreted as f (0) = f' (0) = 0, it follows that[5]

$$f f' = f' - \eta f''. \tag{9.17}$$

---

[4]Here, the chain rule is employed. Thus

$$\frac{\partial}{\partial r} = \frac{d}{d\eta} \times \frac{\partial \eta}{\partial r} = \frac{C}{x}\frac{d}{d\eta} \quad \text{and} \quad \frac{\partial}{\partial x} = \frac{d}{d\eta} \times \frac{d\eta}{dx} = -\frac{C r}{x^2}\frac{d}{d\eta} = -\frac{\eta}{x}\frac{d}{d\eta}.$$

[5]Note that $(f'/\eta)_{\eta \to \infty} = 0$ and $f'' (\infty) \to 0$. Therefore, the constant of integration is zero.

The solution of this equation is

$$f = \frac{\eta^2}{1 + \eta^2/4},$$ (9.18)

$$f' = \frac{2\,\eta}{(1 + \eta^2/4)^2}, \quad \text{and}$$ (9.19)

$$f'' = \frac{2\,(1 - 3\,\eta^2/4)}{(1 + \eta^2/4)^3}.$$ (9.20)

Thus, from Eqs. 9.10 and 9.11, it can be shown that

$$u = \frac{C^2\,\nu_m}{x} \left[\frac{2}{(1 + \eta^2/4)^2} \right], \quad \text{and}$$ (9.21)

$$v = \frac{C\,\nu_m}{x} \left[\frac{\eta - \eta^3/4}{(1 + \eta^2/4)^2} \right].$$ (9.22)

Determination of C

To determine C, we multiply Eq. 9.7 throughout by r and, after using Eq. 9.6, integrate from r = 0 to r = ∞. Then

$$\frac{\partial}{\partial x} \left[\int_0^\infty \rho_m\,u^2\,r\,dr \right] + \{ \rho_m\,v\,u\,r\,|_\infty - \rho_m\,v\,u\,r\,|_0 \}$$
$$= \left\{ \mu_m\,r\,\frac{\partial u}{\partial r}\,|_\infty - \mu_m\,r\,\frac{\partial u}{\partial r}\,|_0 \right\}.$$ (9.23)

Applying the boundary conditions at r = 0 and ∞, it can be shown that all terms in the two curly brackets are zero. Thus,

$$\frac{\partial}{\partial x} \left[\int_0^\infty \rho_m\,u^2\,r\,dr \right] = 0.$$ (9.24)

This implies that $2\,\pi \int_0^\infty \rho_m\,u^2\,r\,dr$, which represents the *momentum of the jet* (J, say), is *independent* of x and also equals the momentum at the exit from the nozzle at x = 0. Thus,

$$2\,\pi \int_0^\infty \rho_m\,u^2\,r\,dr = \rho_0\,U_0^2\,(\frac{\pi}{4}\,D^2) = \text{constant} = J.$$ (9.25)

Substituting for u from Eq. 9.21 and carrying out integration, it can be derived that

$$J = \frac{16}{3}\,\pi\,\rho_m\,\nu_m^2\,C^2 = \rho_0\,U_0^2\,\frac{\pi}{4}\,D^2.$$ (9.26)

Thus, C derives to

$$C = \frac{\sqrt{3}}{8} \, Re \left(\frac{\rho_0}{\rho_m}\right)^{0.5} = \frac{1}{4 \, \nu_m} \sqrt{\frac{3 \, J}{\pi \, \rho_m}} \, , \tag{9.27}$$

where the jet Reynolds number Re is given by

$$Re = \frac{U_0 \, D}{\nu_m} \, . \tag{9.28}$$

Thus, C can be expressed in terms of Re or J. Hence, the final solutions can be expressed as

$$u^* = \frac{u \, x}{\nu_m} = \frac{3}{32} \, Re^2 \, [1 + \eta^2/4]^{-2} \left(\frac{\rho_0}{\rho_m}\right)$$

$$= \frac{3}{8 \, \pi} \left(\frac{J}{\mu_m \, \nu_m}\right) [1 + \eta^2/4]^{-2} \, , \tag{9.29}$$

$$v^* = \frac{v \, x}{\nu_m} = \frac{\sqrt{3}}{8} \, Re \left[\frac{\eta - \eta^3/4}{(1 + \eta^2/4)^2}\right] \left(\frac{\rho_0}{\rho_m}\right)^{0.5}$$

$$= \left[\left(\frac{3}{16 \, \pi}\right) \left(\frac{J}{\mu_m \, \nu_m}\right)\right]^{0.5} \left[\frac{\eta - \eta^3/4}{(1 + \eta^2/4)^2}\right] \, , \tag{9.30}$$

$$\text{and } \eta = \frac{\sqrt{3}}{8} \left(\frac{\rho_0}{\rho_m}\right)^{0.5} \times Re \times \frac{r}{x} \, . \tag{9.31}$$

These solutions show that u and v are *inversely* proportional to x. But, at x = 0, $u = U_0$ and v = 0. As such, the solutions are applicable beyond a sufficiently large x only. Second, $u = u_{max}$ at $\eta = 0$ (or, at the jet axis). Thus,

$$\frac{u}{u_{max}} = \frac{1}{(1 + \eta^2/4)^2} \, . \tag{9.32}$$

Because at any x, $u \to 0$ as $r \to \infty$, it is difficult to identify the location of the edge of the jet exactly. Hence, by convention, the *jet half-width* ($r_{1/2}$) is characterized by value of $\eta_{1/2}$ corresponding to $u/u_{max} = 0.5$. Thus, from Eq. 9.32, $\eta_{1/2} = 1.287$, or

$$\frac{r_{1/2}}{x} = \frac{\eta_{1/2}}{C} = 1.287 \times \frac{8}{\sqrt{3} \, Re} \left(\frac{\rho_0}{\rho_m}\right)^{-0.5} = \frac{5.945}{Re} \left(\frac{\rho_0}{\rho_m}\right)^{-0.5} = \tan \alpha \, . \tag{9.33}$$

This equation shows that the cone angle (α) of the laminar jet is inversely proportional to Re.

Jet Entrainment

Similar to jet momentum J, the mass flow rate $\dot{m}_{jet}$ at any section x of the jet is

$$\dot{m}_{jet} = 2\,\pi \int_0^\infty \rho_m\, u\, r\, dr \,. \tag{9.34}$$

Substituting for u from Eq. 9.29 and carrying out integration, it can be shown that

$$\dot{m}_{jet} = 8\,\pi\,\mu_m\, x \,. \tag{9.35}$$

This equation shows that unlike J, which is constant, $\dot{m}_{jet}$ increases linearly with x and, further, it is independent of U_0 and D. This increase in the mass of the jet can occur only because of *entrainment* of air from the stagnant surroundings; hence, the entrainment rate is given by

$$\dot{m}_{ent} = \frac{d}{dx}\,(\dot{m}_{jet}) = 8\,\pi\,\mu_m = -2\,\pi\,(\rho_m\, r\, v)_{ent}. \tag{9.36}$$

Incidentally, from Eq. 9.30, we note that v = 0 for $\eta = 2$; v is positive for $\eta < 2$ but negative for $\eta > 2$. The negative v at large radii, of course, indicates flow *into the jet* from the surroundings.

9.2.2 Flame Length and Shape Prediction

To predict L_f and $r_f(x)$, mass transfer and energy equations 9.3, 9.4, and 9.5 must be solved, in which u and v are prescribed by Eqs. 9.29 and 9.30, respectively. As mentioned before, this is permissible because uniform properties have been assumed in our simplified analysis. As such, the velocity distributions are not affected by the combustion process. A further simplifying assumption pertains to the *magnitude* of properties. Thus, we assume that $\mu_m = \rho_m D = k_m/Cp_m = \Gamma_m$. This implies that Pr = Sc, and Le = 1. Typically, for gaseous mixtures, $Pr \sim 0.7$ and $Sc \simeq 0.67$. Thus, our assumption of equal magnitude of transport coefficients is not too far from reality.

For an SCR, the mass fraction and the energy equations can be generalized in the usual manner (see Chaps. 7 and 8) by defining a conserved scalar Ψ. Thus,

$$\frac{\partial}{\partial x}\,(\rho_m\, u\, r\, \Psi) + \frac{\partial}{\partial r}\,(\rho_m\, v\, r\, \Psi) = \frac{\partial}{\partial r}\left[\Gamma_m\, r\, \frac{\partial \Psi}{\partial r}\right], \tag{9.37}$$

where $\Psi = \omega_{fu} - \omega_{ox}/R_{st} = h_m + \Delta H_c\,\omega_{fu}$. In fact, we may also write $\Psi = u/U_0$ because Eq. 9.2 also has a zero source.

To *locate* the flame, it is customary to define another conserved scalar f as

$$f = \frac{\Psi - \Psi_A}{\Psi_F - \Psi_A} = \frac{(\omega_{fu} - \omega_{ox}/R_{st}) - (\omega_{fu} - \omega_{ox}/R_{st})_A}{(\omega_{fu} - \omega_{ox}/R_{st})_F - (\omega_{fu} - \omega_{ox}/R_{st})_A} , \tag{9.38}$$

where the subscript A refers to the air stream and F to the fuel stream. Thus, at the exit from the nozzle (x = 0), where only fuel is found ($\omega_{ox} = 0$), f = 1 because $(\omega_{fu} - \omega_{ox}/R_{st})_F = (\omega_{fu})_F = 1$. By the same reasoning, $f_\infty = 0$ because no fuel is to be found in the stagnant surroundings ($\omega_{fu,\infty} = 0$).

In this sense, the concept of f can be generalized for a mixing process of two segregated streams as follows. For such a process, Eq. 9.38 can be written as

$$\Psi = f\,\Psi_F + (1 - f)\,\Psi_A , \tag{9.39}$$

where $0 \le f \le 1$. This relation shows that when f kg (or kg/s) of the *F-stream* material is mixed with $(1 - f)$ kg (or kg/s) of the *A-stream* material, 1 kg (or kg/s) of the *mixture stream* material is formed. In our special case, the F- stream comprises fuel only (although, in general, the F-stream could have some dilutants) and the A-stream comprises air only. Hence, Eq. 9.38 can also be written as

$$\begin{aligned} f &= \frac{(\omega_{fu} - \omega_{ox}/R_{st}) + \omega_{ox,A}/R_{st}}{\omega_{fu,F} + \omega_{ox,A}/R_{st}} \\ &= \frac{(\omega_{fu} - \omega_{ox}/R_{st}) + \omega_{ox,\infty}/R_{st}}{\omega_{fu,F} + \omega_{ox,\infty}/R_{st}} \\ &= \frac{(\omega_{fu} - \omega_{ox}/R_{st}) + 1/R_{st}}{1 + 1/R_{st}} \qquad \omega_{ox,\infty} = \omega_{fu,F} = 1. \end{aligned} \tag{9.40}$$

Thus, although f is a fraction, it equals neither ω_{fu} nor the fuel/air ratio ($1/R_{st}$). The value of f simply characterizes the state of the mixture at a point in terms of the amounts of fuel and oxidant (and, by inference, the amount of product, because in an SCR, $\omega_{pr} = 1 - \omega_{fu} - \omega_{ox}$) at that point. It is for this reason that f is called the *mixture fraction*. Finally, it is, of course, trivial to prove that distributions of f will also be governed by Eq. 9.37.

Now, for an SCR, the flame can be regarded as being a surface located where fuel and oxygen are in *stoichiometric* proportions. That is, $f = f_{stoich}$ or, where

$$\omega_{fu} - \frac{\omega_{ox}}{R_{st}} = 0 . \tag{9.41}$$

and, from Eq. 9.38 or 9.40,

$$f_{stoich} = \frac{1}{1 + R_{st}} . \tag{9.42}$$

It is obvious that *outside* the flame, $0 \le f < f_{stoich}$ because fuel will have zero concentration there, whereas *inside* the flame, $f_{stoich} < f \le 1$ because no oxygen

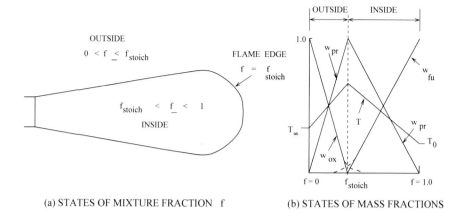

(a) STATES OF MIXTURE FRACTION f   (b) STATES OF MASS FRACTIONS

Fig. 9.2 States of laminar diffusion flame

will be found there. Thus, using Eq. 9.40, we have

$$f \leq f_{stoich} : f = -\frac{\omega_{ox}/R_{st}}{1 + 1/R_{st}} + f_{stoich} , \tag{9.43}$$

$$f \geq f_{stoich} : f = \frac{\omega_{fu}}{1 + 1/R_{st}} + f_{stoich} , \quad \text{and} \tag{9.44}$$

$$f = f_{stoich} : f = \frac{1}{1 + R_{st}} . \tag{9.45}$$

Figure 9.2(a) shows the schematic representation of the f-states of the diffusion flame. The corresponding mass fraction states are graphically plotted in Fig. 9.2(b). When $f = f_{stoich}, \omega_{fu} = \omega_{ox} = 0$. This is a consequence of the fast SCR. However, condition Eq. 9.41 is still satisfied.[6]

Similar to the mixture fraction f, we can also define another variable h^*, as

$$\begin{aligned} h^* &= \frac{(h_m + \Delta H_c \, \omega_{fu}) - (h_m + \Delta H_c \, \omega_{fu})_A}{(h_m + \Delta H_c \, \omega_{fu})_F - (h_m + \Delta H_c \, \omega_{fu})_A} \\ &= \frac{h_m - h_{m,\infty} + \Delta H_c \, \omega_{fu}}{h_{m,0} - h_{m,\infty} + \Delta H_c} \end{aligned} \tag{9.46}$$

$$= \frac{cp_m \, (T - T_\infty) + \Delta H_c \, \omega_{fu}}{cp_m \, (T_0 - T_\infty) + \Delta H_c} . \tag{9.47}$$

---

[6]In reality, the reaction rates are finite but still large. In this case, variations of ω_{fu} and ω_{ox} overlap slightly near $f = f_{stoich}$, the flame surface will have a finite (though small) width, and condition Eq. 9.41 will be satisfied over this width. This is indicated by dotted curves near $f = f_{stoich}$ in Fig. 9.2(b).

Again, h^* will also satisfy Eq. 9.37. Thus, the boundary and initial conditions for $\Psi = f, h^*$ and u/U_0 are

$$\Psi(x, \infty) = 0, \qquad \frac{\partial \Psi}{\partial r}|_{x,r=0} = 0, \quad \text{and} \qquad (9.48)$$

$$\Psi_{0,r} = 1. \qquad (9.49)$$

Because the governing equations and the boundary conditions are the same, the solutions to the three variables are also the same. These solutions can, therefore, be derived from Eq. 9.29:

$$\Psi = \frac{u}{U_0} = f = h^* = \frac{3}{32}\left(\frac{D}{x}\right)\left(\frac{\rho_0}{\rho_m}\right) Re\,(1 + \eta^2/4)^{-2}. \qquad (9.50)$$

This solution[7] shows that Ψ decreases with x at all radii. Ψ is maximum at the axis of the jet and is given by

$$\Psi_{max} = \frac{3}{32}\left(\frac{D}{x}\right)\left(\frac{\rho_0}{\rho_m}\right) Re. \qquad (9.51)$$

Using Eq. 9.50, the value of η_f which represents dimensionless flame radius, can now be evaluated corresponding to $f = f_{stoich}$. Thus, using Eq. 9.31, the physical flame radius r_f will evaluate to

$$\frac{r_f}{x} = \frac{16}{\sqrt{3}} Re^{-1}\left(\frac{\rho_0}{\rho_m}\right)^{-0.5}\left[\left\{\frac{3}{32}\left(\frac{D}{x}\right)\left(\frac{\rho_0}{\rho_m}\right)\frac{Re}{f_{stoich}}\right\}^{0.5} - 1\right]^{0.5} \qquad (9.52)$$

The flame length will obviously correspond to $r_f = 0$. Hence, the value of $x = L_f$ corresponding to $r_f = 0$ is given by

$$\frac{L_f}{D} = \frac{3}{32}\left(\frac{\rho_0}{\rho_m}\right)\frac{Re}{f_{stoich}} = \frac{3}{32} Re\,(1 + R_{st})\left(\frac{\rho_0}{\rho_m}\right). \qquad (9.53)$$

This expression shows that $L_f \propto (Re \times D) \propto \rho_0 U_0 D^2$, but the individual values of ρ_0, U_0, and D have no influence. Also, L_f increases linearly with Re (or the mass flow). In furnaces, designers wish to have a long flame to achieve uniformity of heat transfer over the furnace length. This can be achieved by maintaining a high mass flow and also by having a large D (or, for the same flow rate, using a single nozzle of large D instead of several nozzles of small D). In combustion chambers of gas turbines and in rocket motors, in contrast, small flame lengths are desired, to contain the flame inside the chambers. Short flame lengths can be achieved by having a large number of small-diameter nozzles and dividing the total mass flow between them.

---

[7]Because $f = h^*$, the variation of temperature T with f can be derived from Eq. 9.47. This variation is shown in Fig. 9.2(b). As expected, the maximum flame temperature occurs at $f = f_{stoich}$.

Finally, the location of the maximum flame radius can be found by differentiating r_f with respect to x and equating to zero. The result is

$$x\,(r_{f,max}) = \frac{9}{16}\,L_f\,. \tag{9.54}$$

Substituting this x in Eq. 9.52 and using Eq. 9.53, it can be shown that

$$\frac{r_{f,max}}{L_f} = \frac{3}{Re} \times \left(\frac{\rho_0}{\rho_m}\right)^{-0.5}. \tag{9.55}$$

This equation shows that the maximum flame width is considerably smaller than the flame length. A jet typically remains laminar for $Re < 2000$.

Influence of Temperature-Dependent Properties:

The values of μ_m, D, and k_m increase with temperature, whereas the density ρ_m decreases with temperature. The enhanced transport properties tend to increase the air entrainment into the jet, whereas the reduced density widens the jet and thus reduces the gradients of u, h, and ω across the jet width. The net effect of these cannot be easily estimated unless detailed numerical computations are performed. Turns [122] based on numerical computations of Fay [37], has suggested that for typical hydrocarbon flames, the net effect of variable properties is to increase L_f (by a factor of 2 to 3) over its constant-property estimate provided by Eq. 9.53.

Instead of using the SCR, several authors have obtained numerical solutions using extended reaction mechanisms involving a large number of species and by allowing for effects of property variations and radiation from the flame in wide stagnant surroundings, as well as in confined configurations. Similarly, when flames are issued from noncircular nozzles (square or rectangular slot), the flames are three dimensional; this requires solution of the complete transport equations introduced in Chap. 6. These equations can only be solved using CFD.

9.2.3 Correlations

For use in practical design, experimentally determined empirical correlations are useful. Roper [95] has developed solutions coupled with experimental verifications [96] of diffusion flame lengths produced from different burner geometries in different flow regimes. As indicated earlier, flames in stagnant surroundings can also be influenced by buoyancy when the jet momentum is small. In fact, different magnitudes of a dimensionless parameter called the *Froude number* indicates *momentum-controlled* regime at one extreme, a *buoyancy-controlled* regime at the other extreme, and *transitional* regime in-between. The Froude number (Fr) is defined as

$$Fr = \left[\frac{U_0^2}{g\,L_f}\right]\left[\frac{J}{\dot{m}\,U_0}\right]^2\left[\frac{f_{stoich}^2}{0.6\,(T_f/T_\infty - 1)}\right]. \tag{9.56}$$

The dimensionless ratio in the first bracket represents the conventional Froude number. The ratio inside the second bracket will equal 1 if the velocity profile at the nozzle exit is uniform. Typically, the profile will depend on the wall curvature near the nozzle exit, which in turn will determine the shape of the exit velocity profile. For example, in a rectangular slot burner with large aspect ratio, $u_{max}/U_0 = 1.5$. Then the ratio inside the second bracket will equal 1.5 (Definitions 9.25 and 9.34). T_f in the denominator of the third bracket represents the *mean* flame temperature. Roper [96], for example, uses $T_f = 1500\,K$ for hydrocarbon flames allowing for radiation effects.

The flow regimes are established according to the following criteria:

$$Fr \gg 1 \qquad \text{momentum-controlled,} \qquad (9.57)$$

$$Fr \simeq 1 \qquad \text{transitional, and} \qquad (9.58)$$

$$Fr \ll 1 \qquad \text{buoyancy-controlled.} \qquad (9.59)$$

Roper's correlations[8] are listed in Table 9.1. In this table, the *molar stoichiometric ratio* S is defined as

$$S \equiv \left[\frac{\text{moles of ambient fluid}}{\text{moles of nozzle fluid}}\right]_{\text{stoich}}. \qquad (9.60)$$

Thus, for a pure hydrocarbon fuel ($C_m H_n$) burning in pure air,

$$S = 4.76 \times \left(m + \frac{n}{4}\right). \qquad (9.61)$$

For circular and square ports, the correlations are applicable to both momentum-controlled and buoyancy-controlled situations. For all correlations, the flame length in the transition region $L_{f,tr}$ is predicted [95, 96] by

$$\frac{L_{f,tr}}{L_{f,mom}} = \frac{4}{9} \left(\frac{L_{f,bo}}{L_{f,mo}}\right)^3 \left[\left\{1 + 3.38 \left(\frac{L_{f,mo}}{L_{f,bo}}\right)^3\right\}^{2/3} - 1\right], \qquad (9.62)$$

where $L_{f,bo}$ and $L_{f,mom}$ are flame lengths for buoyancy- and momentum-controlled situations, respectively. Thus, for circular and square ports for which $L_{f,bo} = L_{f,mom}$, $L_{f,tr} = 0.745\,L_{f,mom}$.

9.2.4 Solved Problems

Problem 9.1 A propane (C_3H_8)-air laminar diffusion flame is formed with fuel at 300 K issuing from a 10 mm diameter nozzle and uniform velocity $U_0 = 5.5\,cm/s$ in

---

[8]In all experimental correlations involving $L_{f,exp}$, the constants are dimensional.

stagnant air at 1 atm and 300 K. Estimate L_f using Eq. 9.53 and Roper's correlations. Also determine the rate of heat released from the flame.

For propane, $S = 4.76 \times (3 + 8/4) = 23.8$, $\mu_\infty = \mu_{air}(300) = 18.46 \times 10^{-6}$ N-s/m^2, and $\rho_\infty = 1.01325 \times 10^5/(8314/29)/300 = 1.178$ kg/m^3. Therefore, $Re_\infty = 1.178 \times 0.055 \times 0.01/18.46 \times 10^{-6} = 35.1$. Thus, using circular port correlations,

$$L_{f,th} = 0.01 \times \left(\frac{35.1}{16}\right) \frac{(300/300) \times (300/1500)^{0.67}}{\ln(1 + 1/23.8)} = 0.1815 \text{ m} = 181.5 \text{ mm, and}$$

$$L_{f,exp} = 0.01 \times 1044.58 \times 0.055 \times 0.01 \frac{(300/300)}{\ln(1 + 1/23.8)} = 0.14 \text{ m} = 140 \text{ mm.}$$

In Eq. 9.53, $R_{st} = (3 + 8/4)(32 + 3.76 \times 28)/44 = 15.6$. Also, $\rho_{fu} = (M_{fu}/M_{air}) \times \rho_{air} = 1.787$. To evaluate Re_m, we evaluate properties at $T_f \simeq 1500$ K. Then, $\nu_m = 239.2 \times 10^{-6}$ m^2/s. Thus, $Re_m = 0.055 \times 0.01/239.2 \times 10^{-6} = 2.293$. Hence,

$$L_f = 0.01 \times \frac{3}{32} \times 2.293 \times (1 + 15.6) \times \left(\frac{1500}{300}\right) = 0.178 \text{ m} = 178.0 \text{ mm}$$

These results show that Roper's theory predicts about 30% longer flame length than the length predicted by the experimental correlation. However, Eq. 9.53 predicts a value close to Roper's theoretical correlation.

For this burning jet flame, the Froude number is

$$Fr = \left[\frac{0.055^2}{9.81 \times 0.14}\right] \times (1)^2 \times \left[\frac{16.6^{-2}}{0.6 (1500/300 - 1)}\right] = 3.33 \times 10^{-6} \ll 1$$

Therefore, the flame is buoyancy-controlled. Nonetheless, since the port is circular, both $L_{f,exp}$ and $L_{f,th}$ evaluations are directly applicable.

Finally, the rate of heat released ($\dot{Q}$) from the flame is evaluated as

$$\dot{m}_{jet} = 1.787 \times \frac{\pi}{4} \times 0.01^2 \times 0.055 = 7.72 \times 10^{-6} \text{ kg/s}$$

$$\dot{Q} = \dot{m}_{jet} \, \Delta H_c = 7.72 \times 10^{-6} \times 46.36 \times 10^6 = 357.87 \text{ Watts.}$$

Problem 9.2 In the previous problem, if a rectangular slot burner (b = 2 mm and a = 15 mm) is used, with all other quantities remaining same, estimate $L_{f,exp}$ from Roper's correlation.

Table 9.1 Correlations for laminar diffusion flame length Definitions: inverf [erf $(\Phi)] = \Phi$, b = short side, a = long side, g = 9.81 m/s^2—Diffusion coefficient $D_\infty = \nu_\infty$ is assumed

Burner geometry	Flame length	Remarks
Circular port	$\frac{L_f}{D}\|_{th} = \frac{1}{16}\left(\frac{U_0 D}{\nu_\infty}\right) \times \frac{(T_\infty/T_0)\times(T_\infty/T_f)^{0.67}}{\ln(1+1/S)}$	All
	$\frac{L_f}{D}\|_{exp} = 1044.58\,(U_0\,D)\,\frac{(T_\infty/T_0)}{\ln(1+1/S)}$	Regimes
Square port (a = b)	$\frac{L_f}{a}\|_{th} = \frac{1}{16}\left(\frac{U_0 a}{\nu_\infty}\right)\frac{(T_\infty/T_0)\times(T_\infty/T_f)^{0.67}}{[\text{inverf}\{(1+S)^{-0.5}\}]^2}$	All
	$\frac{L_f}{a}\|_{exp} = 1045\,(U_0\,a)\,\frac{(T_\infty/T_0)}{[\text{inverf}\{(1+S)^{-0.5}\}]^2}$	Regimes
Rectangular (a > b)	$\frac{L_f}{b}\|_{th} =$ $\frac{1}{16}\left(\frac{U_0 b}{\nu_\infty}\right)\left(\frac{\dot{m}\,U_0}{J}\right)\frac{f_{stoich}^{-1}\times(T_\infty/T_0)^2\times(T_\infty/T_f)^{0.33}}{[\text{inverf}\{(1+S)^{-1}\}]^2}$	$Fr \gg 1$
	$\frac{L_f}{b}\|_{exp} = 5375\,(U_0\,b)\left(\frac{\dot{m}\,U_0}{J}\right)\frac{f_{stoich}^{-1}\times(T_\infty/T_0)^2}{[\text{inverf}\{(1+S)^{-1}\}]^2}$	$Fr \gg 1$
Rectangular (a > b)	$\frac{L_f}{b}\|_{th} = 1.233\left[\frac{\nu_\infty^2}{g\,(T_f/T_\infty-1)}\right]^{1/3} A^{4/3}\left(\frac{T_f}{T_\infty}\right)^{2/9}$	$Fr \ll 1$
	$\frac{L_f}{b}\|_{exp} = 2371.26\left[\frac{\nu_\infty^4}{g\,(T_f/T_\infty-1)}\right]^{1/3} A^{4/3}$	$Fr \ll 1$
	$A = \left(\frac{U_0 b}{\nu_\infty}\right)\left(\frac{T_\infty}{T_0}\right)\left[4\,\text{inverf}\{(1+S)^{-1}\}\right]^{-1}$	

In this case, since the flame is buoyancy-controlled, we first evaluate A in Table 9.1 where (inverf $\{(1+23.8)^{-1}\} = 0.03546$).

$$A = \frac{1}{4}\left(\frac{0.055\times0.002}{15.67\times10^{-6}}\right)\left(\frac{300}{300}\right)\left[\text{inverf}\{(1+23.8)^{-1}\}\right]^{-1} = \frac{1.755}{0.03546} = 49.5$$

Now, $L_{f,bo}$ is evaluated as

$$L_{f,bo,exp} = 2371.26\left[\frac{(15.89\times10^{-6})^4}{9.81\,(1500/300-1)}\right]^{1/3} \times 49.5^{4/3}$$

$$= 2371.26 \times (11.46\times10^{-8}) \times 181.72$$

$$= 0.0493\text{ m} = 49.3\text{ mm}$$

$$\dot{m}_{jet} = 1.787\times0.002\times0.015\times0.055 = 2.95\times10^{-6}\text{ kg/s}$$

$$\dot{Q} = \dot{m}_{jet}\,\Delta H_c = 2.95\times10^{-6}\times46.36\times10^6 = 136.7\text{ Watts}$$

Thus, largely due to smaller fuel flow rate, the flame from the rectangular slot is shorter than that from the circular nozzle.

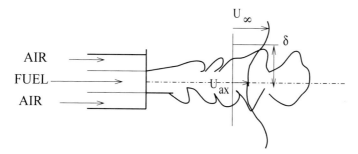

Fig. 9.3 Turbulent diffusion flame in coflowing surroundings

9.3 Turbulent Diffusion Flames

A turbulent jet is formed when the Reynolds number is high.[9] As shown in Fig. 8.2(b), a turbulent jet flame has a nearly constant width. The radial distribution of the axial velocity, however, is nearly uniform over the greater part of the width δ (see Fig. 9.3). Turbulent jets occur more frequently than laminar ones not only in combustion equipment, but also in environmental flows (liquid discharges in rivers or ponds, for example).

For turbulent jets, Eqs. 9.1 to 9.5 again apply. To solve them, the effective values of viscosity, diffusivity, and conductivity must be known. The latter scalar-transport properties are modeled as

$$\rho_m \, D_{eff} \simeq \frac{\mu_{eff}}{Sc_t} \qquad \frac{k_{m,eff}}{C p_m} \simeq \frac{\mu_{eff}}{Pr_t} \, , \qquad (9.63)$$

where the turbulent Schmidt and Prandtl numbers are taken to be $Sc_t = Pr_t \simeq 0.9$. Thus, our equation set is closed when μ_{eff} is prescribed.

9.3.1 Velocity Prediction

The simplest model of effective viscosity assumes that μ_{eff} is related to the *time-mean* velocity and length scale characterizing the jet. Then, a dimensionally correct formula [113] is given by

$$\mu_{eff} = C \, \rho_m \, |u_\infty - u_{ax}| \, \delta \, , \qquad (9.64)$$

---

[9] $Re > 16000$ is typical. However, much depends on the geometry of the burner and the curvature of the burner bounding wall at the exit. For $Re > 30000$, the jet is certainly turbulent.

where u_∞ is the velocity of the coflowing stream, u_{ax} is the velocity at the jet axis, δ is the flow width at any x, and $C \simeq 0.01$. If the surroundings were stagnant, then $u_\infty = 0$ and $u_{ax} = u_{max}$. As seen from Fig. 9.3, the velocity variation in the vicinity of $r = \delta$ is asymptotic in the radial direction. As such, it becomes difficult to estimate δ exactly. However, as in the case of a laminar jet, $r_{1/2}$ can be determined precisely (see Eq. 9.33). Typically, $\delta/r_{1/2} \simeq 2.5$. Therefore, for *stagnant* surroundings, which are of interest here, the effective viscosity is taken as

$$\mu_{eff} = 0.0256 \, \rho_m \, u_{max} \, r_{1/2} \, . \tag{9.65}$$

This equation shows that μ_{eff} is independent of local radius r. Hence, μ_{eff} can be treated as a constant[10] with respect to our set of transport equations. The problem of prediction of turbulent jet is thus the same as that of a laminar jet and all previous solutions apply. It is only required to change μ to μ_{eff}.

Thus, the jet half-width $r_{1/2}$, for example, can be estimated from Eq. 9.33

$$\frac{r_{1/2}}{x} = \frac{5.945}{Re_{turb}} \left(\frac{\rho_0}{\rho_m} \right)^{-0.5} = 5.945 \times \left[\frac{0.0256 \, \rho_m \, u_{max} \, r_{1/2}}{\rho_m \, U_0 \, D} \right] \left(\frac{\rho_0}{\rho_m} \right)^{-0.5} .$$

Canceling $r_{1/2}$, we have[11]

$$\frac{u_{max}}{U_0} = 6.57 \left(\frac{D}{x} \right) \left(\frac{\rho_0}{\rho_m} \right)^{0.5} \tag{9.66}$$

where, for a constant-density jet $\rho_0 = \rho_m$. From Eq. 9.29, at $\eta = 0$, $u_{max}^* = (3/32) \, Re_{turb}^2 \, (\rho_0/\rho_m)$. Therefore, using Eq. 9.65, it can be shown that

$$\left(\frac{u_{max}}{U_0} \right)^2 = 3.662 \left(\frac{D^2}{r_{1/2} \, x} \right) \left(\frac{\rho_0}{\rho_m} \right) \tag{9.67}$$

Eliminating D from this equation and Eq. 9.66, we have

$$\frac{r_{1/2}}{x} = \frac{3.662}{(6.57)^2} = 0.0848 \, . \tag{9.68}$$

The prediction from this equation agrees well with experiments for (x/D) > 6.5. Equation 9.68 shows that unlike the laminar jet, $r_{1/2}/x$ is independent of Re in a turbulent jet. Thus, a turbulent jet becomes really self-similar only after a certain distance from the nozzle exit.

---

[10]The more refined turbulence models discussed in Chap. 6, when executed through CFD, also predict nearly constant μ_{eff} away from the nozzle exit.

[11]Experimentally, it is observed that $u_{max} = U_0$ for $x < 6.5 \, D$. The axial decay indicated in Eq. 9.66 is verified only for $x > 6.5 \, D$.

Replacing $r_{1/2}$ and u_{max} in Eq. 9.65,

$$\mu_{eff} = 0.0256 \, \rho_m \times 6.57 \, U_0 \left(\frac{D}{x}\right) \left(\frac{\rho_0}{\rho_m}\right)^{0.5} \times (0.0848 \, x)$$

$$= 0.01426 \, (\rho_0 \, \rho_m)^{0.5} \, U_0 \, D . \tag{9.69}$$

Thus, for a constant-density jet, μ_{eff} is indeed constant. From Eq. 9.31, $\eta = (\sqrt{3}/8) \, Re_{turb} \, (\rho_0/\rho_m)^{0.5} \, (r/x)$. Therefore, using Eqs. 9.65 and 9.67, it can be shown that the similarity variable for the turbulent jet is given by

$$\eta = 1.287 \left(\frac{r}{r_{1/2}}\right), \tag{9.70}$$

and, from Eq. 9.32,

$$\frac{u}{u_{max}} = \left[1 + 0.414 \left(\frac{r}{r_{1/2}}\right)^2\right]^{-2}. \tag{9.71}$$

9.3.2 Flame Length and Shape Prediction

In laminar flames, it was found that the flame length increases with the jet Reynolds number (or the nozzle velocity U_0). However, when the nozzle velocity is increased further, the flame length actually decreases at first, owing to enhanced mixing caused by turbulence. On further increase, it is found that the flame length remains nearly constant with U_0. Figure 9.4 shows this typical behavior of turbulent diffusion flames first revealed by measurements made by Hawthorne et al. [44]. Further, in this region, in which L_f is independent of U_0, different fuels have significantly different values of L_f/D, indicating dependence on air–fuel ratio R_{st}.

When the nozzle velocity is very high, small holes appear just downstream of the nozzle exit; they grow in number with an increase in U_0. Eventually, the flame detaches from the nozzle. This is called *liftoff*. If U_0 is increased further, then the flame *blows out*. Our interest here is in the attached flame region when L_f is nearly constant.

Remember though, that estimation of the flame length and radius is not easy. This is because a turbulent flame is essentially *unsteady* and its edges are jagged, as shown in Fig. 9.3. Fragments of gas intermittently detach from the main body of the flame and flare, diminishing in size. Turbulence affects not only L_f but also the entire reaction zone near the edge of the flame. Compared with a laminar flame, this zone is also much thicker. This implies that if the time-averaged values $\overline{\omega}_{fu}$ and $\overline{\omega}_{ox}$ are plotted with radius r, then the two profiles show considerable overlap around the crossover point near $f = f_{stoich}$. The situation is similar to but more pronounced than for a laminar diffusion flame (see Fig. 9.2(b)). Unlike this overlap, which is caused

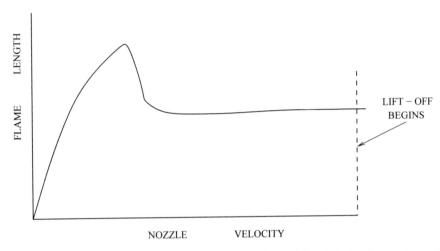

Fig. 9.4 Typical variation of L_f with nozzle velocity U_0 in a turbulent diffusion flame—stagnant surroundings

by finite chemical kinetic rates, however, in turbulent diffusion flames, the overlap is caused by turbulence. Measured profiles of time-averaged mass fractions, therefore, can be misleading because they give the impression that different mass fractions of fuel and oxygen at a given point are present at the *same time*. In fact, in the presence of turbulence, the fuel consumption rates actually experienced are not as high as those estimated from time-averaged mass fractions from $R_{fu} \propto \overline{\omega}^x_{fu} \overline{\omega}^y_{ox}$. This is because the fuel and oxidant at a point are present at *different times*. Thus, fluctuations in mass fractions cause the instantaneous value of one reactant (say, fuel) mass fraction at one instant to be in excess of stoichiometric requirement and the other reactant (air) to be in excess at another instant at the *same point*. A theoretical estimate of L_f, therefore, requires that this probabilistic phenomenon must be properly accounted for.[12]

Because of the assumption of radially independent effective viscosity, Eq. 9.37 and its solution, Eq. 9.50 or 9.71, are again applicable. Here, the solution for the time-averaged quantities is written by combining Eq. 9.71 with Eqs. 9.66 and 9.68, to read as

$$\Psi = \frac{u}{U_0} = f = h^* = 6.57 \left(\frac{D}{x}\right) \left(\frac{\rho_0}{\rho_m}\right) \left[1 + 57.6 \left(\frac{r}{x}\right)^2\right]^{-2}. \qquad (9.72)$$

---

[12]Complete treatment of this phenomenon requires construction of transport equations for scalar fluctuations $\overline{(T')^2}$, $\overline{(\omega'_k)^2}$ and $\overline{(f')^2}$. These equations comprise convection, diffusion, and source terms. The sources comprise *production* and *dissipation* of these scalar fluctuations. Spalding [111], however, argued that the fuel consumption rate R_{fu} is related to the process of dissipation of mass fractions and proposed that $R_{fu} \propto -\rho_m \left(\epsilon/k\right) \sqrt{\overline{(\omega'_{fu})^2}}$. This is called the *Eddy break-up* (EBU) model. Other approaches for dealing with this probabilistic phenomenon are discussed by Kuo [68].

Accounting for Fluctuations

Suppose that the value of the instantaneous mixture fraction f truly fluctuates between an instantaneous low value $\hat{f}_l$ and a high value $\hat{f}_h$. The manner of variation of f with time between these two extremes is unknown, so an approximation is required. The simplest assumption is that the fluid spends *equal* times at the two extremes and sharply moves from one extreme to the other (that is, without spending any time to move from one extreme to the other). Then, this time variation will be as shown in Fig. 9.5(a). From this figure, we deduce that the time-averaged $(\overline{f})$ and fluctuating mixture fractions (f') will be given by

$$\overline{f} = \frac{1}{2}(\hat{f}_h + \hat{f}_l) \quad \text{and} \quad f' = \frac{1}{2}(\hat{f}_h - \hat{f}_l). \tag{9.73}$$

In Fig. 9.5(b), variations of ω_{fu}, ω_{ox}, and T are plotted with f. A few observations are pertinent.

1. Time averaged $\overline{\omega}_{fu} = 0.5\,(\hat{\omega}_{fu,l} + \hat{\omega}_{fu,h})$ (shown by by filled circle) is greater than ω_{fu} (shown by open circle) corresponding to time-averaged $\overline{f} > f_{stoich}$.
2. Likewise, time averaged $\overline{\omega}_{ox} = 0.5\,(\hat{\omega}_{ox,l} + \hat{\omega}_{ox,h})$ also exceeds ω_{ox} (which is zero) corresponding to time-averaged $\overline{f} > f_{stoich}$.
3. Time averaged $\overline{T} = 0.5\,(\hat{T}_l + \hat{T}_h)$ is *lower* than T corresponding to time-averaged $\overline{f} > f_{stoich}$.
4. The above observations will also apply when $\overline{f} < f_{stoich}$.
5. If f_{stoich} does not lie between $\hat{f}_l$ and $\hat{f}_h$ then $\overline{T}$, $\overline{\omega}_{ox}$ and $\overline{\omega}_{fu}$ will of course correspond to $\overline{f}$.

Figure 9.5(b) thus explains why it is possible to find finite amounts of fuel and oxidant when $\overline{f} = f_{stoich}$ in a turbulent flame. The flame shape can be visualized as shown in Fig. 9.6. The reaction zone will be a finite volume enclosed by the $\hat{f}_l = \overline{f} - f' = f_{stoich}$ and $\hat{f}_h = \overline{f} + f' = f_{stoich}$ surfaces. The latter will form the *outer* surface because the mixture there is mainly fuel-lean, $(\omega_{fu} < \overline{\omega}_{fu})$ but the reaction itself occurs only when mixture has its maximum fuel content. The true $\overline{f} = f_{stoich}$ surface will reside between the two surfaces.

To determine outer $(r_{f,out})$ and inner $(r_{f,in})$ flame radii, we recast Eq. 9.72 and insert appropriate values of f. Thus,

$$\frac{r_{f,out}}{x} = \left[\frac{1}{57.6} \left\{ \sqrt{\left(\frac{6.57}{f_{stoich} - f'} \right) \left(\frac{D}{x} \right) \left(\frac{\rho_0}{\rho_m} \right)} - 1 \right\} \right]^{0.5}, \tag{9.74}$$

$$\frac{r_{f,in}}{x} = \left[\frac{1}{57.6} \left\{ \sqrt{\left(\frac{6.57}{f_{stoich} + f'} \right) \left(\frac{D}{x} \right) \left(\frac{\rho_0}{\rho_m} \right)} - 1 \right\} \right]^{0.5}, \tag{9.75}$$

$$\frac{r_{f,stoich}}{x} = \left[\frac{1}{57.6} \left\{ \sqrt{\left(\frac{6.57}{f_{stoich}} \right) \left(\frac{D}{x} \right) \left(\frac{\rho_0}{\rho_m} \right)} - 1 \right\} \right]^{0.5}. \tag{9.76}$$

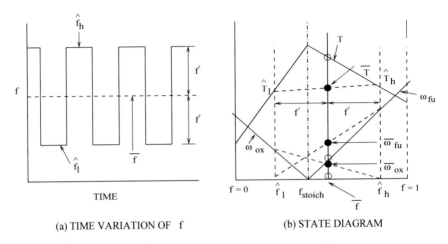

(a) TIME VARIATION OF f   (b) STATE DIAGRAM

Fig. 9.5 Accounting for turbulent fluctuations in a turbulent diffusion flame

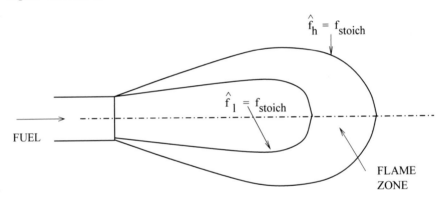

Fig. 9.6 Flame zone a turbulent diffusion flame—stagnant surroundings

The last equation conforms with the definition of r_f used for a laminar flame (Eq. 9.52). To deduce the outer and inner radii, f' corresponding to $r_{f,stoich}$ must be known. Spalding [113] recommends

$$f' \simeq l_m \times |\frac{\partial \overline{f}}{\partial r}|_{stoich} = (0.1875 \times r_{1/2}) \times |\frac{\partial \overline{f}}{\partial r}|_{stoich}$$

$$= 24.07 \left(\frac{\rho_0}{\rho_m}\right) \left(\frac{D}{x}\right) \left(\frac{r_{f,stoich}}{x}\right) \left[1 + 57.6 \left(\frac{r_{f,stoich}}{x}\right)^2\right]^{-3} \quad (9.77)$$

where the mixture-fraction gradient is evaluated from Eq. 9.72 and the mixing length for the round jet is specified using Eq. 6.91. Finally, the flame lengths can be determined by setting $r_f = 0$ in Eqs. 9.74, 9.75 and 9.76. Thus,

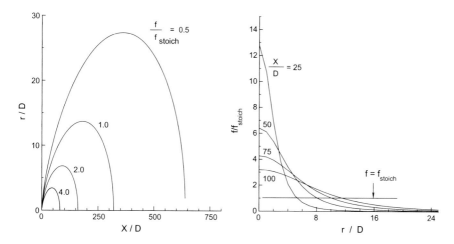

Fig. 9.7 Profiles of f for CH₄ ($f_{stoich} = 0.05507$) turbulent diffusion flame—stagnant surroundings. ($T_o = 300$ K, $T_m = 1500$ K)

$$\frac{L_{f,out}}{D} = \frac{6.57}{f_{stoich} - f'}\left(\frac{\rho_0}{\rho_m}\right) \tag{9.78}$$

$$\frac{L_{f,in}}{D} = \frac{6.57}{f_{stoich} + f'}\left(\frac{\rho_0}{\rho_m}\right) \tag{9.79}$$

$$\frac{L_{f,stoich}}{D} = \frac{6.57}{f_{stoich}}\left(\frac{\rho_0}{\rho_m}\right) = 6.57\,(1 + R_{st})\left(\frac{\rho_0}{\rho_m}\right). \tag{9.80}$$

Notwithstanding the fact that flame lengths are difficult to determine exactly, the last equation nonetheless confirms the experimentally observed dependence of L_f on f_{stoich} or R_{st} and D; no dependence on Re (or, U_0) as observed in Fig. 9.4. Some experimental correlations demonstrate that L_f is influenced by ρ_0/ρ_∞ as well as by ρ_0/ρ_m (see next subsection)

Figure 9.7 shows profiles of the mixture fraction f evaluated according to Eq. 9.72. The fuel is methane (CH₄, $f_{stoich} = (1 + R_{st})^{-1} = 0.055$). The figure on the left shows that the profiles are bulgy because property variations are not accounted for. If $f = f_{stoich}$ is regarded as the indicator of flame length, then $L_f/D \simeq 330$ for methane and its radius at x /d = 50, for example, will be $r_f/D \simeq 8$, as seen from the figure on the right. This implies that $L_f/r_f(x/D = 50) \simeq 41$, compared with $L_f/r_{f,max} = Re/3$ in a laminar flame (see Eq. 9.55).

Figure 9.8 shows the variations of inner, stoichiometric, and outer radii, evaluated using Eqs. 9.74, 9.75, and 9.76 in which f' is evaluated from Eq. 9.77. Over the greater part of the flame length, $f' \simeq 0.2\,f_{stoich}$. This figure confirms our expectation, as shown in Fig. 9.6.

Fig. 9.8 Variation of $r_{f,in}$, $r_{f,out}$ and $r_{f,stoich}$ for CH$_4$ ($f_{stoich} = 0.05507$) turbulent diffusion flame—stagnant surroundings. ($T_o = 300$ K, $T_m = 1500$ K)

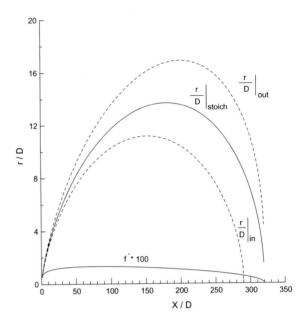

9.3.3 Correlations for L_f

Several authors (see, for example, Refs. [8, 18, 32]) have proposed correlations for L_f of turbulent diffusion flames with and without the presence of coflowing streams. Based on the experimental data[13] of Delichatsois [32], Turns [122] recommends the following correlations for stagnant surroundings ($U_\infty = 0$),

$$\frac{L_f}{D} = \left[\frac{23}{f_{stoich}}\right]\left[\frac{\rho_0}{\rho_\infty}\right]^{0.5} \quad Fr \geq 5 \tag{9.81}$$

$$= \left[\frac{13.5}{f_{stoich}}\right]\left[\frac{\rho_0}{\rho_\infty}\right]^{0.5}\left[\frac{Fr^{0.4}}{(1 + 0.07\,Fr^2)^{0.2}}\right] \quad Fr < 5\,, \tag{9.82}$$

where, for turbulent flames, the Froude number is defined as

$$Fr \equiv \left[\frac{U_0^2}{gD}\right]^{0.5}\left[\frac{f_{stoich}^3}{(T_f/T_\infty - 1)}\right]^{0.5}\left[\frac{\rho_\infty}{\rho_0}\right]^{0.25}. \tag{9.83}$$

Here, $T_f \simeq T_{ad}$.

A correlation that is valid for subsonic jets, for all Froude numbers and presence of a coflowing airstream is given by Kazantsev [63]. In this correlation, L_f is obtained from solution of a transcendental equation

---

[13] Fuels used were hydrogen, methane, and propane, and lengths were estimated from video frames.

$$LR^5 - K_t (1 - LR^2) = 0, \qquad (9.84)$$

where

$$LR = \frac{L_f}{L_{f,mom}} \qquad (9.85)$$

$$K_t = \left(\frac{1 + D_1}{D_2}\right) \left(\frac{U_0^2}{g D}\right) \left[\frac{L_f}{D}\Big|_{mom}\right]^{-3}, \qquad (9.86)$$

$$\frac{L_f}{D}\Big|_{mom} = \left[\frac{11.2}{f_{stoich}} - 2.14\right] \left[\frac{\rho_0/\rho_m}{1 + D_1}\right]^{0.5} \qquad (9.87)$$

$$D_1 = R_{st} \left[\frac{U_\infty}{U_0}\right], \qquad (9.88)$$

$$D_2 = 14.8 \times 10^{-4} \left(\frac{\rho_\infty}{\rho_0}\right) \left(\frac{\rho_\infty}{\rho_m} - 1\right), \quad \text{and} \qquad (9.89)$$

$$\frac{\rho_0}{\rho_m} = \frac{0.2304 \, \rho_0 \, R_{st} \, (h_0 + R_{st} \, h_\infty + \Delta H_c)}{[11.2/f_{stoich} - 2.14]^2}. \qquad (9.90)$$

In Eq. 9.90, the sensible enthalpies[14] and the heat of combustion are evaluated in kJ/kg. Kazantsev [63] shows that when the Froude number is large (momentum-controlled $K_t > 20$), $LR \to 1$. Similarly, if the surroundings are stagnant, $D_1 = 0$.

Sometimes, external air flow is directed at an angle θ with respect to the jet axis. Then, Kazantsev [63] recommends that L_f predicted by Eq. 9.84 be multiplied by $[1 + D_1 (1 - \cos \theta)]^{-0.5}$.

9.3.4 Correlations for Liftoff and Blowout

As mentioned earlier, when the jet velocity is sufficiently high, the base of the flame is separated from the nozzle; the separation distance being occupied by a large number of hole-like structures. This separation distance is called the *liftoff height*, H_{lift}. With a further increase in velocity, the flame detaches and blows out.

Based on the dimensional analysis of the liftoff phenomenon in CH_4, C_3H_8, and C_2H_6 flames, Kalghatgi [57] proposes a correlation for H_{lift}. Experimental data on flames of several hydrocarbons were considered to derive a curve-fit correlation, which reads as follows:

$$\frac{S_{l,max} \, H_{lift}}{\nu_0} = 50 \left(\frac{U_0}{S_{l,max}}\right) \left(\frac{\rho_0}{\rho_\infty}\right)^{1.5}, \qquad (9.91)$$

where $S_{l,max}$ occurs at an equivalence ratio slightly greater than 1 (see Chap. 8).

---

[14]If $T_0 = T_\infty = T_{ref}$ then $h_0 = h_\infty = 0$.

The blowout condition is characterized by height H_{blow}, which is defined as

$$\frac{H_{blow}}{D} = 4 \left[\frac{f_0}{f_{stoich}} \left(\frac{\rho_0}{\rho_\infty} \right)^{0.5} - 5.8 \right]. \tag{9.92}$$

Then, defining $Re_{blow} \equiv S_{l,max} \times H_{blow}/\nu_0$, Kalghatgi [57] proposes following correlation for blowout velocity.

$$\frac{U_{0,blow}}{S_{l,max}} = 0.017 \, Re_{blow} \left[1 - 3.5 \times 10^{-6} \, Re_{blow} \right] \left(\frac{\rho_\infty}{\rho_0} \right)^{1.5}. \tag{9.93}$$

The theory for liftoff and blowout can be developed from several standpoints. Pitts [92] has provided a simple theory based on experimental data for non-reacting jet flows that predicts the blowout experimental data well. It is of interest to determine conditions under which the *spark-ignition* of lifted turbulent jet flames is successful. Such probable conditions have been determined in Ref. [2].

9.4 Solved Problems

Problem 9.3 In Problem 9.1, let the nozzle diameter be 6.5 mm and uniform velocity $U_0 = 70.0$ m/s. All other conditions are the same. Estimate L_f for methane and propane using Eqs. 9.80 and the correlations by Delichatsios [32] and Kazantsev [63].

For **methane**, $\rho_0 = 1.01324 \times 10^5/(8314/16)/300 = 0.65$ kg/m^3, $\rho_\infty = 0.65 \times 29/16 = 1.178$ kg/m^3 and corresponding to $T_f = 1500$ K, $\rho_m \simeq \rho_{air} = 1.01324 \times 10^5/(8314/29)/1500 = 0.2356$ kg/m^3. Also $R_{st} = 17.16$ and $f_{stoich} = 1/18.16 = 0.055$. Then, using Eq. 9.80,

$$L_{f,stoich} = 0.0065 \times 6.57 \, (1 + 17.16) \left(\frac{0.65}{0.2356} \right) = 2.139 \text{ m} = 2139 \text{ mm}$$

To use correlation from [32], we first evaluate the Froude number with $T_f = T_{ad} = 2226$ K.

$$Fr = \left[\frac{70^2}{9.81 \times 0.0065} \right]^{0.5} \left[\frac{0.055^3}{(2226/300 - 1)} \right]^{0.5} \left[\frac{1.178}{0.65} \right]^{0.25} = 1.637 < 5.$$

Hence, using Eq. 9.82,

$$L_f = 0.0065 \times \left[\frac{13.5}{0.055} \right] \left[\frac{0.65}{1.178} \right]^{0.5} \left[\frac{1.637^{0.4}}{(1 + 0.07 \times 1.637^2)^{0.2}} \right] = 1.395 \text{ m} = 1395 \text{ mm}.$$

Finally, using Eq. 9.84 with $U_\infty = 0$ (or, $D = 0$), $h_0 = h_\infty = 0$, and $\Delta H_c = 50.016 \times 10^3$ kJ/kg,

$$\frac{\rho_0}{\rho_m} = \frac{0.2304 \times 0.65 \times 17.16 \times 50.016 \times 10^3)}{[11.2\,(1 + 17.16) - 2.14]^2} = 3.1735 .$$

$$D_2 = 14.8 \times 10^{-4} \left(\frac{1.178}{0.65}\right) \left(\frac{1.178}{0.65} \times 3.1735 - 1\right) = 0.01274 .$$

$$\frac{L_f}{D}\bigg|_{mom} = \left[\frac{11.2}{f_{stoich}} - 2.14\right](3.1735)^{0.5} = 358.92 .$$

$$K_t = \frac{1}{0.01274} \times \left(\frac{70^2}{9.81 \times 0.0065}\right)[358.92]^{-3} = 0.13 .$$

Because $K_t \to 0$ (very small), the flame is dominated by buoyancy. Solving the transcendental Eq. 9.84, we have $LR \simeq 0.6067$, or

$$L_f = D \times LR \times \frac{L_f}{D}\bigg|_{mom} = 0.0065 \times 0.6067 \times 358.92 = 1.415\,\text{m} = 1415\,\text{mm}$$

Similar calculations for **propane** ($\Delta H_c = 46.35$ MJ, $T_{ad} = 2267$ K, $R_{st} = 15.6$, $\rho_0 = 1.787$ kg/m^3, L_f/D (mom) = 545.66, $K_t = 0.1036$, LR = 0.5845) have been carried out, and results shown in Table 9.2. The table shows that the estimates from Eq. 9.80 and the two correlations differ. This only further confirms the experimental difficulties associated with estimation of correct flame length.

Problem 9.4 A turbulent diffusion flame is formed while burning propane at 300 K in stagnant environment at 1 atm and 300 K . The nozzle diameter is 10 mm. Estimate the blowout velocity for this flame.

For propane $\rho_0 = 1.787$ kg/m^3 and $\mu_0 = 8.256 \times 10^{-6}$ kg/m-s (from Eq. C.2 and Table C.3). Also, $f_0 = 1$ and $f_{stoich} = 1/16.6 = 0.06024$. Hence, from Eq. 9.92, the characteristic height is

$$H_{blow} = 4 \times 0.01 \times \left[\left(\frac{1}{0.06024}\right) \times \left(\frac{1.787}{1.178}\right)^{0.5} - 5.8\right] = 0.586\,\text{m}$$

Table 9.2 Predicted L_f (mm) for Problem 9.3

Equations	Methane	Propane
(9.80)	2139	3558
(9.82)	1395	2170
(9.84)	1415	2073

From Chap. 8, correlation Eq. 8.33 shows that for propane, $S_{l,max} \simeq 34.22$ cm/s. Then,

$$Re_{blow} = 0.3422 \times 0.586 \times \frac{1.787}{8.256 \times 10^{-6}} = 4.355 \times 10^4$$

Hence, from Eq. 9.93, ,

$$U_{0,blow} = 0.3422 \times 0.017 \times 43550 \times \left[1 - 3.5 \times 10^{-6} \times 43550\right] \left(\frac{1.178}{1.787}\right)^{1.5}$$

$$= 174.3 \, \text{m/s}.$$

Equation 9.93 shows that $U_{0,blow}$ increases with diameter. So, flames from large diameter burners are difficult to blow out.

9.5 Burner Design

Gas burners are designed to obtain stable clean blue flame without flashback and liftoff. This requires that the *Port-Loading* L_{port} (W/mm^2) must be properly constrained over a range of values. Figure 9.9 shows a schematic of a typical domestic burner [53]. Burners are designed to deliver a specified heating power (P = $\dot{m}_{fu} \times \Delta H_c$) with a specified flame height (L_f).

Fuel gas, under pressure, issues from the fuel manifold at the rate $\dot{m}_{fu}$ through an orifice (diameter d_{orf}) into a convergent–divergent mixing tube with a throat (diameter d_{th}). Because of high fuel velocity, primary air $\dot{m}_{a,pr}$ is entrained into the tube. Thus, a rich premixed fuel + air mixture issues through the burner port (diameter d_{port}) with a unburned velocity V_u and equivalence ratio $\Phi > 1$. The flame is formed to the right of the port where secondary air $\dot{m}_{a,sec}$ is also drawn. Burner flames are thus partly premixed, and partly, diffusion flames. For safe and stable operation, the various parameters are selected in the ranges given below.

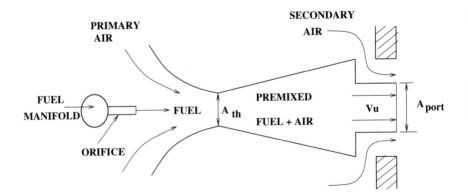

Fig. 9.9 Schematic of a domestic gas burner

$$\left(6.8 < L_{port} \left(\frac{W}{mm^2} \right) < 12.7 \right) , \qquad \left(2 < \frac{V_u}{S_l} < 6 \right) ,$$

$$\left(1.25 < \frac{A_{port}}{A_{th}} < 1.5 \right) , \qquad \left(50 < \frac{A_{th}}{A_{orf}} < 100 \right) .$$

Following definitions are used in design.

$$P = \dot{m}_{fu} \times \Delta H_c \ (\Phi = 1) , \tag{9.94}$$

$$L_{port} = \frac{P}{A_{port}} , \tag{9.95}$$

$$\dot{m}_{a,stoich} = \dot{m}_{fu} \times \left(\frac{A}{F} \right)_{stoich} , \tag{9.96}$$

$$\beta_{a,pr} = \frac{\dot{m}_{a,pr}}{\dot{m}_{a,stoich}} = \frac{\dot{n}_{a,pr}}{\dot{n}_{a,stoich}} = \frac{1}{\Phi} , \tag{9.97}$$

where typically, $0.4 < \beta_{a,pr} < 0.6$, and $\beta_{a,sec} = 1 - \beta_{a,pr}$. $\beta_{a,pr}$ is called the *primary aeration*.

Further, when flame length is predicted by using correlations in Table 9.1, effective molar stoichiometric ratio S_{eff} must be used in place of S. Thus,

$$S_{eff} = \frac{\dot{n}_{a,sec}}{\dot{n}_{fu} + \dot{n}_{a,pr}}$$

$$= \frac{S \times (1 - \beta_{a,pr})}{1 + S \, \beta_{a,pr}} \tag{9.98}$$

$$= \frac{S \times (1 - 1/\Phi)}{1 + S/\Phi} , \tag{9.99}$$

where S is defined in Eq. 9.61. Because $\beta_{a,pr} < 1$ (or, $\Phi > 1$) and $S \gg 1$ for hydrocarbons, $S_{eff} \ll S$ and $S_{eff} \rightarrow (\Phi - 1)$.

Exercises

1. Show that Eqs. 9.9 satisfy the continuity Eq. 9.6.
2. Derive Eqs. 9.10 to 9.16.
3. Show that for constant properties, and Le $= 1$, Eq. 9.37 is identical to Equation 9.7 when $\Psi = u/U_0$. Hence, using Eq. 9.29, derive Eqs. 9.50 and 9.52.
4. Show that for a turbulent round jet, Eqs. 9.70 and 9.71 are valid.
5. Using Eq. 9.71 determine $\dot{m}_{jet}$ for a turbulent round jet. Hence, show that

$$\dot{m}_{jet} = 8 \pi \mu_{eff} x$$

$$\frac{d \dot{m}_{jet}}{dx} = 0.404 \sqrt{\rho_m J}$$

$$\frac{v_{ent}}{U_0} = - \left(\frac{0.865}{\eta} \right) \left(\frac{\rho_0}{\rho_m} \right)^{0.5} \left(\frac{D}{x} \right).$$

6. A laminar jet diffusion flame of ethylene (C_2H_4) is formed from a 10 mm diameter nozzle. The jet velocity is 12 cm/s. Both the fuel and the stagnant surroundings are at 300 K and 1 atm. Using Roper's correlations in Table 9.1, estimate the value of L_f/D. Hence, using Eqs. 9.54 and 9.55, estimate $r_{f,max}$ for the jet and the radial profiles of u, T, and ω_{fu} at the axial location where $r_{f,max}$ occurs. Also, estimate the cone angle α of the jet. Take $cp_{mix} = 1230$ J/kg-K.

7. In the previous problem, if the jet velocity is reduced to 6 cm/s, what should be the nozzle diameter so that $r_{f,max}$ occurs at the same axial location as before? What will be the value of new $r_{f,max}$?

8. In Problem 9.1 in the text, if, instead of pure air, the ambient fluid was pure oxygen, what will be the value of S? Hence, determine the flame lengths using Roper's correlations and comment on the result. Also determine the heat release rate.

9. A burner with N holes of diameter d_{port} is to be designed to deliver $P = 50$ kW and flame length $L_f = 50$ mm. The fuel is propane (C_3H_8). The fuel and primary air ($\Phi = 1.525$) issue out with velocity V_u, at temperature $T_u = 300$ K and pressure $p = 1$ atm. It is assumed that the thermal port loading $L_{port} = 12$ W/mm^2 port area, and $V_u/S_l = 4$. Also, the inter-port spacing is assumed to be sufficiently high, so each port delivers a standalone flame. Determine

   (a) $\dot{m}_{fu}$ and $\dot{m}_{a,sec}$ and S_{eff} for the burner,
   (b) N and d_{port},
   (c) If $A_{port}/A_{th} = 1.4$ and $A_{th}/A_{orf} = 70$, estimate d_{th} and d_{orf},
   (d) Minimum inter-port spacing and $Re_{port} = V_u\, d_{port}/v_u$.
       (Hint: Use Roper's correlation for a circular port $L_{f,exp}$ from Table 9.1 and Eq. 9.55 to evaluate $r_{f,max}$.)

10. Consider turbulent diffusion flame of hydrogen burning in stagnant air. Both fuel and air are at 300 K. The nozzle diameter is 1 mm. What is the minimum fuel velocity required to achieve momentum-dominated flame? Estimate the flame length for the minimum velocity. Also, determine the heat release rate.

11. Using the minimum velocity U_0 determined in the previous problem, estimate the flame length and the heat release rate with and without coflowing air velocity— $U_\infty = 0.4 \times U_0$. Use correlations due to Kazantsev [63].

12. An ethane–air turbulent jet flame issues from a 4 mm diameter nozzle with $U_0 = 55$ m/s. Both fuel and stagnant air are at 300 K and 1 atm. Estimate H_{lift}/L_f, H_{blow}, and U_{blow} for this flame. Take $\mu_0 = 9.7 \times 10^{-6}$ kg-m/s. (Hint: You need to determine $S_{l,max}$ using correlation Eq. 8.35).

13. Using computed profiles for methane shown in Fig. 9.7, estimate L_f/r_f at (x/D) $= 25, 75$, and 100 as well as location of $L_f/r_{f,max}$. Comment on the result.

Chapter 10
Combustion of Particles and Droplets

10.1 Introduction

So far we have been concerned with the combustion of gaseous fuels. However, solid (coal, wood, agricultural residues, or metals) and liquid (diesel, kerosene, ethanol, etc.) fuels are routinely used in a variety of applications. Such fuels burn at a relatively slow flux rate, expressed in kg/m^2-s. To achieve high heat release rates (J/s), it becomes necessary to increase the surface area of the fuel. This is achieved by atomizing liquid fuels and by pulverizing solid fuels. Liquid fuel atomization is achieved by employing high-pressure fuel injectors. Pulverized solid fuel is achieved by crushing/grinding operations to obtain tiny particles (typical diameter $<250\,\mu m$, in the case of coal). This operation becomes increasingly expensive as the size of the particles is reduced.

Solid fuels are burned in a wide variety of ways. The most primitive systems burn fuels (typically wooden logs) by placing them on the ground and replacing (or feeding) fuel intermittently in a batch mode. More advanced versions include a fixed grate, as shown in Fig. 10.1(a). The combustion air is drawn in by natural convection and the airflow rate is controlled by dampers. Such systems are used in small furnaces or cookstoves. These systems were considerably improved in power generation applications by having a fixed (sloping) or traveling endless *grate* on which solid fuel is fed continuously via a *stoker* and the combustion air is supplied by means of a fan/blower. There are many types of grate-and-stoker arrangements. Figure 10.1(b) shows the schematic of a traveling grate over-feed spreader stoker. In this type, coal particle sizes range up to 4 to 5 cm. The grate speed is adjusted so the deposited coal (10 to 20 cm thick) burns fully by the end of its travel. Because the fuel bed is relatively thin, the plenum air easily passes through the fuel bed with a very small pressure drop.[1] This grate is also suitable for wood chips, agro-residues, and the like.

---

[1] Very fine coal particles are, in fact, carried away upward and burnt in suspension.

© Springer Nature Singapore Pte Ltd. 2020
A. W. Date, *Analytic Combustion*,
https://doi.org/10.1007/978-981-15-1853-9_10

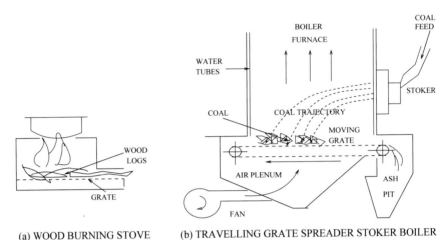

(a) WOOD BURNING STOVE   (b) TRAVELLING GRATE SPREADER STOKER BOILER

Fig. 10.1 Solid fuel burning

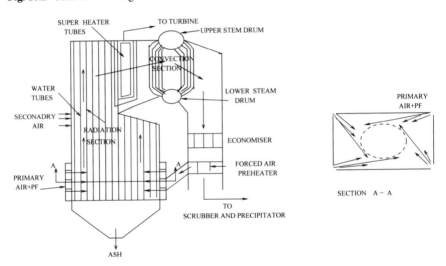

Fig. 10.2 Schematic of a pulverized fuel furnace

In contrast to the fixed-bed combustion, in large modern power stations (>100 MW), solid fuels are burned in suspension. This is done in two ways: (a) pulverized fuel combustion (PFC) and (b) fluidized bed combustion (FBC).

Figure 10.2 shows the schematic of a typical PFC furnace (approximately 50 to 60 m high and 10 to 15 m width and breadth). Preheated air and pulverized coal are injected/blown through a battery of burners/nozzles. These nozzles are located in a variety of ways; the most common location is at the corners of the furnace as shown (see section A-A). The line of injection is tangent to a circle at the center of the furnace. Such tangential injection sets up a vortex, which causes abundant

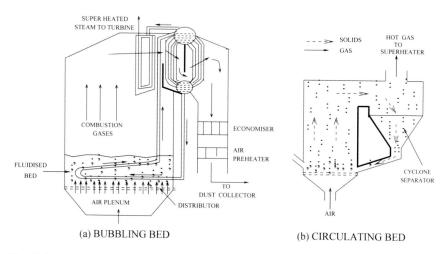

Fig. 10.3 Schematic of a fluidized bed furnace

mixing between fuel and air. Above the primary air + fuel injection, secondary air is often added to reduce NOx emissions. Thus, whereas the hot combustion products travel upward, the ash falls to the bottom, from which it is removed intermittently. The heat of combustion is transmitted by convection and radiation to the water tubes lining the walls of the furnace as well as to the superheater tubes, whereas ash is collected intermittently from the bottom hopper. The burned products (called flue gases) are led through the economizer (for boiler feedwater heating), preheater (for air preheating), wet scrubber, baghouse (for removal of larger particles), and electrostatic precipitator (for removal of fine particulates). Finally, the gases are exhausted through the chimney.

Fluidized bed combustion is a relatively new technology. This technology is particularly suited for coals with high ash and moisture content, and controls SOx and NOx emissions more effectively than PF combustion technology. Figure 10.3(a) shows the schematic of a fluidized bed furnace. In this, coal particles (about 2 to 4 cm) are continuously fed (from the top) on a distributor plate. The holes in the plate are in the form of nozzles through which high-velocity air is blown. The velocity is just high enough to suspend the particles, but not so high that it blows the particles out of the bed. When the bed is sustained in this way, it takes the appearance[2] of a fluid with a wavy free surface. Within the bed, the particles are continuously churned, so the bed appears to be bubbling. During combustion, the bed temperature is considerably low and uniform (typically between 800 and 950 °C); thus avoiding peak temperatures. This results in lower NOx emissions and prevents ash agglomeration and slagging, as well as achieving reduced corrosion. The bed can also be mixed with limestone to reduce SOx emissions. The disadvantage is that there is considerably higher ash

---

[2]The bed also assumes many properties of a real fluid. For example, when tilted, the free surface of the bed remains horizontal.

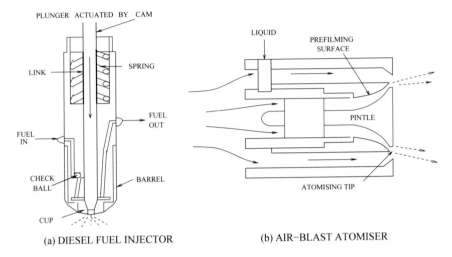

(a) DIESEL FUEL INJECTOR   (b) AIR–BLAST ATOMISER

Fig. 10.4 Fuel atomization

carryover in the free-board region above the bed. The very fine coal particles also burn in the free-board region.[3] The water tubes embedded in the bed pick up the heat of combustion to form steam. Because of the bubbling nature of the bed, the heat transfer coefficient between the bed and the tubes is also very high. However, the continuous bombardment of particles may also cause tube erosion, requiring frequent replacement.

Liquid fuels again require atomization to achieve enhanced surface area. Such atomized fuel is burned in diesel engines, gas turbines, oil-fired furnaces, and rocket motor combustors [83]. The atomized fuel in the form of droplets (typically $<70\,\mu m$) first vaporize and then burn on mixing with the surrounding gas. Figure 10.4 shows the two commonly used injectors for fuel atomization. In a diesel engine, the fuel is injected just before the end of the compression stroke when the cylinder pressure is high. As such, the fuel is injected under pressure (typically $>300\,atm$) by means of a plunger through tiny holes, as shown.[4] In oil-fired furnaces and gas turbines, the pressure is low; therefore, air-blast-type injectors suffice. For low-viscosity fuels, the liquid film becomes unstable under the shear of airflow past a curved surface.

---

[3]Largely, to prevent this carryover burning and the associated loss of unburned carbon, an alternative concept of a circulating fluidized bed is now practiced. In this, the air velocity through the distributor is high enough to physically transport the particles upward to the full chamber height, as shown in Fig. 10.3(b). The particles are then returned via the downward leg of a cyclone separator. This type of particle burning achieves high mixing rates and there is no clearly defined bed. Also, there are no embedded tubes in the combustion zone. The burned gases are led directly to the superheater tubes and the convection section. The circulating fluidized bed is particularly suitable for high-ash-content coals and other coarse solid fuels.

[4]Fuel is also atomized in spark ignition engines by using a carburetor. However, the atomization is achieved by air turbulence near the point of fuel entrainment, rather than by a large pressure difference across a nozzle.

This forms liquid ligaments which break up farther downstream; thus giving larger drop sizes [74]. Sometimes, swirling co-flowing air is also provided to enhance shear in the after-injector region. When fuel atomization is carried out, not only the droplet diameter but penetration and dispersion (that is, number density and mass distribution) are important.

In the sections to follow, we develop simple expressions for droplet evaporation and solid particle drying times, as well as for their burning times.

10.2 Governing Equations

Droplet evaporation, particle drying, and burning are mass transfer phenomena. In different situations, mass transfer problems can be classified into three types:

1. Inert mass transfer without heat transfer.
2. Inert mass transfer with heat transfer.
3. Heat and mass transfer with chemical reaction.

In the first two types, which are inert, the mass generation rate $R_k = 0$ for all species involved. A typical example of the first type is evaporation of water (or any liquid) in air, when both the air and water are at the same temperature. As such, heat transfer between air and water is absent, but mass transfer nonetheless takes place because of the difference in mass fractions of water vapor in water and air. In the second type, the water temperature may be higher or lower than the air temperature and, mass transfer (evaporation) will be accompanied by heat transfer. Mass transfer between air and water is particularly important in air-conditioning applications.[5]

The third type involves a chemical reaction and, therefore, directly relevant to combustion. The process of combustion of liquid and solid fuels is different, however. In liquid burning, the fuel substance is simply converted to vapor by latent heat transfer *without a change in chemical composition*. The fuel vapor substance then burns in the gas phase. In the gas phase, chemical reactions are called *homogeneous* reactions of the type already considered when dealing with gaseous fuel combustion. On the other hand, solid fuels comprise volatiles (essentially, mixture of hydrocarbons, CO, H_2, HCN, hydroxyls, etc.), char (essentially, carbon), and ash (mineral matter). As such, after initiation of combustion, volatiles evolve quickly from the solid surface. This evolution is assisted by heat transfer from the burning gas phase. The volatiles burn in the gaseous phase through homogeneous reactions. In contrast, the char is left behind and burns at the solid surface. This surface burning occurs through *heterogeneous* chemical reactions.[6] The products of these heterogeneous reactions again yield a combustible gas (comprising mainly CO and H_2) that

---

[5]The science of psychrometry deals with the properties of air–water mixtures.

[6]There are many situations, other than combustion involving heterogeneous reactions. For example, in catalysis, one species from the gaseous phase is preferentially adsorbed at the surface via a heterogeneous reaction. Surfaces with catalytic agents are often used in pollution abatement and other applications.

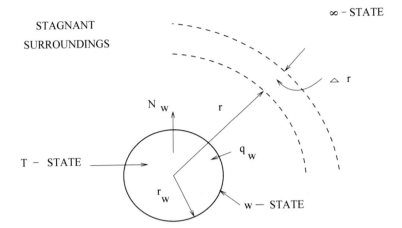

Fig. 10.5 Radial mass transfer from spherical droplet/particle

burns through homogeneous reactions. Thus, whereas liquid fuels burn principally through homogeneous reactions coupled with latent heat transfer, solid fuels burn with a combination of heterogeneous and homogeneous reactions.

We consider any of the aforementioned situations to occur in stagnant gaseous surroundings. We further assume that the particle/droplet is perfectly spherical. As such, the Stefan flow model (see Chap. 6) must be developed in spherical coordinates in which only radial heat/mass transfer is permitted. Figure 10.5 shows the model situation. The total solid/liquid mass transfer flux N_w is radially outward. Following mass transfer, the particle/droplet radius will diminish from its original value $r_{w,i}$ to zero. As such, the domain length $(r_\infty - r_w)$ will be a function of time. But, since the initial radius $r_{w,i} \ll r_\infty$, we may assume that the domain length remains constant; hence, we have a (quasi-) steady-state mass transfer situation. Thus, the instantaneous mass transfer rate $\dot{m}_w$ (kg/s) will be given by

$$\frac{d}{dt}\left[\frac{4\,\pi}{3}\,\rho_p\,r_w^3\right] = 4\,\pi\,\rho_p\,r_w^2\,\frac{d\,r_w}{dt} = -\dot{m}_w \;, \tag{10.1}$$

where ρ_p is the particle or droplet density.

The main task now is to determine $\dot{m}_w$. To do this, we set up the mass transfer equation. Thus, for any species k, consider the spherical shell of length Δr located at radius r from the center of the particle/droplet. Then, the mass balance over volume $\Delta V = 4\,\pi\,r^2\,\Delta r$ will give

$$N_k\,4\,\pi\,r^2\,|_r - N_k\,4\,\pi\,r^2\,|_{r+\Delta r} + R_k\,4\,\pi\,r^2\,\Delta r = 0 \;,$$

or, dividing by ΔV and letting $\Delta r \to 0$,

$$\frac{1}{4\,\pi\,r^2}\,\frac{d}{dr}\left[N_k\,4\,\pi\,r^2\right] = R_k \;, \tag{10.2}$$

where the species radial mass flux N_k with convecting radial velocity V_r is given by

$$N_k = \rho_k \, V_r + m_k'' = \rho_m \, V_r \, \omega_k - \rho_m \, D \, \frac{d \, \omega_k}{dr} \tag{10.3}$$

It will be recognized that Eq. 10.2 is same as Eq. 6.55 if area A in the latter is interpreted as $4 \pi r^2$ and $dy \equiv dr$.

Similarly, the energy equation can be derived as

$$\frac{1}{4 \pi r^2} \frac{d}{dr} \left[4 \pi r^2 \left\{ \rho_m \, cp_m \, V_r \, T - k_m \, \frac{d \, T}{dr} \right\} \right] = - \frac{1}{4 \pi r^2} \frac{d}{dr} \left[4 \pi r^2 \sum m_k'' \, h_k \right]$$
$$+ \dot{Q}_{chem} + \dot{Q}_{rad} \, , \tag{10.4}$$

where in a SCR, $\dot{Q}_{chem} = \Delta H_c \, |R_{fu}|$ is the heat generation from chemical reaction.

Equations 10.2 and 10.4 coupled with Eq. 10.1 are the main equations governing heat/mass transfer in stagnant surroundings. Their solutions for the three mass transfer situations mentioned earlier require different boundary conditions.

10.3 Droplet Evaporation

10.3.1 Inert Mass Transfer without Heat Transfer

Consider a pure single-component liquid droplet (without any dissolved air) whose temperature T_T equals that of the surrounding air, $T_T = T_\infty$. Then, in the absence of any heat transfer, only Eq. 10.2 is relevant with $R_k = 0$, and we must consider two species: liquid vapor and air. Writing this equation for ω_v (vapor) and ω_a (air) and adding the two equations, we have

$$(N_v + N_a) \, 4 \pi r^2 = \dot{m}_v + \dot{m}_a$$
$$= \left\{ \rho_m \, V_r \, (\omega_v + \omega_a) - \rho_m \, D \, \frac{d \, (\omega_v + \omega_a)}{dr} \right\} \, 4 \pi r^2$$
$$= \rho_m \, V_r \, 4 \pi r^2 = \text{constant}, \tag{10.5}$$

where $\omega_v + \omega_a = 1$. Thus, the total mass transfer is *independent* of radius r and therefore must equal $\dot{m}_w = 4 \pi r_w^2 \, N_w$ at the surface. However, because the surroundings are stagnant, $N_a \, 4 \pi r_w^2 = \dot{m}_a = 0$. Thus, Eq. 10.5 can be written as

$$\dot{m}_w = \dot{m}_v = \left\{ \rho_m \, V_r \, \omega_v - \rho_m \, D \, \frac{d \, \omega_v}{dr} \right\} \, 4 \pi r^2 = \rho_m \, V_r \, 4 \pi r^2 \, , \tag{10.6}$$

or, on rearrangement,

$$\frac{d\,\omega_v}{\omega_v - 1} = \left[\frac{\dot{m}_w}{4\,\pi\,\rho_m\,D}\right]\frac{dr}{r^2}. \tag{10.7}$$

Integrating this equation once gives

$$\ln(\omega_v - 1) = -\left[\frac{\dot{m}_w}{4\,\pi\,\rho_m\,D}\right]\frac{1}{r} + C, \tag{10.8}$$

where C is the constant of integration. Its value is determined from the boundary condition $\omega_v = \omega_{v,\infty}$ at $r = \infty$. Thus, $C = \ln(\omega_{v,\infty} - 1)$. Setting the second boundary condition $\omega_v = \omega_{v,w}$ at $r = r_w$ gives

$$\dot{m}_w = 4\,\pi\,\rho_m\,D\,r_w\,\ln\left[\frac{\omega_{v,\infty} - 1}{\omega_{v,w} - 1}\right]$$

$$= 4\,\pi\,\rho_m\,D\,r_w\,\ln[1 + B] \quad \text{and} \tag{10.9}$$

$$B = \frac{\omega_{v,\infty} - \omega_{v,w}}{\omega_{v,w} - 1}, \tag{10.10}$$

where B is the Spalding number, and the surface mass fraction $\omega_{v,w}$ is determined by assuming that saturation conditions prevail at the surface. Thus, the partial pressure of the vapor is given by $p_{v,w} = p_{sat}(T_w = T_T)$. Then[7]

----

[7]In psychrometry, it is customary to define the specific humidity W = kg of water vapor/kg of dry air = m_v/m_a. However, using the ideal gas relation, $p_v\,V = m_v\,R_v\,T$ and $p_a\,V = m_a\,R_a\,T$. Therefore, taking the ratio

$$W = \frac{m_v}{m_a} = \frac{R_a}{R_v} \times \frac{p_v}{p_a} = \frac{M_v}{M_a} \times \frac{p_v}{p_a} \simeq 0.622 \times \frac{p_v}{p_{tot} - p_v},$$

and, the mass fraction can be deduced as

$$\omega_v = \frac{\text{kg of water vapor}}{\text{kg of air} + \text{water vapor}} = \frac{W}{1 + W}.$$

Thus, in the w state, W_w and $\omega_{v,w}$ can be calculated using $p_v = p_{sat}(T_w)$. Also, the ∞-state is prescribed as $p_{v,\infty} = \Phi \times p_{sat}(T_\infty)$. Therefore, W_∞ and $\omega_{v,\infty}$ can be calculated knowing the relative humidity Φ.

For computer applications, the saturation conditions (corresponding to $\Phi = 1$) for the range $-20 < T_w\,(C) < 100$ can be correlated as

$$\omega_{v,w} \simeq 3.416 \times 10^{-3} + (2.7308 \times 10^{-4})\,T_w + (1.372 \times 10^{-5})\,T_w^2$$
$$+ (8.2516 \times 10^{-8})\,T_w^3 - (6.9092 \times 10^{-9})\,T_w^4 + (3.5313 \times 10^{-10})\,T_w^5$$
$$- (3.7037 \times 10^{-12})\,T_w^6 + (6.1923 \times 10^{-15})\,T_w^7$$
$$+ (9.9349 \times 10^{-17})\,T_w^8$$

$$\omega_{v,w} = \left(\frac{p_{sat} (T_w)}{p_{tot}} \right) \times \left(\frac{M_v}{M_{mix}} \right) = x_{v,w} \left(\frac{M_v}{M_{mix}} \right) , \tag{10.11}$$

where p_{tot} is the total pressure. Using Eqs. 10.1 and 9.9, it is now possible to evaluate the total evaporation time t_{eva}. Thus, setting $\rho_p = \rho_l$ (liquid density),

$$4 \pi \rho_l r_w^2 \frac{d r_w}{dt} = - 4 \pi r_w \rho_m D \ln [1 + B] ,$$

or, canceling $4 \pi r_w$,

$$\frac{\rho_l}{2} \frac{d r_w^2}{dt} = - \rho_m D \ln [1 + B] .$$

Integrating this equation from t $= 0$ $(r_w = r_{w,i} = D_{w,i}/2)$ to t $= t_{eva}$ $(r_w = 0)$, we have

$$t_{eva} = \frac{\rho_l D_{w,i}^2}{8 \rho_m D \ln [1 + B]} . \tag{10.12}$$

Comments on Eqs. 10.9 and 10.12

1. Equation 10.9 can also be written as

$$N_w = \frac{\dot{m}_w}{4 \pi r_w^2} = \left(\frac{\rho_m D}{r_w} \right) \ln [1 + B] .$$

This equation is same as Eq. 6.56 if it is recognized that $\rho_m D = \Gamma_m$ and that, for a spherical system,

$$L_{eff} = 4 \pi r_w^2 \int_0^L \frac{dr}{4 \pi r^2} = r_w .$$

The instantaneous evaporation rate $\dot{m}_w$ is thus directly proportional to the droplet radius r_w but the instantaneous mass flux is inversely proportional to r_w.
2. Because $\ln [1 + B] \to B$ as $B \to 0$, Eq. 10.9 will read as

$$\dot{m}_w (B \to 0) = 4 \pi r_w \rho_m D B = 4 \pi r_w \left(\frac{\rho_m D}{\omega_{v,w} - 1} \right) (\omega_{v,w} - \omega_{v,\infty}) .$$

This solution shows that for small B (or $\dot{m}_w$), the diffusion mass transfer rate is directly proportional to $(\omega_{v,w} - \omega_{v,\infty})$, but with *effective* diffusivity D/$(1 - \omega_{v,w})$. Therefore, the mass transfer solution bears similarity with the heat conduction solution in which heat transfer rate is directly proportional to $(T_w - T_\infty)$.
3. Equation 10.12 shows that the evaporation time is directly proportional to the square of the initial droplet diameter. For rapid evaporation, atomization of the liquid into very small diameter droplets will be preferred.

4. If the droplet were placed in *moving surroundings*, then mass transfer would be governed by convection. For this case, to a first approximation, the Reynolds Flow model (see Chap. 6) can be invoked. Thus,

$$\frac{\dot{m}_{w,conv}}{\dot{m}_{w,diff}} = \frac{g^* \, 4 \, \pi \, r_w^2 \, \ln \, [1 + B]}{\rho_m \, D \, 4 \, \pi \, r_w \, \ln \, [1 + B]} = \frac{1}{2} \left[\frac{g^* \, D_w}{\rho_m \, D} \right] = \frac{Sh}{2} \, , \tag{10.13}$$

where Sh is called the Sherwood number analogous to the Nusselt number in heat transfer, and g^* is estimated from Eq. 6.62. To carry out this evaluation, the heat transfer coefficient α must be evaluated from a correlation for flow of a gas over a sphere. This correlation [51] reads as

$$Nu = \frac{\alpha \, D_w}{k_m} = 2 + 0.6 \, Re^{0.5} \, Pr^{1/3} \text{ and} \tag{10.14}$$

$$Re = \frac{|u_g - u_p| \, D_w}{\nu_m} \, , \tag{10.15}$$

where u_g is the velocity of the surrounding gas, u_p is the velocity of the droplet, and $D_w = 2 \, r_w$. Thus, $|u_g - u_p|$ represents the relative velocity between particle and gas.[8]

5. Substituting Eq. 10.13 in Eq. 10.1, equation for the rate of change of droplet radius can be set up. It reads

$$\frac{d}{dt} \left[\frac{4 \, \pi}{3} \, \rho_l \, r_w^3 \right] = - \, \dot{m}_{w,conv}$$

$$4 \, \pi \, r_w^2 \, \rho_l \, \frac{d \, r_w}{dt} = - \, 4 \, \pi \, r_w^2 \, g^* \, \ln \, [1 + B]$$

$$\rho_l \, \frac{d \, r_w}{dt} = - \, g^* \, \ln \, [1 + B] \, . \tag{10.18}$$

---

[8]In cooling towers and spray dryers, atomized liquid droplets fall under the action of gravity. The *free-fall or terminal velocity* of the droplet is given by

$$u_{p,t} = \sqrt{\frac{4 \, g \, D_p}{3 \, C_d} \left(\frac{\rho_p}{\rho_{air}} - 1 \right)} \tag{10.16}$$

where g is the acceleration due to gravity and C_d is the drag coefficient. The latter is correlated with Reynolds number Re as

$$\begin{aligned} C_d &= \frac{24}{Re} & Re < 1 \\ &= \frac{18.5}{Re^{0.6}} & 1 < Re < 1000 \\ &= 0.44 & Re > 1000 \end{aligned} \tag{10.17}$$

Replacing g^* from Eq. 6.62 and using Eqs. 10.14 and 10.15, it can be shown that

$$\rho_l \frac{d\,r_w}{dt} = - \left(\frac{\alpha}{cp_m}\right) P F_c \ln\,[1+B]$$

$$= - \left(\frac{k_m}{cp_m\,D_w}\right) (2+0.6\,Re^{0.5}\,Pr^{1/3})\,P F_c \ln\,[1+B]\,,$$

$$r_w \frac{d\,r_w}{dt} = - \left(\frac{\Gamma_h}{\rho_l}\right) (1+0.3\,Re^{0.5}\,Pr^{1/3})\,P F_c \ln\,[1+B]\,,$$

$$r_w \frac{d\,r_w}{dt} = - \left(\frac{\Gamma_h}{\rho_l}\right) (1+C\,\sqrt{r_w})\,P F_c \ln\,[1+B]\,,\quad \text{and} \qquad (10.19)$$

$$C = 0.3 \times Pr^{1/3} \times \sqrt{\frac{2\,|\,u_g - u_p\,|}{\nu_m}}\,,$$

where $\Gamma_h = k_m/cp_m$. Integration of Eq. 10.19 gives the evaporation time

$$t_{eva} = \frac{\rho_l}{\Gamma_h\,P F_c \ln\,(1+B)} \int_0^{r_{w,i}} \frac{r_w\,d\,r_w}{1+C\,\sqrt{r_w}}\,. \qquad (10.20)$$

This solution requires numerical integration.

6. In applying the preceding relations, the mixture properties are evaluated using following rules:
First, the mean temperature is determined as $T_m = 0.5\,(T_w + T_\infty)$. The values of $k_{v,m}, k_{a,m}, \mu_{a,m}, \mu_{v,m}, cp_{a,m}, cp_{v,m}$, and D are evaluated at temperature T_m. Then, the mixture properties are determined knowing mole fraction $x_v = p_v/p_{tot}$, where p_v is the vapor pressure and p_{tot} is the total pressure.

$$x_{v,m} = 0.5\,(x_{v,w} + x_{v,\infty})\,, \qquad (10.21)$$

$$M_{mix} = x_{v,m}\,M_v + (1 - x_{v,m})\,M_a\,, \qquad (10.22)$$

$$\rho_m = \frac{p_{tot}\,M_{mix}}{R_u\,T_m}\,, \qquad (10.23)$$

$$k_m = x_{v,m}\,k_v + (1 - x_{v,m})\,k_a\,, \qquad (10.24)$$

$$\mu_m = x_{v,m}\,\mu_v + (1 - x_{v,m})\,\mu_a\,, \qquad (10.25)$$

$$cp_m = w_{v,m}\,cp_v + (1 - w_{v,m})\,cp_a\,, \qquad (10.26)$$

$$\text{and}\quad D_m = D_{ref} \times \left(\frac{T_m}{T_0}\right)^{1.5} \times \left(\frac{p_0}{p_{tot}}\right)\,, \qquad (10.27)$$

where $w_{v,m} = x_{v,m}\,(M_v/M_{mix})$, $T_0 = 273\,K$, and $p_0 = 1$ atm and values of D_{ref} (m^2/s) are given in Appendix F for several binary mixtures.

Problem 10.1 A water droplet at $25\,°C$ evaporates in dry air at $25\,°C$ and 1 atm pressure. Its initial diameter is 0.1 mm. Calculate the evaporation time.

Because the droplet is evaporating in dry air, $\omega_{v,\infty} = x_{v,\infty} = 0$. From steam tables, $p_{sat,25} = 0.03169$ bar. Therefore,

$$W_w = 0.622 \times \frac{0.03169}{1.01325 - 0.03169} = 0.02035 \qquad \omega_{v,w} = \frac{0.02035}{1 + 0.02035} = 0.02,$$

and $x_{v,w} = p_w/p_{tot} = 0.03169/1.01325 = 0.03127$ or $x_{v,m} = 0.5\,(0 + 0.03127) = 0.01564$.

Hence,

$$B = \frac{0 - 0.02}{0.02 - 1} = 0.0203, \qquad \text{and } \ln\,(1 + B) = 0.020145.$$

Also, $M_{mix} = (1 - 0.01564) \times 29 + 0.01564 \times 18 = 28.834$. Therefore,

$$\rho_m = \frac{1.01325 \times 10^5 \times 28.83}{8314 \times 298} = 1.172\,\frac{\text{kg}}{\text{m}^3}.$$

Finally, for air–water vapor system (see Appendix F), with $p_{tot} = p_0$,

$$D_m = 24.0 \times 10^{-6} \times \left(\frac{298}{300}\right)^{1.5} = 2.376 \times 10^{-5}\,\frac{\text{m}^2}{\text{s}}.$$

Therefore, from Eq. 10.12, taking $\rho_l = 1000\,\text{kg/m}^3$,

$$t_{eva} = \frac{1000 \times (0.1 \times 10^{-3})^2}{8 \times 1.172 \times 2.376 \times 10^{-5} \times 0.020145} = 2.23\,\text{s}.$$

Problem 10.2 In the previous problem, if the relative humidity of air is 25% and the relative velocity between the droplet and air is 5 m/s, estimate the evaporation time. Take Sc = 0.6 and Pr = 0.714.

In the ∞-state, $p_{v,\infty} = \Phi \times p_{sat}\,(T_\infty) = 0.25 \times 0.03169 = 0.00792$ bar. Therefore, $x_{v,\infty} = 0.00792/1.01325 = 0.007816$ and

$$W_\infty = 0.622 \times \frac{0.007816}{1 - 0.007816} = 0.007878,$$

giving $\omega_{v,\infty} = 0.007878/(1 + 0.007878) = 0.007816$.

Also, from Problem 10.1, $\omega_{v,w} = 0.02$. Therefore,

$$B = \frac{0.007816 - 0.02}{0.02 - 1} = 0.01237.$$

Similarly, $x_{v,m} = 0.5\,(0.03127 + 0.007816) = 0.01954$. Therefore, in appropriate units, $M_{mix} = 28.79$, $\rho_m = 1.177$, $cp_m = 1015.5$, $D_m = 2.376 \times 10^{-5}$, and $\nu_m = D_m \times Sc = 1.42 \times 10^{-5}$.

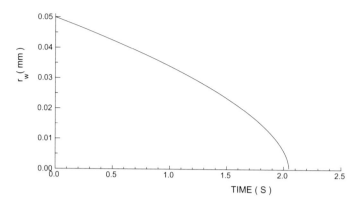

Fig. 10.6 Time variation of water droplet radius (Problem 10.2)

Hence, C is given by

$$C = 0.3 \times \sqrt{2} \times 0.714^{1/3} \times \left(\frac{5}{1.42 \times 10^{-5}} \right)^{0.5} = 225.3 \,.$$

For an air–water vapor system, $PF_c \simeq 1$. Thus, taking $\Gamma_h = \rho_m \, \nu_m / Pr = (1.177 \times 1.42 \times 10^{-5})/0.714 = 2.34 \times 10^{-5}$, and carrying out numerical integration of Eq. 10.20, the evaporation time at $r_w = 0$ is 2.045 s. This can be confirmed from Fig. 10.6, in which the time variation of r_w is shown.[9]

10.3.2 Inert Mass Transfer with Heat Transfer

We consider evaporation of a cold droplet at temperature T_l, say, in hot surroundings at $T_\infty > T_l$, so evaporation will take place in the presence of heat transfer from the surroundings to the droplet. As a result, T_l will be a function of time and, in fact, there may also be temperature gradients within the droplet. That is, T_l will be nonuniform inside the droplet. The temperature gradients can set up density gradients in the droplet resulting in natural convection within the droplet. Analysis of this transient phenomenon with effects of variable properties, radiation, and liquid circulation is quite complex and has been a subject of considerable research (see, for example, Refs. [1, 35, 45, 71]).

---

[9] An approximate answer can be developed if r_w in the denominator of Eq. 10.20 is replaced by the mean radius $0.5 \times r_{w,i}$. Then, analytical integration gives

$$t_{eva} = \rho_l \, r_{w,i}^2 \times \left[2 \Gamma_h \ln (1 + B) (1 + C \sqrt{0.5 \times r_{w,i}}) \right]^{-1} = 2.005 \text{ s.}$$

With respect to highly volatile fuels in which vaporization rates are high, the heat conducted into the liquid may penetrate only a short distance from the surface because of the high Prandtl number of such fuels. As such, the core of the droplet will remain at nearly constant temperature $T_T = T_l$, whereas the surface temperature T_w will be somewhat greater than T_l. In this scenario, the droplet evaporation will be quasi-steady. We assume this to be the case and follow the analysis of Spalding [113]. Assuming no chemical reaction and radiation and assuming that all species in the gas phase have the same temperature, the right-hand side of Eq. 10.4 is zero. Therefore, integrating this equation with respect to r and using Eq. 10.6, we have

$$\frac{\dot{m}_w}{4\pi} T - \Gamma_h r^2 \frac{\partial T}{\partial r} = \text{constant} = \frac{\dot{m}_w}{4\pi} T_w - \Gamma_h r_w^2 \frac{\partial T}{\partial r}\Big|_{r_w}, \qquad (10.28)$$

where $\Gamma_h = k_m/c_{p_m}$. We now define *inward* heat transfer (see Fig. 10.5) as

$$Q_w = 4\pi r_w^2 q_w = -\left(-4\pi r_w^2 k_m \frac{\partial T}{\partial r}\Big|_{r_w}\right). \qquad (10.29)$$

Substituting this equation in Eq. 10.28, we have

$$\frac{dT}{T - T_w + Q_w/(\dot{m}_w c_{p_m})} = \left(\frac{\dot{m}_w}{4\pi\Gamma_h}\right)\frac{dr}{r^2},$$

or, upon integration,

$$\ln\left[T - T_w + \frac{Q_w}{\dot{m}_w c_{p_m}}\right] = -\left(\frac{\dot{m}_w}{4\pi\Gamma_h}\right)\frac{1}{r} + C_1.$$

Using the boundary condition $T = T_\infty$ at $r = r_\infty$, the constant of integration C_1 can be determined to give

$$\ln\left[\frac{T - T_w + Q_w/(\dot{m}_w c_{p_m})}{T_\infty - T_w + Q_w/(\dot{m}_w c_{p_m})}\right] = -\left(\frac{\dot{m}_w}{4\pi\Gamma_h}\right)\frac{1}{r}, \qquad (10.30)$$

or, setting $T = T_w$ at $r = r_w$ and rearranging,

$$\dot{m}_w = 4\pi\Gamma_h r_w \ln\left[1 + \frac{\dot{m}_w c_{p_m}(T_\infty - T_w)}{Q_w}\right]. \qquad (10.31)$$

This equation is implicit in $\dot{m}_w$ in which Q_w and T_w are unknown. Further progress can be made by making assumptions. These are described under the headings *Quasi-steady Evaporation* and *Equilibrium Evaporation*.

Quasi-steady Evaporation

To calculate the instantaneous mass transfer rate, we need to execute the following algorithm, in which $r_{w,i}$ is known and T_∞ and $\omega_{v,\infty}$ are also treated as known constants.

1. We assume $T_w = T_l$ for all times during evaporation and hence, evaluate $\omega_{v,w}$ using Clausius–Clapeyron equation 2.83,

$$x_{v,w} = \frac{p_{v,w}}{p_1} = \exp\left\{-\frac{h_{fg}}{R_g}\left(\frac{1}{T_w} - \frac{1}{T_{bp}}\right)\right\}, \qquad (10.32)$$

where $p_1 = 1$ atm and $T_{bp} =$ boiling point of the liquid.[10] Knowing $p_{v,w}$, $\omega_{v,w}$ can be evaluated using Eq. 10.11.
2. Evaluate $\dot{m}_w$ from Eqs. 10.9 and 10.10.
3. Evaluate dr_w/dt (and, hence, r_w) from Eq. 10.1.
4. Hence, evaluate Q_w from Eq. 10.31 as

$$Q_w = \frac{\dot{m}_w\, cp_m\, (T_\infty - T_w)}{\exp\left(\dot{m}_w/4\,\pi\,\Gamma_h\,r_w\right) - 1} \qquad (10.33)$$

Note that a positive $\dot{m}_w$ reduces Q_w.[11]
5. The heat transfer to the droplet will be used in two ways: firstly, to cause vaporization ($\dot{m}_w\, h_{fg}$), and secondly, to cause increase in the liquid temperature (due to conduction heat transfer Q_l into the drop). Thus,

$$Q_l = \rho_l \left(\frac{4\,\pi}{3} r_w^3\right) cp_l \frac{dT_l}{dt} = Q_w - \dot{m}_w\, h_{fg}. \qquad (10.34)$$

This model assumes that the near-surface layer in the droplet, through which conduction heat transfer takes place, is very small compared to the droplet radius. In another model, known as the *onion skin* model, the instantaneous value of Q_l is expressed as

$$Q_l = \dot{m}_w\, cp_l\, (T_w - T_{l,i}) = Q_w - \dot{m}_w\, h_{fg}, \qquad (10.35)$$

where $T_{l,i}$ is the *initial* temperature of the droplet at $t = 0$. This model assumes that the core of the droplet remains at its initial temperature throughout its lifetime and is particularly suited to high-Prandtl-number oil droplets.
6. Using Eqs. 10.34 or 10.35, T_l at the new time step is determined. Then, setting $T_w = T_l$ (see step 1), the procedure is repeated until $r_w = 0$. One can thus obtain a complete time history.
7. Execution of this algorithm requires numerical solution of two simultaneous first-order equations for dr_w/dt and dT_l/dt on a computer.

Equilibrium Evaporation:

This is defined as a process in which $Q_l = 0$ at all times and, therefore, $Q_w = \dot{m}_w\, h_{fg}$. Thus, according to Eqs. 10.35, $T_l = T_{l,i} = T_w$ and the droplet remains at

---

[10]Boiling points of fuels are given in Appendix C.
[11]This fact is used in *transpiration cooling* of solid surfaces, in which a coolant gas is injected through the surface.

a constant and uniform temperature with respect to time. However, the task is to evaluate its value, and hence that of $\omega_{v,w}$, so $\dot{m}_w$ can be evaluated. Because there are two unknowns, we need two equations.

1. The first can be obtained by recognizing that because $Q_w = \dot{m}_w h_{fg}$, Eq. 10.31 can be written as

$$\dot{m}_w = 4\pi\,\Gamma_h\,r_w\,\ln\left[1 + \frac{c p_m\,(T_\infty - T_w)}{h_{fg}}\right]. \tag{10.36}$$

Then, eliminating $\dot{m}_w$ between this equation and Eq. 10.9 and rearranging, we have

$$\frac{1 - \omega_{v,\infty}}{1 - \omega_{v,w}} = \left[1 + \frac{c p_m\,(T_\infty - T_w)}{h_{fg}}\right]^{Le}, \tag{10.37}$$

where the Lewis number Le $= \Gamma_h/(\rho_m\,D)$. In air–water evaporation, its value is nearly 1. In liquid fuel evaporation, Le may differ from 1 (see Problem 10.3(b)).
2. The second equation is obtained by invoking Eq. 10.32 to evaluate $x_{v,w} = p_v(T_w)/p_{tot}$. The mass fraction at the droplet surface $\omega_{v,w}$ can now be evaluated from Eq. 10.11.
3. Simultaneous solution of Eqs. 10.32 and 10.37 must be carried out iteratively in general. However, in a typical combustion environment, $T_\infty \gg T_{bp}$, and, therefore, $\omega_{v,w} \to 1$, implying that $T_w \to T_{bp}$. This is often the case with highly volatile fuels that have a large B. For such a case,

$$\dot{m}_w = 4\pi\,\Gamma_h\,r_w\,\ln\left[1 + \frac{c p_m\,(T_\infty - T_{bp})}{h_{fg}}\right]. \tag{10.38}$$

Thus, for large B, $\dot{m}_w$ can be determined without iteration.
4. At the other extreme, when T_∞ is very low and $\omega_{v,\infty} \to 0$, from Eq. 10.37, we notice that T_w must be close to T_∞. In this case, Eq. 10.31 is unhelpful. Therefore, $\dot{m}_w$ must be determined from Eqs. 10.9 and 10.10 in which $\omega_{v,w}$ is determined from $x_{v,w}$ (see Eq. 10.11)

$$x_{v,w} = \frac{p_v(T_w)}{p_{tot}} \simeq \frac{p_v(T_\infty)}{p_{tot}} = \exp\left\{-\frac{h_{fg}}{R_g}\left(\frac{1}{T_\infty} - \frac{1}{T_{bp}}\right)\right\}. \tag{10.39}$$

Again, noniterative evaluation of $\dot{m}_w$ becomes possible.
5. In the general case in which T_∞ is neither very low nor very high, however, Eq. 10.36 applies and T_w and must be determined iteratively.
6. Thus, using Eqs. 10.6, 10.36, or 10.38, the evaporation time can be deduced from Eqs. 10.12 or 10.20, as applicable, in which $\rho_m\,D$ is replaced by Γ_h.

Problem 10.3 A $50\,\mu$m liquid n-hexane (C_6H_{14}) droplet evaporates in a stationary pure nitrogen environment at 1 atm and 850 K. Calculate the evaporation time (a) using $T_w = T_{bp}$ and (b) using $T_w = T_l$ and an equilibrium evaporation assumption.

From Appendix C (Table C.1), the fuel properties are: $\rho_l = 659 \, \text{kg/m}^3$, $h_{fg} = 335 \, \text{kJ/kg}$, $T_{bp} = 273 + 69 = 342 \, \text{K}$.

(a) Solution without iteration

Here, we assume $T_l = T_w = T_{bp}$. Therefore, Eq. 10.38 is applicable. We evaluate properties at $T_m = 0.5 \, (342 + 850) = 596 \, \text{K}$. Then, as $x_{v,\infty} = 0$ and $x_{v,w} = 1$, $x_{v,m} = 0.5$ and, hence, $k_m = 0.5 \, (k_{fu} + k_{N_2})$. Similarly, cp_m is determined using Eq. 10.26. Thus, from Appendix C (Table C.2), k_{fu} (596) = 0.0497 W/m-K, cp_{fu} (596) = 2.879 kJ/kg-K (from Table B.2), k_{N_2} (596) = 0.0445 W/m-K, cp_{N_2} (596) = 1.074 kJ/kg-K (from Table D.2).

Now, $M_{mix} = 0.5 \, (86 + 28) = 57$. Thus,

$$k_m = 0.5 \, (0.04967 + 0.0445) = 0.0478 \, \text{W/m-K} ,$$

and

$$cp_m = 0.5 \left(2.877 \times \frac{86}{57} + 1.074 \times \frac{28}{57} \right) = 2.434 \, \text{kJ/kg-K} .$$

Hence, $\Gamma_h = 0.0478/2434 = 1.963 \times 10^{-5}$ and, from Eq. 10.38

$$B = \frac{2.434 \times (850 - 342)}{335} = 3.691.$$

Therefore, the initial burning rate will be

$$\dot{m}_w = (4 \, \pi \, 25 \times 10^{-6}) \, 1.963 \times 10^{-5} \, \ln(4.691) = 0.953 \times 10^{-8} \, \text{kg/s},$$

and, from Eq. 10.12,

$$t_{eva} = \frac{659 \times (50 \times 10^{-6})^2}{8 \times 1.963 \times 10^{-5} \times \ln (1 + 3.691)} = 0.00678 \, \text{s}.$$

(b) Solution by iteration

Here, we assume that under equilibrium, $T_l = T_w < T_{bp}$. Then, the solution is obtained by iteration between Eqs. 10.37 and 10.32. The procedure is as follows:

1. Assume T_w.
2. Evaluate $x_{v,w}$ from

$$x_{v,w} = \frac{p_v(T_w)}{p_{tot}} = \exp - \left\{ \frac{h_{fg} \, M_v}{R_u} \left(\frac{1}{T_w} - \frac{1}{T_{bp}} \right) \right\}. \qquad (10.40)$$

3. Evaluate $M_{mix,w}$ from $x_{v,w}$. Therefore, the first estimate is

$$\omega_1 = x_{v,w} \times \frac{M_v}{M_{mix,w}} .$$

4. Evaluate $T_m = 0.5\,(T_w + T_\infty)$. Hence, evaluate cp_v, k_v, μ_v at T_m. The properties of N_2 are evaluated from

$$\mu_{N_2} = (136 + 0.258 \times T) \times 10^{-7}\ \frac{N\text{-}s}{m^2}$$

$$cp_{N_2} = 963 + 0.1867 \times T\ \frac{J}{kg\text{-}K}\,,\quad \text{and}$$

$$k_{N_2} = 0.0134 + 5.2 \times 10^{-5} \times T\ \frac{W}{m\text{-}K}\,.$$

Thus, mean properties cp_m, k_m, μ_m can be evaluated. Here M_{mix} is evaluated from $x_{v,m} = 0.5 \times x_{v,w}$. Also, $D_m = 0.757 \times 10^{-5} \times (T_m/288)^{1.5}\ m^2/s$.

5. Evaluate Pr, Sc, and Le.
6. Using Eq. 10.37, evaluate

$$\omega_2 = 1 - \left[1 + \frac{cp_{v,T_m}\,(T_\infty - T_w)}{h_{fg}}\right]^{-Le}.$$

7. If $\omega_2 = \omega_1$, accept the value of T_w. If not, go to step 1, assuming a new value of T_w.

The results are given in Table 10.1. It is seen that convergence is obtained at $T_w = 322.1\,K$ (or, $49.1\,°C < T_{bp}$). Thus, B from energy considerations evaluates to

$$B_h = \frac{2.009 \times (850 - 322.1)}{335} = 3.167\,.$$

This value of B is less than that evaluated in part (a). It is interesting to note that the value of B from mass fractions (see Eq. 10.10) evaluates to

$$B_m = \frac{0 - 0.7791}{0.7791 - 1} = 3.527\,.$$

This difference between B_h and B_m arises because Le $= 0.888\,(\neq 1)$. Finally, the evaporation time is determined ($\Gamma_h = 0.045/2009 = 2.24 \times 10^{-5}$) as

$$t_{eva} = \frac{659 \times (50 \times 10^{-6})^2}{8 \times 2.24 \times 10^{-5} \times \ln\,(1 + 3.167)} = 0.00645\,s\,.$$

These results show that our assumption of $T_w = T_{bp}$ in part (a) is reasonable, although the droplet temperature is now nearly $20\,°C$ less than the boiling point.

Problem 10.4 Repeat the preceding problem by assuming $T_\infty = 850, 1000, 1500,$ and $2000\,K$,

In this problem, the procedures of Problem 10.3 are followed for each T_∞. The results are given in Table 10.2. The results show that the droplet temperature and the associated B increase with T_∞ with a consequent reduction in evaporation time. The

Table 10.1 C_6H_{14} droplet evaporation in N_2 at 850 K (Problem 10.3(b))

T_w	cp_m	ρ_m	k_m	μ_m	Le	M_{mix}	ω_1	ω_2
338	2350	1.102	0.0464	2.85E-05	1.25	53.72	0.96	0.878
333	2243	1.031	0.0459	2.88E-05	1.12	50.05	0.907	0.85
328	2135	0.969	0.0454	2.88E-05	1.007	46.81	0.85	0.819
323	2029	0.914	0.0450	2.87E-05	0.906	43.97	0.79	0.786
322.1	2009.5	0.904	0.0450	2.87E-05	0.888	43.5	0.7792	0.7796

Table 10.2 C_6H_{14} Droplet evaporation in N_2—effect of T_∞ (Problem 10.4)

T_∞	B (a)	t_{eva} (a)		B (b)	t_{eva} (b)	T_w
K		sec			sec	°C
850	3.693	0.00702		3.167	0.00645	49.15
1000	5.145	0.00544		4.657	0.00528	56.78
1500	10.59	0.00295		9.981	0.00299	62.40
2000	17.88	0.00192		16.967	0.00197	63.60

last column shows the value of $T_w = T_l$ predicted at convergence. The values of t_{eva} confirm that the $T_w = T_{bp}$ assumption becomes increasingly more accurate at large values of T_∞. For this reason, this assumption is employed to estimate t_{eva} and B in combustion environments.

10.4 Droplet Combustion

We now consider a single fuel droplet in an oxidizing environment, such as air. Figure 10.7(a) shows what happens when the droplet burns. Owing to the relative motion between the droplet and the air (upward because of natural convection), the flame front appears egg-shaped (that is, stretched upward). The analysis of this type of combustion requires the solution of three-dimensional equations. To simplify matters, however, we assume a perfectly spherical flame front around the droplet, as shown in Fig. 10.7(b).

We further assume that the gas phase is quasi-steady and that the distances between the droplets are much larger than the droplet diameter, so a single droplet analysis suffices. Further, the gas-phase combustion is assumed to take place according to the SCR, so mass transfer equations for fuel, oxygen (in air), and products and energy equations must be invoked. Assuming equality of diffusion coefficients and a unity Lewis number, these equations can be transformed into a single equation for conserved property Φ. This equation reads

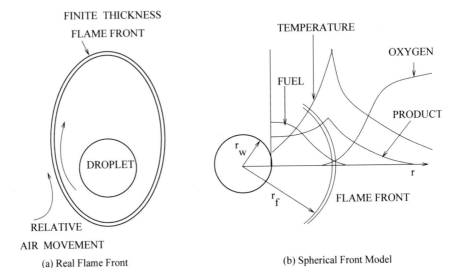

Fig. 10.7 Liquid droplet burning

$$\frac{1}{4\,\pi\,r^2}\,\frac{d}{dr}\left[4\,\pi\,r^2\,\left\{\rho_m\,V_r\,\Phi - \Gamma\,\frac{d\,\Phi}{dr}\right\}\right] = 0\,,\qquad(10.41)$$

where

$$\Phi = \omega_{fu} - \frac{\omega_{O_2}}{r_{st}} = \omega_{N_2}\,,$$

$$= cp_m\,T + \omega_{fu}\,\Delta H_c = cp_m\,T + \frac{\omega_{O_2}}{r_{st}}\,\Delta H_c\,.\qquad(10.42)$$

N_2 is inert during combustion. Also, the burn rate $\dot{m}_w = 4\,\pi\,r^2\,\rho_m\,V_r = $ constant.

10.4.1 Droplet Burn Rate

Let Φ_T represent the value of Φ *inside* the droplet and Φ_w at the droplet surface
$r = r_w$. Then, integrating Eq. 10.41 once,

$$\frac{\dot{m}_w}{4\,\pi}\,\Phi - \Gamma\,r^2\,\frac{d\,\Phi}{dr} = \text{constant} = \frac{\dot{m}_w}{4\,\pi}\,\Phi_w - \Gamma\,r_w^2\,\frac{d\,\Phi}{dr}\,|_{r_w}\,,\qquad(10.43)$$

or, setting $r = 0$ where $\Phi = \Phi_T$,

$$\frac{\dot{m}_w}{4\,\pi}\,(\Phi_T - \Phi_w) = -\,\Gamma\,r_w^2\frac{d\,\Phi}{dr}\,|_{r_w}\,, \qquad (10.44)$$

or, eliminating Φ_w from the last two equations, we have

$$\frac{\dot{m}_w}{4\,\pi}\,(\Phi - \Phi_T) - \Gamma\,r^2\frac{d\,\Phi}{dr} = 0\,. \qquad (10.45)$$

The solution to this equation is

$$\ln\,(\Phi - \Phi_T) = -\,\left(\frac{\dot{m}_w}{4\,\pi\,\Gamma}\right)\frac{1}{r} + C_1\,.$$

Setting $\Phi = \Phi_\infty$ at $r = \infty$, the solution can be written as

$$\Phi^* = \frac{\Phi - \Phi_T}{\Phi_\infty - \Phi_T} = \exp\left(-\frac{\dot{m}_w}{4\,\pi\,\Gamma\,r}\right), \qquad (10.46)$$

and, further, by setting $\Phi = \Phi_w$ at $r = r_w$, the instantaneous burn rate can be deduced as

$$\dot{m}_w = 4\,\pi\,\Gamma\,r_w\,\ln\,[1 + B] \qquad (10.47)$$

$$\text{where }\;B = \frac{\Phi_\infty - \Phi_w}{\Phi_w - \Phi_T}\,. \qquad (10.48)$$

Equation 10.47 for the burning rate has the same form as that of inert mass transfer Eq. 10.9. The Spalding number B, however, must be properly interpreted. Also, in the presence of relative velocity between the droplet and the surrounding air, Eq. 10.47 must be multiplied by Nu/2 (see Eq. 10.14).

10.4.2 Interpretation of B

Depending on the definition of Φ, several deductions can now be made.

1. If $\Phi = \omega_{fu} - \omega_{O_2}/r_{st}$, then, in the ∞-state, $\Phi_\infty = -\,\omega_{O_2,\infty}/r_{st}$. Similarly, in the T state, in which only fuel (without any dissolved gases) is found, $\Phi_T = \omega_{fu,T} = 1$. In the w-state, there can be no oxygen (owing to to the SCR assumption). As such, $\Phi_w = \omega_{fu,w}$. Hence,

$$B = \frac{\omega_{fu,w} + \omega_{O_2,\infty}/r_{st}}{1 - \omega_{fu_w}}\,. \qquad (10.49)$$

For the inert mass transfer, we assume that the droplet is at uniform temperature $T_l = T_T$. However, in a combustion environment, T_∞ is very large and, therefore, $T_w \to T_{bp}$. If this is the case, then $\omega_{fu,w} \to 1$ and Eq. 10.49 does not prove useful because $B \to \infty$ (see Problem 10.4).

Incidentally, from Eq. 10.44, it can be shown that

$$\Phi_w - \Phi_T = \omega_{fu,w} - 1 = \frac{4\pi \Gamma r_w^2}{\dot{m}_w} \frac{d\,\omega_{fu}}{dr}\,|_w . \tag{10.50}$$

This identity is used below.

2. If $\Phi = cp_m\,T + \omega_{fu}\,\Delta H_c$, then, using Eqs. 10.29, 10.44, and 10.50

$$\Phi_\infty - \Phi_w = cp_m\,(T_\infty - T_w) - \omega_{fu,w}\,\Delta H_c \quad \text{and} \tag{10.51}$$

$$\Phi_w - \Phi_T = \left(\frac{4\pi \Gamma r_w^2}{\dot{m}_w}\right)\left(cp_m \frac{d\,T}{dr}\,|_w + \Delta H_c \frac{d\,\omega_{fu}}{dr}\,|_w\right)$$

$$= \frac{Q_w}{\dot{m}_w} + (\omega_{fu,w} - 1)\,\Delta H_c . \tag{10.52}$$

Therefore,

$$B = \frac{cp_m\,(T_\infty - T_w) - \omega_{fu,w}\,\Delta H_c}{Q_w/\dot{m}_w + (\omega_{fu,w} - 1)\,\Delta H_c} . \tag{10.53}$$

If $T_w \neq T_{bp}$, then $\omega_{fu,w}$ and T_w must be iteratively determined using equilibrium relation 10.32. Further, using Eq. 10.35,

$$\frac{Q_w}{\dot{m}_w} = h_{fg} + cp_l\,(T_w - T_{l,i})$$

and $T_{l,i}$ is the temperature at which the droplet enters the combustion space. If, however, $T_{l,i} = T_{bp}$ then $\omega_{fu,w} = 1$, and[12]

$$B = \frac{cp_m\,(T_\infty - T_w) - \Delta H_c}{h_{fg}} . \tag{10.54}$$

3. If $\Phi = cp_m\,T + \omega_{O_2}/r_{st}\,\Delta H_c$, then, $\omega_{O_2,T} = \omega_{O_2,w} = 0$. Further, no oxygen is present between the droplet surface and the flame front (see Fig. 10.7). As such, $d\,\omega_{O_2}/dr\,|_w = 0$ and hence,

---

[12]Equation 10.54 is valid as long as pressures are low. For hydrocarbon fuels, critical pressures vary from 10 to 45 atm. If the pressure is high, $h_{fg} \to 0$ and $B \to \infty$. As such, the quasi-steady assumption must break down and one must adopt unsteady analysis, in which Q_l is evaluated from Eq. 10.34. High pressures are encountered in IC engines and in modern gas turbine plants.

$$\Phi_\infty - \Phi_w = cp_m \, (T_\infty - T_w) + \left(\frac{\omega_{O_2,\infty}}{r_{st}}\right) \Delta H_c \quad \text{and} \tag{10.55}$$

$$\Phi_w - \Phi_T = \left(\frac{4 \pi \Gamma r_w^2}{\dot{m}_w}\right) \left(cp_m \, \frac{dT}{dr} \Big|_w + \frac{\Delta H_c}{r_{st}} \frac{d\omega_{O_2}}{dr} \Big|_w\right)$$

$$= \frac{Q_w}{\dot{m}_w} \, . \tag{10.56}$$

Thus,

$$B = \frac{cp_m \, (T_\infty - T_w) + (\omega_{O_2,\infty}/r_{st}) \, \Delta H_c}{Q_w/\dot{m}_w} \, . \tag{10.57}$$

Again, $Q_w/\dot{m}_w$ for $T_w \neq T_{bp}$ and $T_w = T_{bp}$ must be evaluated in the manner described above.

4. The droplet burn time can be estimated using Eqs. 10.12 or 10.20, as appropriate. Usually $\omega_{O_2,\infty}$ is known and $T_w = T_{bp}$ is a reasonable approximation in combustion environment. Therefore, Eq. 10.57 applies with $Q_w/\dot{m}_w = h_{fg}$. As such, using Eq. 10.12, for example, the droplet burn time in a stagnant environment will read as

$$t_{burn} = \frac{\rho_l \, D_{w,i}^2}{8 \, \Gamma_m \, \ln (1 + B)} \tag{10.58}$$

$$\text{and } B = \frac{cp_m \, (T_\infty - T_{bp}) + (\omega_{O_2,\infty}/r_{st}) \, \Delta H_c}{h_{fg}} \, . \tag{10.59}$$

This so-called D^2 law ($t_{burn} \propto D_{w,i}^2$) has been experimentally verified by many investigators for droplet burning as well (see, for example, Ref. [87])

5. For most hydrocarbons, $r_{st} \simeq 3.5$. Thus, if we consider droplet burning in air ($\omega_{O_2,\infty} = 0.232$) at ambient temperature ($T_\infty = 300\,\text{K}$), then using the values of h_{fg}, ΔH_c, and T_{bp} from Appendix C, it can be shown from Eq. 10.59 that Spalding number B $\simeq 8$ to 10. If the droplet is injected at $T < T_{bp}$, B $\simeq 4$ to 8 from Eq. 10.54.

10.4.3 Flame Front Radius and Temperature

From Eq. 10.46

$$\Phi = \Phi_T + (\Phi_\infty - \Phi_T) \exp\left(-\frac{\dot{m}_w}{4 \pi \Gamma r}\right) \, . \tag{10.60}$$

This equation shows that starting from $\Phi = \Phi_T$ for $0 < r < r_w$, the value of Φ exponentially decays with increasing radius r.

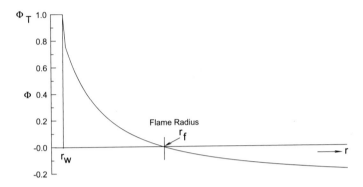

Fig. 10.8 Variation of Φ with radius r—droplet burning

Flame Radius

If we take $\Phi = \omega_{fu} - \omega_{O_2}/r_{st}$, then Eq. 10.60 will read as

$$\Phi = \omega_{fu} - \frac{\omega_{O_2}}{r_{st}} = 1 - \left(\frac{\omega_{O_2,\infty}}{r_{st}} + 1 \right) \exp \left(- \frac{\dot{m}_w}{4\,\pi\,\Gamma\,r} \right). \tag{10.61}$$

This variation with r is shown in Fig. 10.8. From Chap. 9, we recall that the flame front is a surface at which fuel and oxygen are in stoichiometric proportions. Thus, the flame front $(r = r_f)$ will occur where the left-hand side of Eq. 10.61 is zero (see Eq. 9.41). Thus, the expression for the flame radius is deduced as

$$r_f = \left(\frac{\dot{m}_w}{4\,\pi\,\Gamma} \right) \left[\ln \left(1 + \frac{\omega_{O_2,\infty}}{r_{st}} \right) \right]^{-1}. \tag{10.62}$$

However, from Eq. 10.47,

$$r_w = \left(\frac{\dot{m}_w}{4\,\pi\,\Gamma} \right) [\ln\,(1 + B)]^{-1}. \tag{10.63}$$

Therefore,

$$\frac{r_f}{r_w} = \frac{\ln\,(1 + B)}{\ln\,(1 + \omega_{O_2,\infty}/r_{st})}. \tag{10.64}$$

For hydrocarbon fuels, typically $\omega_{O_2,\infty}/r_{st} = 0.232/3.5 \simeq 0.065$ and the denominator in this equation tends to $\omega_{O_2,\infty}/r_{st}$. Thus,

$$\frac{r_f}{r_w} \simeq \frac{r_{st}}{\omega_{O_2,\infty}} \ln\,(1 + B). \tag{10.65}$$

Using practically encountered values of B (8 to 10) for hydrocarbon fuels, therefore, $r_f/r_w \simeq 30$. In reality, during the transient burning process, the ratio varies between 7 and 17. Nonetheless, the preceding estimate confirms the fact that the flame radius is considerably greater than the droplet radius [122].

Flame Temperature

If $\Phi = cp_m \, T + w_{O_2}/r_{st} \, \Delta H_c$, then, adding Eqs. 10.55 and 10.56, and further using Eq. 10.60, we have

$$cp_m \, T + \frac{w_{O_2}}{r_{st}} \, \Delta H_c = \frac{Q_w}{\dot{m}_w} \left[(1+B) \, \exp\left(-\frac{\dot{m}_w}{4\,\pi\,\Gamma\,r} \right) - 1 \right]$$
$$+ \, cp_m \, T_w \, . \tag{10.66}$$

Setting $T = T_f$ at $r = r_f$ and noting that $w_{O_2,f} = 0$, we have

$$T_f = \frac{Q_w}{\dot{m}_w \, cp_m} \left[(1+B) \, \exp\left(-\frac{\dot{m}_w}{4\,\pi\,\Gamma\,r_f} \right) - 1 \right] + T_w \, . \tag{10.67}$$

However, from Eq. 10.62,

$$\exp\left(-\frac{\dot{m}_w}{4\,\pi\,\Gamma\,r_f} \right) = (1 + w_{O_2,\infty}/r_{st})^{-1} \, .$$

Substituting this equation in Eq. 10.67 and noting that $B \gg w_{O_2,\infty}/r_{st}$, it follows that

$$T_f \simeq T_w + \left(\frac{Q_w}{\dot{m}_w \, cp_m} \right) B \, . \tag{10.68}$$

Further, if $T_w = T_{bp}$ then $Q_w/\dot{m}_w = h_{fg}$. Hence,

$$T_f \simeq T_{bp} + \left(\frac{h_{fg}}{cp_m} \right) B \, . \tag{10.69}$$

Gas-Phase Profiles

Using Eqs. 10.61, 10.62 and 10.66, it is now possible to construct variations of mass fractions and temperature with respect to r. This variation is shown in Fig. 10.9 for n-heptane ($T_\infty = 298$ K and $p = 1$ atm) as an example. It is seen that the fuel mass fraction falls to zero at $r = r_f$. As $r \to \infty$, $w_{O_2} \to 0.232$ and $w_{pr} \to w_{N_2} = 0.768$. The temperature peaks at $r = r_f$ and nearly equals T_{ad}.

Problem 10.5 Consider combustion of an n-hexane (C_6H_{14}) droplet (diameter $= 50\,\mu$m) in air at 1 atm and 300 K. Estimate (a) the initial burning rate, (b) the ratio of the flame radius to the droplet radius, (c) burning time, and (d) flame temperature. Assume $T_l = T_w = T_{bp}$.

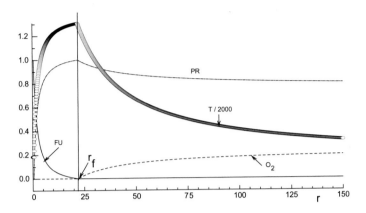

Fig. 10.9 Variation of ω_{fu}, ω_{O_2}, and T with radius r—n-heptane droplet burning

From Appendix C, for n-hexane, $T_{bp} = 69 + 273 = 342$ K, $T_{ad} = 2277$ K, ΔH_c $= 45.1 \times 10^6$ J/kg, $h_{fg} = 335340$ J/kg, and $\rho_l = 659.4$ kg/m^3. Further, for n-hexane, it can be shown that $r_{st} = 3.535$.

(a) Initial burning rate To evaluate this rate, we need to know Γ_m and B. It will be recognized from Fig. 10.9 that the temperature varies considerably in the gas phase and therefore, Spalding [113] recommends that to account for this variation, Γ_m be evaluated as

$$\Gamma_m = \left[r_w \int_0^\infty \frac{dr}{\Gamma(r)\, r^2} \right]^{-1}.$$

However, such an evaluation will require an iterative solution. To simplify matters, we assume $T_f = T_{ad}$ and evaluate properties at $T_m = 0.5\,(2277 + 342) = 1309.5$ K. Then, from Appendices B (Table B.4) and C (Table C.2),

$$cp_{fu}\,(T_m) = 4487.8\,\frac{J}{\text{kg-K}} \qquad k_{fu}\,(T_m) = 0.2477\,\frac{W}{\text{m-K}}$$

$$cp_a\,(T_m) = 1190\,\frac{J}{\text{kg-K}} \qquad k_a\,(T_m) = 0.083\,\frac{W}{\text{m-K}}.$$

Therefore $cp_m = 0.5\,(4487.8 + 1190) = 2839$, $k_m = 0.5\,(0.2477 + 0.083) = 0.1645$, and, hence, $\Gamma_m = 0.1654/2839 = 5.83 \times 10^{-5}$. Therefore, from Eq. 10.59,

$$B = \frac{2839 \times (300 - 342) + (0.232/3.535) \times 45.1 \times 10^6}{3.35 \times 10^5} = 8.48$$

Thus, the initial burn rate is

$$\dot{m}_w = 4 \times \pi \times \Gamma_h \times r_{w,i} \times \ln(1 + B)$$

$$= 4 \times \pi \times 5.83 \times 10^{-5} \times \frac{50}{2} \times 10^{-6} \times \ln(1 + 8.48) = 4.12 \times 10^{-8} \text{ kg/s.}$$

This rate is nearly 4 times greater than that evaluated for droplet evaporation in Problem 10.3.

(b) Ratio of flame radius to drop radius From Eq. 10.64,

$$\frac{r_f}{r_w} = \frac{\ln(1 + 8.48)}{\ln(1 + 0.232/3.535)} = 35.83 .$$

This estimate is greater than the expected value of about 20.

(c) Burning Time Using Eq. 10.58,

$$t_{burn} = \frac{659.4 \times (50 \times 10^{-6})^2}{8 \times 5.83 \times 10^{-5} \times \ln(1 + 8.48)} = 0.00157 \text{ s.}$$

This time is again nearly 4 times smaller than that evaluated for droplet evaporation in Problem 10.3.

(d) Flame Temperature From Eq. 10.69,

$$T_f \simeq 342 + \left(\frac{3.35 \times 10^5}{2839} \right) \times 8.48 = 1343 \text{ K.}$$

This estimate is considerably less than the expected $T_f \simeq T_{ad}$. The difficulty is associated with the value of cp_m which is too large. For most hydrocarbons, if we take $cp_{m,eff} = 0.5 (cp_a + cp_m)$, the flame temperature prediction turns out to be more reasonable. In the present case, $cp_{m,eff} = 0.5 (1190 + 2839) = 2014.5$ gives $T_f = 1752$ K, which is reasonably accurate for real flames in which radiation loss is present.

Problem 10.6 For the conditions of Problem 10.5, determine flame temperature by allowing for flame radiation effects. Assume droplet surface emissivity $\epsilon = 1.0$ and $T_w \rightarrow T_{bp}$.

When flame radiation is accounted for, the expression for gas-phase *inward* heat transfer will read as

$$Q_w = \dot{m}_w h_{fg} - 4 \pi r_w^2 \epsilon \sigma (T_f^4 - T_w^4) . \tag{10.70}$$

Therefore, from Eq. 10.57

$$B_h = \frac{cp_m (T_\infty - T_w) + (w_{O_2,\infty}/r_{st}) \Delta H_c}{h_{fg} - 4 \pi r_w^2 \epsilon \sigma (T_f^4 - T_w^4)/\dot{m}_w} , \tag{10.71}$$

where $\dot{m}_w$ must be determined from Eq. 10.47 and the associated B from Eq. 10.49.

Table 10.3 Determination of T_f for different T_w (Problem 10.6)

T_w (°C)	T_f (K)	$\Gamma_m \times 10^5$	$\Gamma_h \times 10^5$	B	B_h	$\omega_{v,w}$	$\dot{m}_w \times 10^8$
68.0	2612.2	5.567	5.235	108.6	146.52	0.992	8.217
67.0	2384.1	5.267	4.660	53.36	89.87	0.9804	6.610
66.5	2297.3	5.140	4.450	42.31	76.41	0.9764	6.093
66.0	2220.0	5.03	4.270	34.95	66.68	0.9703	5.665

This shows that T_f must be determined iteratively, along with the equilibrium relation Eq. 10.40, to determine $\omega_{fu,w}$. Thus, we choose T_w close to $T_{bp} = 69$ C. Calculations are now carried out for the initial radius $r_w = 25\,\mu$m only. Further steps are as follows:

1. Assume T_f. Hence, evaluate cp_m, k_m, ρ_m, and D_m at $T_m = 0.5\,(T_f + T_w)$ as described in part (b) of Problem 10.3.
2. Evaluate B (see Eq. 10.49) and $\dot{m}_{w1} = 4\,\pi\,r_w\,\rho_m\,D_m\,\ln\,(1 + B)$ where $D_m = D_{CO_2-N_2}(T_m) = 11 \times 10^{-6}\,(T_m/300)^{1.5}$.
3. Knowing $\dot{m}_{w1}$, evaluate B_h (Eq. 10.71).
4. Evaluate $\dot{m}_{w2} = 4\,\pi\,r_w\,(k_m/cp_m)\,\ln\,(1 + B_h)$.
5. If $\dot{m}_{w2} = \dot{m}_{w1}$, accept T_f. If not, go to step 1.

Converged values of T_f for each assumed T_w are shown in Table 10.3. The table shows that T_f increases with T_w. From Appendix C, $T_{ad} = 2277$ K for C_6H_{14}. It is seen that $T_f \rightarrow T_{ad}$ when $T_w \simeq 66.4\,°$C. For this value, the initial burning rate is predicted at $\simeq 6 \times 10^{-8}$ kg/s. This is higher than that of Problem 10.5 where radiation is neglected. This is because the value of B_h is now much greater than the 8.48 obtained in Problem 10.5. This shows the effect of radiation on the driving force. Finally, the burn time can be estimated through integration as

$$t_{burn} = \frac{\rho_l}{\Gamma_h} \int_0^{r_{w,i}} \frac{dr_w^2}{1 + B_h(r_w^2)} \tag{10.72}$$

Numerical integration is required to account for B_h as a function of droplet radius r_w. Overall, this manner of estimating T_f and T_w is procedurally much more secure than the simplifying assumption made in Eq. 10.69 which requires adjustments to value of cp_m.

10.5 Solid Particle Combustion

10.5.1 Stages of Combustion

As is well known, coals are formed from buried vegetation over a long period of time. The different stages of coal formation are plant debris $\rightarrow$ peat $\rightarrow$ lignite (brown coal) $\rightarrow$ bituminous coal $\rightarrow$ anthracite $\rightarrow$ graphite. The *proximate analysis* of coal is given in terms of percentage moisture, volatile matter, fixed carbon (or char), and ash. As one moves down the chain, ash and volatile matter decrease and the fixed carbon percentage increases, with an accompanied increase in calorific value ($\sim$28 MJ/kg). Coal used in PF or fluidized bed boilers comprises volatile matter (typically, 30% to 40%) and ash (15% to 40%), and the rest is char (also called coke). Anthracite is pure coal with very small amounts of volatiles and ash. The typical density of char particles is about 1300 kg/m^3 and that of ash about 2500 to 5500 kg/m^3. The *ultimate analysis* is given in terms of the proportions of chemical elements C, H, S, N, and O. The chemical formula for coal can be taken as $CH_{0.8} S_x N_y O_z$ where x, y, and z vary but are each <0.1.

Woody biomass, on the other hand, is much lighter (about 600 to 750 kg/m^3). The typical proximate analysis of wood is volatile matter (70% to 85%), ash (0.5% to 2%), and fixed carbon (15% to 30%). Typical chemical analysis of wood yields C (50%), H (6%), O (40%), Ash (2%), N and S (<0.2%) by mass. The LHV of wood[13] is about 16 MJ/kg. The chemical formula for a typical wood may be taken as $C_{4.42} H_{6.1} O_{2.42} N_{0.01} (H_2O)_n$, where $n = 5.555 \, f/(1 - f)$ and f is percentage moisture content in wood on a dry basis.

Solid particles typically burn in three stages: (1) drying, (2) devolatilization, and (3) combustion. All three processes occur simultaneously layer by layer, starting from the surface of the particle. Here, they will be assumed to be sequential for the particle as a whole. This assumption is particularly justified for small particles. Although our interest is in combustion, the analysis of drying and devolatilization is briefly mentioned below.

Drying

Neglecting internal resistance to moisture removal from the particle, the energy balance for the particle may be written as

$$\frac{dU}{dt} = -\dot{m}_w h_{fg} + \alpha A_p (T_F - T_p) + \epsilon \sigma A_p (T_F^4 - T_p^4), \qquad (10.73)$$

where T_F refers to furnace temperature, A_p to particle surface area, and the particle internal energy U is given by

$$U = (m_w c_w + m_{dp} c_{dp}) T_p. \qquad (10.74)$$

In this equation, m_w refers to mass of water, m_{dp} refers to mass of dry particle and cv_w and cv_{dp} are specific heats of water and dry particles. Assuming water evaporates

---

[13]For more than 200 species of biomass, Channiwala [23] has correlated the HHV (MJ/kg) as

$$HHV = 0.3491 \times \%C + 1.1783 \times \%H - 0.1034 \times \%O$$
$$- 0.0211 \times \%A + 0.1005 \times \%S - 0.0151 \times \%N,$$

where A is Ash. All percentages are on a mass basis.

at $T_{bp} = 100\,°C = 373\,K$ ($h_{fg} = 2432\,kJ/kg$), Eq. 10.73 is integrated[14] to yield drying time t_{dry}.

$$t_{dry} = \frac{m_{w,i}\, h_{fg} + (m_{w,i}\, cv_w + m_{dp}\, cv_{dp})\,(T_{bp} - T_{p,i})}{A_p\left\{\alpha\,(T_F - \overline{T}_p) + \epsilon\,\sigma\,(T_F^4 - \overline{T}_p^4)\right\}}$$ (10.75)

where $\overline{T}_p \simeq 0.5\,(T_{bp} + T_{pi})$.

Problem 10.7 A $100\,\mu m$ coal particle, initially at $300\,K$ and 15% moisture (dry basis), is injected into a furnace at $1500\,K$. The dry particle density is $1200\,kg/m^3$. Estimate the drying time. Take $\epsilon = 1$ and $cv_{dp} = 1300\,J/kgK$.

In this problem, $T_m = 0.5\,(1500 + 300) = 900\,K$ and the average particle temperature $\overline{T}_p = 0.5\,(300 + 373) = 336.5\,K$. Taking furnace gas properties corresponding to those of air, $k_g(T_m) = 0.062\,W/m\text{-}K$ (see Table D.1). For stagnant surroundings, $Nu_d = 2$. Therefore, $\alpha = 2 \times 0.062/(100 \times 10^{-6}) = 1240\,W/m^2\text{-}K$. Also, $h_{fg} = 2432\,kJ/kg$ and the dry particle mass evaluates to $m_{dp} = \rho_p\,\pi\,d_p^3/6 = 6.286 \times 10^{-10}\,kg$. Therefore, $m_{w,i} = 0.15 \times m_{dp} = 0.942 \times 10^{-10}\,kg$ and $A_p = \pi \times (100 \times 10^{-6})^2 = 3.1416 \times 10^{-8}\,m^2$. Using these data, Eq. 10.75 gives $t_{dry} = 0.00585\,s$ or $5.85\,ms$.

Devolatilization

As the particle temperature is increased, volatiles flow out through the porous structure of the particle. The volatiles burn in the gas phase and provide heat for further heating of the particle, which enhances the volatilization rate further. Volatiles comprise CO, CO_2, hydrocarbons $C_m H_n$, H_2, and tar. Volatile evolution begins at about $250\,°C$ to $300\,°C$ and is almost completed at approximately $450\,°C$ in wood particles and at $600\,°C$ to $900\,°C$ for different types of coals. On completion of volatilization, char and ash are left behind in the particle. In different temperature ranges, the rates of volatilization are different. On the average, however, the rate of devolatilization can be expressed as

$$\frac{d\,m_{vol}}{dt} = -\,m_{vol}\,A_{vol}\,\exp\left(-\frac{E_{vol}}{R_u\,T_p}\right),$$ (10.76)

where m_{vol} is the mass of volatiles given by

$$m_{vol} = m_{dp} - m_{char} - m_{ash}\,.$$ (10.77)

Typical values of A_{vol} and E_{vol} are given in Table 10.4. If the particle temperature is assumed to be nearly constant during devolatilization, then Eq. 10.76 can be integrated to yield time for devolatilization. Thus,

---

[14]During drying, the temperature difference $(T_F - T_p)$ is large. Hence, the variation of T_p with time is ignored.

Table 10.4 Devolatilization constants in Eq. 10.76 [19]

Fuel	A_{vol} (s^{-1})	E_{vol} (kJ/kmol)
Lignite	280	47.32×10^3
Bituminous coal	700	49.4×10^3
Wood	7×10^7	130×10^3

$$t_{vol} = \left[A_{vol} \exp\left(-\frac{E_{vol}}{R_u\, T_p} \right) \right]^{-1} \times \ln\left(\frac{m_{vol,i}}{m_{vol,f}} \right), \tag{10.78}$$

where $m_{vol,f}$ is the final volatile mass in the particle.

It is emphasized again that both drying and devolatilization are extremely complex phenomena that depend on the physical structure of the particle under consideration [19]. Several thermochemical processes involving phase change occur inside the particle, resulting in structural and irreversible chemical change. For example, bituminous coal particles swell during devolatilization. Similarly, if the volatile gases were to be condensed by lowering their temperature, they would not constitute the same coal substance. Fortunately, both drying and devolatilization are much more rapid processes than the burning rate of char that is left behind. As such, estimation of particle burning time based purely on the burning rate of char (or fixed carbon or coke) suffices for most practical applications. Besides, in more purer forms of coals, such as anthracite or graphite, which contain very small amounts of volatiles, such an estimate is directly applicable.

10.5.2 Char Burning

Surface Reactions

Unlike liquid droplet burning, which was *physically controlled*, char burning is both physically and kinetically controlled. Char first burns through heterogeneous surface reactions that release combustible gas (principally CO and some H_2 and CH_4), which then burns in the gas phase through homogeneous reactions. Oxygen from the gas phase reacts with the char surface to produce principally CO and CO_2. The CO_2 may also react further with the surface to produce CO. The latter may then burn in the gas phase. The surface reaction rates are represented by Arrhenius-type expressions

$$k = k_0 \exp\left(-\frac{E}{R_u\, T_w} \right),$$

where T_w is the surface temperature. Although several reactions take place, only those involving CO and CO_2 are considered adequate for representing the burning phenomenon [19, 108, 122]. These reactions along with the rate constants and asso-

Table 10.5 Surface reactions in char combustion

Reaction	k_i (m/s))	ΔH_i (MJ/kg)
R1. $C^* + O_2 \rightarrow CO_2$	593.83 T_g exp $(-18000/T_w)$ for $T_w < 1650\,\text{K}$ $(2.632 \times 10^{-5}\, T_w - 0.03353)\, T_g$ for $T_w > 1650\,\text{K}$	32.76
R2. $C^* + 0.5\,O_2 \rightarrow CO$	1.5×10^5 exp $(-17966/T_w)$	9.2
R3. $C^* + CO_2 \rightarrow 2\,CO$	4.016×10^8 exp $(-29790/T_w)$	-14.4

ciated heats of combustion ΔH_i (MJ/kg of C) are given in Table 10.5. Reaction R3 is endothermic and hence ΔH_3 is negative. Similarly, in reaction R1, T_g represents near-surface gas temperature. In these reactions, C^* represents fixed carbon or char. The corresponding burning rate (kg/s) expressions are given by

$$\dot{m}_{w1} = 4\,\pi\, r_w^2\, \rho_w\, k_1\, \frac{M_C}{M_{O_2}}\, \omega_{O_2,w}\,, \tag{10.79}$$

$$\dot{m}_{w2} = 4\,\pi\, r_w^2\, \rho_w\, k_2\, \frac{M_C}{M_{O_2}}\, \omega_{O_2,w}\,, \tag{10.80}$$

$$\text{and }\dot{m}_{w3} = 4\,\pi\, r_w^2\, \rho_w\, k_3\, \frac{M_C}{M_{CO_2}}\, \omega_{CO_2,w}\,. \tag{10.81}$$

Gas-Phase Reactions

The CO produced at the surface burns in the gas-phase via the following homogeneous reaction.

$$R4.\quad CO + \frac{1}{2}\,O_2 \rightarrow CO_2 \quad \Delta H_4 = 10.1\,\text{MJ/kg of CO}$$

$$k_4 = 2.24 \times 10^{12}\, \exp\left(-\frac{20134.7}{T}\right)$$

$$R_{CO} = -\,\rho_m^{1.75}\, k_4 \left(\frac{M_{CO}}{M_{H_2O}^{0.5}\, M_{O_2}^{0.25}}\right) \omega_{CO}\, \omega_{O_2}^{0.25}\, \omega_{H_2O}^{0.5}\, \frac{\text{kg}}{\text{m}^3\text{-s}}\,, \tag{10.82}$$

where ω_{H_2O} is treated as inert and does not take part in the burning reactions, its magnitude being very small[15]

---

[15] For example, water vapour can heterogeneously react with surface char to produce CO and H_2. The latter can further react with the surface to produce CH_4. As such, two additional reactions are

$$R5:\ C^* + H_2O \rightarrow CO + H_2 \quad \text{and} \quad R6:\ C^* + H_2 \rightarrow CH_4\,,$$

with $k_5 \simeq 2\,k_3$ and $k_6 = 0.035\, \exp\,(-\,17900/T_w)$.

Preliminary analysis of char burning is carried out via two reaction models. In model 1, only reaction R1, which produces CO_2 is considered. As such, in this model, gas-phase reactions are absent. Reaction R1 typically dominates when the particle temperature is low (say, below 1300 K). In model 2, R2 and R3 are considered along with the gas-phase reaction (10.82). This combination of reactions is applicable when the particle temperature is high. The presence of an endothermic reaction R3 implies that the heat release will be smaller than that of model 1. Implications of these models are developed below.

Model 1

In this model, the fuel (carbon) is nonvolatile. Therefore, $w_{fu} = 0$ everywhere and we may consider diffusion of only O_2 and/or CO_2 in the gas-phase. O_2 flows to the surface, CO_2 flows away from the surface, and the heat of combustion is partly conducted and partly radiated to the surroundings. For this reaction R1, the stoichio-metric ratio $r_{st1} = 32/12 = 2.667$. Considering burning in stagnant surroundings, Eqs. 10.47 and 10.48 are applicable. The aim of the model is to evaluate char burning rate when the particle temperature is known.

Burning Rate from mass transfer equation: T_w known:

If we take $\Phi = w_{fu} - w_{O_2}/r_{st1} = -w_{O_2}/r_{st1}$, then

$$\Phi_\infty - \Phi_w = \frac{w_{O_2,w} - w_{O_2,\infty}}{r_{st1}} \, ,$$

and, from Eq. 10.44,

$$\Phi_w - \Phi_T = -\frac{4\pi \Gamma_m r_w^2}{\dot{m}_w r_{st1}} \frac{d w_{O_2}}{dr} \Big|_w \, ,$$

where $\Gamma_m = \rho_m D$ and $\dot{m}_w$ is the total burn rate of carbon. But, from Eq. 10.6, it follows that

$$\dot{m}_{O_2} = -r_{st1} \dot{m}_w = \dot{m}_w w_{O_2,w} - 4\pi \Gamma_m r_w^2 \frac{d w_{O_2}}{dr} \Big|_w \, , \tag{10.83}$$

where the negative sign indicates that the flow of oxygen is *inward to the surface*. Using this equation, we have

$$\Phi_w - \Phi_T = -\left(\frac{r_{st1} + w_{O_2,w}}{r_{st1}}\right) \, ,$$

and, from Eq. 10.48

$$B_m = \frac{\Phi_\infty - \Phi_w}{\Phi_w - \Phi_T} = \frac{w_{O_2,\infty} - w_{O_2,w}}{r_{st1} + w_{O_2,w}}. \tag{10.84}$$

Thus, from Eq. 10.47

$$\dot{m}_w = 4\,\pi\,\Gamma_m\,r_w\,\ln\left[1 + \frac{w_{O_2,\infty} - w_{O_2,w}}{r_{st1} + w_{O_2,w}}\right]. \tag{10.85}$$

This equation, which is derived from accounting for diffusion (that is, physically controlled conditions) considerations, permits evaluation of the char burning rate when $w_{O_2,w}$ is known.[16] The value of $w_{O_2,w}$, however, is fixed by chemical kinetics. Thus, equating kinetic- and diffusion-controlled rates of burning from Eqs. 10.79 and 10.85, respectively, it is possible to evaluate $w_{O_2,w}$. Further, because B_m is usually very small, it can be derived that

$$\frac{w_{O_2,\infty}}{w_{O_2,w}} \simeq 1 + \frac{\rho_{m,w}\,r_w\,k_1}{\Gamma_m}. \tag{10.87}$$

Thus, if the particle temperature T_w is known, then k_1 can be readily evaluated from Table 10.5 and, hence, the burning rate $\dot{m}_w$.

Kinetic and Physical Resistances

When B_m is small, Eq. 10.85 may be written as

$$w_{O_2,\infty} - w_{O_2,w} = \dot{m}_w\,R_{phy} \qquad \rightarrow R_{phy} = \frac{r_{st1} + w_{O_2,w}}{4\,\pi\,\Gamma\,r_w}. \tag{10.88}$$

Similarly, Eq. 10.79 may be written as

$$w_{O_2,w} - 0 = \dot{m}_w\,R_{kin} \qquad \rightarrow R_{kin} = \frac{r_{st1}}{\rho_w\,4\,\pi\,r_w^2\,k_1}, \tag{10.89}$$

where, following an electrical analogy, R_{phy} and R_{kin} may be viewed as physical and kinetic resistances to burning. Adding the last two equations, therefore, the burning rate may be expressed as

$$\dot{m}_w = \frac{w_{O_2,\infty}}{R_{phy} + R_{kin}}. \tag{10.90}$$

---

[16]If char is highly reactive $w_{O_2,w} \rightarrow 0$ and $B_m \rightarrow w_{O_2,\infty}/r_{st1} = 0.232/2.667 = 0.087$ (a small number). Therefore, $\ln(1 + B_m) \rightarrow B_m$ and

$$\dot{m}_w = 4\,\pi\,\Gamma_m\,r_w\left(\frac{w_{O_2,\infty}}{r_{st1}}\right). \tag{10.86}$$

This expression was derived by Nusselt in 1924 [113].

Further, noting that because $r_{st1} \gg w_{O_2,w}$,

$$\frac{R_{kin}}{R_{phy}} \propto \left(\frac{\Gamma_m}{k_1\, r_w}\right)\left(\frac{1}{\rho_w}\right). \tag{10.91}$$

Thus, if $R_{kin} \gg R_{phy}$, particle combustion is said to be kinetically controlled. Such a type of combustion occurs when the particle size is very small (or, toward the end of burning when particle size has been reduced considerably) and when pressure (or density) and temperature are very low.[17] In contrast, physically controlled combustion occurs when T_w is large and $w_{O_2,w} \to 0$.

Particle Burn Time

Using Eqs. 10.1 and 10.90, we have

$$\frac{d}{dt}\left[\frac{4\,\pi}{3}\,\rho_p\, r_w^3\right] = -\dot{m}_w = -\frac{w_{O_2,\infty}}{A_{kin}/r_w^2 + A_{phy}/r_w}, \tag{10.92}$$

where $A_{phy} = R_{phy}\, r_w$ and $A_{kin} = R_{kin}\, r_w^2$. For a given temperature T_w, they are constants; therefore, further integration of Eq. 10.92 gives[18]

$$t_{burn} = 4\,\pi \left(\frac{\rho_p\, r_{w,i}^2}{w_{O_2,\infty}}\right)\left[\frac{A_{phy}}{2} + \frac{A_{kin}}{r_{w,i}}\right]$$

$$= 4\,\pi \left(\frac{\rho_p\, r_{w,i}^3}{w_{O_2,\infty}}\right)\left[\frac{R_{phy}}{2} + R_{kin}\right]. \tag{10.93}$$

Burning rate from energy equation: T_w unknown

When T_w is not known, as in practical combustion, we must invoke $\Phi = cp_m\, T + (w_{O_2}/r_{st1})\, \Delta H_1$. Then,

$$\Phi_\infty - \Phi_w = cp_m\,(T_\infty - T_w) + (w_{O_2,\infty} - w_{O_2,w})\,\frac{\Delta H_1}{r_{st1}}, \tag{10.94}$$

and, using Eqs. 10.45 and 10.83,

$$\Phi_w - \Phi_T = \left(\frac{4\,\pi\,\Gamma\, r_w^2}{\dot{m}_w}\right)\left(cp_m\,\frac{d\,T}{dr}\big|_w + \frac{\Delta H_1}{r_{st1}}\frac{d\,w_{O_2}}{dr}\big|_w\right)$$

$$= \frac{Q_w}{\dot{m}_w} + \left(\frac{w_{O_2,w}}{r_{st1}} + 1\right)\Delta H_1. \tag{10.95}$$

---

[17]When T_w is small, k_1 is also small.

[18]This theory for estimation of burn time assumes that the particle, when fully burned, vanishes in size. In reality, ash remains in the particle and, therefore, the fully burned particle does have a finite radius. Also, during burning, particle swelling may take place.

However, in the presence of radiation, the inward conduction and radiation heat transfer are given by

$$Q_w + 4\pi r_w^2 \epsilon \sigma (T_\infty^4 - T_w^4) = \dot{m}_w \Delta H_1 .$$

Hence, the Spalding number and burn rate are given by

$$B_h = \frac{\Phi_\infty - \Phi_w}{\Phi_w - \Phi_T}$$

$$= \frac{cp_m (T_\infty - T_w) + (\omega_{O_2,\infty} - \omega_{O_2,w}) \Delta H_1/r_{st1}}{\Delta H_1 + 4\pi r_w^2 \epsilon \sigma (T_w^4 - T_\infty^4)/\dot{m}_w + \Delta H_1 (1 + \omega_{O_2,w}/r_{st1})}$$

and $\dot{m}_w = 4\pi \Gamma_m r_w \ln [1 + B_h]$.   (10.96)

Thus, when T_w is not known, Eqs. 10.90 and 10.96 must be solved simultaneously along with Eq. 10.87 to estimate $\omega_{O_2,w}$. Also, if Le $= 1$ is assumed then $B_m = B_h$. Usually, $T_\infty < T_w$ in the combustion space. If this temperature difference is too great, B_h reduces considerably to an extent that the particle may even extinguish.

Model 2

This model is invoked when the particle temperature is high. In this model, reactions R1 and R2 are ignored[19] but reaction R3 ($r_{st3} = 44/12 = 3.667$) is retained, and considered along with the gas-phase reaction R4 ($r_{st4} = 0.5 \times 32/28 = 0.5714$). The simultaneity of reactions R3 and R4 implies that R4 produces CO_2 in the gas-phase, which then diffuses to the char surface, and is consumed producing CO at the surface according to R3. CO, however, is completely consumed at the *flame front* at $r = r_f$. Thus, the overall scenario can be partitioned in 4 zones as follows:

$$r_w \leq r < r_f : \omega_{CO} + \omega_{CO_2} + \omega_{N_2} = 1; \quad \omega_{O_2} \simeq 0 , \qquad (10.97)$$

$$r = r_f : \omega_{CO_2,f} + \omega_{N_2,f} = 1; \quad \omega_{O_2,f} = \omega_{CO,f} = 0 , \qquad (10.98)$$

$$r > r_f : \omega_{O_2} + \omega_{CO_2} + \omega_{N_2} = 1; \quad \omega_{CO} = 0 , \qquad (10.99)$$

$$r \to \infty : \omega_{O_2} + \omega_{N_2} = 1; \quad \omega_{CO_2,\infty} = \omega_{CO,\infty} = 0 . \qquad (10.100)$$

If we let $\Phi = \omega_{N_2}$, then $\Phi_T = 0$ and $\Phi_\infty = 1 - \omega_{O_2,\infty}$. Hence, from Eq. 10.60 it is easy to show that

$$\omega_{N_2} = \omega_{N_2,\infty} \exp \left(-\frac{\dot{m}_w}{4\pi r \Gamma} \right) \quad \text{for all } r > r_w . \qquad (10.101)$$

---

[19]This neglect stems from the high value of T_w which renders $\omega_{O_2,w} \simeq 0$ and hence, $\dot{m}_{O_2,w} = \dot{m}_{w1} = \dot{m}_{w2} \simeq 0$.

Region $r < r_f$

In this region, reaction R4 is fast so that SCR assumption applies, and hence, Eq. 10.41 also applies. For this CO, O_2, and CO_2 system, Φ can be defined as

$$\Phi = \omega_{CO} - \frac{\omega_{O_2}}{r_{st4}} = \frac{\omega_{O_2}}{r_{st4}} + \frac{\omega_{CO_2}}{1 + r_{st4}} = \omega_{N_2} .$$

If we take the second definition for Φ, then recognizing that $\omega_{O_2,w} \simeq 0$ and $\omega_{CO_2,\infty} = 0$,

$$\Phi_\infty - \Phi_w = \frac{\omega_{O_2,\infty}}{r_{st4}} + \frac{\omega_{CO_2,\infty}}{1 + r_{st4}} - \left[\frac{\omega_{O_2,w}}{r_{st4}} + \frac{\omega_{CO_2,w}}{1 + r_{st4}} \right]$$
$$= \frac{\omega_{O_2,\infty}}{r_{st4}} - \frac{\omega_{CO_2,w}}{1 + r_{st4}} . \tag{10.102}$$

Similarly, recognizing that $d\,\omega_{O_2}/dr\,|_w \simeq 0$ and using Eq. 10.44

$$\Phi_w - \Phi_T = \left(\frac{4\pi \Gamma r_w^2}{\dot{m}_w} \right) \left(\frac{1}{1 + r_{st4}} \right) \frac{d\,\omega_{CO_2}}{dr} \Big|_w . \tag{10.103}$$

Further, analogous to Eq. 10.83,

$$\dot{m}_{CO_2} = \dot{m}_w\, \omega_{CO_2,w} - 4\pi \Gamma r_w^2 \frac{d\,\omega_{CO_2}}{dr} \Big|_w . \tag{10.104}$$

However, from reaction R3,[20]

$$\dot{m}_{CO_2} = - r_{st3}\, \dot{m}_w , \tag{10.105}$$

where $\dot{m}_w \simeq \dot{m}_{w3}$. From the last two equations, it can be shown that

$$4\pi \Gamma r_w^2 \frac{d\,\omega_{CO_2}}{dr} \Big|_w = (r_{st3} + \omega_{CO_2,w})\, \dot{m}_w . \tag{10.106}$$

Hence, Eq. 10.103 can be written as

$$\Phi_w - \Phi_T = \frac{r_{st3} + \omega_{CO_2,w}}{1 + r_{st4}} . \tag{10.107}$$

Combining this equation with Eqs. 10.102, we have

$$B_m = \frac{\{(1 + r_{st4})/r_{st4}\}\, \omega_{O_2,\infty} - \omega_{CO_2,w}}{r_{st3} + \omega_{CO_2,w}}$$
$$= \frac{2.75\, \omega_{O_2,\infty} - \omega_{CO_2,w}}{3.667 + \omega_{CO_2,w}} . \tag{10.108}$$

---

[20] Also, $\dot{m}_{CO} = (1 + r_{st_3})\, \dot{m}_w.$

This equation is also derived in Turns [122]. Now, to evaluate $\omega_{CO_2,w}$, we invoke kinetically and physically controlled relations. Thus,

$$\dot{m}_{w,kin} \simeq \dot{m}_{w3} = \rho_w \, 4 \, \pi \, r_w^2 \left(\frac{k_3}{r_{st3}} \right) \omega_{CO_2,w} \tag{10.109}$$

$$\dot{m}_{w,phy} = 4 \, \pi \, r_w \, \Gamma_m \, \ln \, (1 + B_m)$$

$$= 4 \, \pi \, r_w \, \Gamma_m \, \ln \left[1 + \frac{2.75 \, \omega_{O_2,\infty} - \omega_{CO_2,w}}{3.667 + \omega_{CO_2,w}} \right]. \tag{10.110}$$

Recognizing that $\omega_{CO_2,w} << (\, 3.667$ and $2.75 \, \omega_{O_2,\infty})$[21] and equating $\dot{m}_{w,kin}$ with $\dot{m}_{w,phy}$, it can be deduced that

$$\omega_{CO_2,w} \simeq \left(\frac{\Gamma_m}{\rho_w \, r_w} \right) \left(\frac{3.667}{k_3} \right) \, \ln \left[\frac{3.667 + 2.75 \, \omega_{O_2,\infty}}{3.667} \right]. \tag{10.111}$$

Hence, B_m can be evaluated from Eq. 10.108 and the burn time can be estimated as

$$t_{burn} = \frac{\rho_p \, D_{w,i}^2}{8 \, \Gamma_m \, \ln \, (1 + B_m)}. \tag{10.112}$$

Flame Radius $r = r_f$

It is now possible to determine flame radius in Model 2. Thus, writing Eq. 10.46 with $r = r_f$, we have

$$\dot{m}_w = 4 \, \pi \, \Gamma \, r_f \, \ln \left[1 + \frac{\Phi_\infty - \Phi_f}{\Phi_f - \Phi_T} \right]. \tag{10.113}$$

Recognizing that at the flame front, $\omega_{O_2,f} = 0$, we have

$$\Phi_\infty - \Phi_f = \frac{\omega_{O_2,\infty}}{r_{st4}} - \frac{\omega_{CO_2,f}}{1 + r_{st4}} \tag{10.114}$$

$$\Phi_f - \Phi_T = (\Phi_f - \Phi_w) + (\Phi_w - \Phi_T)$$

$$= \left(\frac{\omega_{CO_2,f} - \omega_{CO_2,w}}{1 + r_{st4}} \right) + \left(\frac{r_{st3} + \omega_{CO_2,w}}{1 + r_{st4}} \right)$$

$$= \frac{\omega_{CO_2,f}}{1 + r_{st4}} + \frac{r_{st3}}{1 + r_{st4}}. \tag{10.115}$$

Therefore, it follows that

$$\frac{\Phi_\infty - \Phi_T}{\Phi_f - \Phi_T} = \frac{2.75 \, \omega_{O_2,\infty} + r_{st3}}{\omega_{CO_2,f} + r_{st3}}. \tag{10.116}$$

---

[21] If simplifying assumptions are not made, then $\omega_{CO_2,w}$ must be determined iteratively by equating Eqs. 10.109 and 10.110.

Similarly, if we let $\Phi = w_{N_2}$, then $\Phi_T = 0$, $\Phi_\infty = 1 - w_{O_2,\infty}$ and $\Phi_f = w_{N_2,f} = 1 - w_{CO_2,f}$. Therefore, Eq. 10.116 can also be written as

$$\frac{\Phi_\infty - \Phi_T}{\Phi_f - \Phi_T} = \frac{w_{N_2,\infty}}{1 - w_{CO_2,f}} . \tag{10.117}$$

Thus, equating Eqs. 10.116 and 10.117, $w_{CO_2,f}$ can be determined from the following equation:

$$\frac{w_{CO_2,f}}{w_{O_2,\infty}} = \frac{6.417}{1.75 \, w_{O_2,\infty} + 4.667} . \tag{10.118}$$

Knowing $w_{O_2,\infty}$, r_f can be determined using Eqs. 10.113 and 10.117, where $w_{CO_2,f}$ is determined from Eq. 10.118 and, $\dot{m}_w$ is determined from Eq. 10.109.

Region $r \geq r_f$

For this region, effects of simultaneity of R3 and R4 can be accounted by combining them as follows:

$$R5 = R3 + 2\,R4 : \quad C^* + O_2 = CO_2 . \tag{10.119}$$

Therefore, we can define conserved property as

$$\Phi = w_C - \frac{w_{O_2}}{r_{st5}} = w_C + \frac{w_{CO_2}}{1 + r_{st5}} = c_{p_m} T = w_{N_2} ,$$

where $r_{st5} = 32/12 = 2.667$. In this region, there is no heat of reaction, and hence, $\Phi = h_m = c_{p_m} T$. Taking the second definition, and following the arguments of Model 1, it can be shown that

$$w_{CO_2} = (1 + r_{st5}) \left(1 - \exp\left(-\frac{\dot{m}_w}{4\pi r \Gamma} \right) \right)$$

$$= (1 + r_{st5}) \left(1 - \frac{w_{N_2}}{w_{N_2,\infty}} \right) , \tag{10.120}$$

$$T = T_w + \frac{F_1 - \Delta H_{eff}}{c_{p_m}} , \tag{10.121}$$

$$F_1 = \left[\Delta H_{eff} + c_{p_m} (T_\infty - T_w) \right] \left(1 - \frac{w_{CO_2}}{1 + r_{st5}} , \right)$$

$$\Delta H_{eff} = \Delta H_3 - \frac{4\pi r_w^2 \, \epsilon \, \sigma \, (T_\infty^4 - T_w^4)}{\dot{m}_w} . \tag{10.122}$$

Problem 10.8 Estimate the initial burning rate and burn time of a $100\,\mu m$ carbon particle in still air at $300\,K$ and 1 atm using Model 1 and Model 2. The particle temperature is $1700\,K$. Assume $M_{mix} = 30$, $\rho_p = 1200\,kg/m^3$. Neglect radiation effects.

Model 1: In this model, taking $T_g \simeq T_w > 1650$ K, the kinetic rate coefficient k_1 (see Table 10.5) evaluates to

$$k_1 = (2.632 \times 10^{-5} \times 1700 - 0.03353) \times 1700 = 19.064 \text{ m/s}.$$

We now calculate mass diffusivity for $CO_2 \sim N_2$ pair (Appendix F) as $D_w = 11 \times 10^{-6} \times (1700/300)^{1.5} = 1.484 \times 10^{-4} \text{ m}^2/\text{s}$ and $\rho_w = (101325 \times 30)/(8314 \times 1700) = 0.215 \text{ kg/m}^3$. Therefore, $\Gamma_m = \rho_w D_w = 3.19 \times 10^{-5} \text{ kg/ms}$. Thus, from Eq. 10.87,

$$\frac{w_{O_2,\infty}}{w_{O_2,w}} \simeq 1 + \frac{0.215 \times 50 \times 10^{-6} \times 19.064}{3.19 \times 10^{-5}} = 7.425 \,.$$

Therefore, $w_{O_2,w} = 0.232/7.425 = 0.03124$. Further, from Eqs. 10.89 and 10.88,

$$R_{\text{kin}} = \left[0.215 \times 4\,\pi \times (50 \times 10^{-6})^2 \times 19.06 \times \left(\frac{12}{32} \right) \right]^{-1} = 2.071 \times 10^7 \text{ s/kg},$$

$$R_{\text{phy}} = \frac{2.667 + 0.03124}{4\,\pi \times 3.19 \times 10^{-5} \times 50 \times 10^{-6}} = 13.46 \times 10^7 \text{ s/kg}.$$

Thus, the particle combustion is largely physically controlled. From Eq. 10.90,

$$\dot{m}_w = \frac{0.232}{(13.46 + 2.071) \times 10^7} = 1.494 \times 10^{-9} \text{ kg/s} \,,$$

and, from Eq. 10.93,

$$t_{burn} = 4\,\pi \left(\frac{1200 \times (50 \times 10^{-6})^3}{0.232} \right) \left[\left(\frac{13.46}{2} + 2.071 \right) \times 10^7 \right] = 0.715 \text{ s}.$$

It is interesting to note that from the simpler model of Nusselt (Eq. 10.86), the burn rate is predicted as
$$\dot{m}_w = 4\,\pi\,(3.19 \times 10^{-5})\,(50 \times 10^{-6})\,(0.232/2.667) = 1.64 \times 10^{-9} \text{ kg/s}.$$

Model 2: In this model, the kinetic rate coefficient k_3 evaluates to

$$k_3 = 4.016 \times 10^8 \times \exp\left(-\frac{29790}{1700} \right) = 9.85 \text{ m/s}.$$

Therefore, iterating[22] between Eqs. 10.109 and 10.110, gives $w_{CO_2,w} = 0.137$. Thus, from Eq. 10.109

---

[22] Without iterations, Eq. 10.111 gives $w_{CO_2,w} = 0.177$.

$$\dot{m}_w = 0.215 \times 4\pi \times (50 \times 10^{-6})^2 \times \frac{12}{44} \times 9.85 \times 0.137$$
$$= 2.48 \times 10^{-9} \text{ kg/s}.$$

Further, from Eq. 10.108

$$B_m = \frac{2.75 \times 0.232 - 0.137}{3.667 + 0.137} = 0.1317.$$

Therefore, from Eq. 10.112

$$t_{burn} = \frac{1200 \times (100 \times 10^{-6})^2}{8 \times 3.19 \times 10^{-5} \times \ln(1 + 0.1317)} = 0.38 \text{ s}.$$

Thus, the initial burning rate from Model 2 is 65% greater than that from Model 1, and consequently, the burn time predicted by Model 2 is smaller. Of course, since T_w is large, prediction from Model 2 may be trusted. Estimates such as these help the designer to predict the required residence time for the particle in the furnace. The actual residence times will, of course, be greater for a variety of reasons. For example, in a burning particle cloud issuing from nozzles, $w_{O_2, \infty}$ experienced by each particle will be ≤ 0.232 owing to the presence of combustion products [9, 72, 129].

Finally, in this problem, the value of T_∞ (though given) does not enter the calculations anywhere because radiation has not been considered. The effect of T_∞ and radiation is considered in Problem 10.10. In the next problem, however, we estimate the flame radius.

Problem 10.9 In the previous problem, using Model 2, estimate $w_{CO_2, f}$ and r_f.
$w_{CO_2, f}$ can be determined from Eq. 10.118, or from Eq. 10.120. We take the former equation.

$$w_{CO_2, f} = \frac{6.417 \times 0.232}{4.667 + 1.75 \times 0.232} = 0.293$$

Therefore, from Eq. 10.116

$$\frac{\Phi_\infty - \Phi_T}{\Phi_f - \Phi_T} = \frac{2.75 \times 0.232 + 3.667}{0.293 + 3.667} = 1.087$$

Hence, from Eq. 10.113

$$r_f = \frac{2.48 \times 10^{-9}}{4\pi \times 3.19 \times 10^{-5} \times \ln(1.0871)} = 7.406 \times 10^{-5} \text{ m},$$

or

$$\frac{r_f}{r_{w,i}} = \frac{7.406 \times 10^{-5}}{50 \times 10^{-6}} = 1.481.$$

Table 10.6 Variation of T_w with T_∞ (Problem 10.10)

T_∞	$\dot{m}_w \times 10^9$	$w_{O_2,w}$	B_h	R_{phy}	R_{kin}	T_w
K	kg/s			$\times 10^{-6}$	$\times 10^{-6}$	K
2000	1.82	0.0175	0.0326	117.80	9.60	2193.6
1500	1.50	0.0310	0.0316	134.32	20.49	1704.1
1250	0.562	0.146	0.0134	157.70	254.90	1344.5
1000	9.8E-3	0.230	2.57E-4	188.2	24996	1001.50

Thus, the initial flame radius in carbon burning is much smaller than that for droplet burning ($r_f / r_w \simeq 30$). Carbon monoxide flame is observed as an envelope of pale blue gas surrounding the particle.

Problem 10.10 Using Model 1, determine T_w when $T_\infty = 1000, 1250, 1500$, and $2000\,$K. Assume $r_{w,i} = 50\,\mu$m and allow for radiation with $\epsilon = 1$.

In this problem, Eqs. 10.90 and 10.96 must be solved simultaneously, along with Eq. 10.87. Because these equations are nonlinear, an iterative solution must be obtained. For each T_∞, T_w is guessed until burn rates evaluated from Eqs. 10.90 and 10.96 are in agreement. The results of the computations are shown in Table 10.6. The results show that as T_∞ increases, so does the particle temperature T_w shown in the last column. The middle entry, in particular, shows that to keep T_w near 1700 K (a value assumed in Problem 10.8) T_∞ must be high at 1500 K when radiation is accounted for. Also, the results confirm that at high T_w, combustion is mainly physically controlled, but at low temperatures, it is kinetically controlled. For $T_\infty \leq 1000$ K, extinction condition is almost reached. In this problem, effect of T_∞ on T_w has been studied for the initial particle diameter. But as combustion proceeds, r_w decreases. Recall that radiation heat transfer is $\propto r_w^2$, and therefore, it becomes less important as the radius diminishes. As such, for a given T_∞, particle temperature T_w first increases, but then decreases, with $w_{O_2,w} \rightarrow w_{O_2,\infty}$, as r_w diminishes, and thus, leading to near-extinction of combustion. Study of effect of r_w on T_w is left as an exercise.

Exercises

1. With reference to Fig. 10.5, derive energy Eq. 10.4 by applying First law of thermodynamics to a spherical shell.
2. Consider the free-fall of a 1 mm water droplet in stagnant dry air starting from rest. Both air and water are at $25\,^\circ$C. Calculate the time for evaporation and the height of fall before complete evaporation. (Hint: As the droplet evaporates, its diameter decreases, and hence, the terminal velocity and the Reynolds number. Therefore, make quasi-static assumption. That is (see Eq. 10.18), replace (d/dt) $= u_p \times$ (d/dx) where x is the vertical coordinate.)
3. Water droplet 0.5 mm in dia evaporates in 25% RH air at $40\,^\circ$C and 1 atm. Determine the initial evaporation rate and temperature and vapor mass fraction at the droplet surface. Assume Le $= 1$, $cp_{v,m} = 1.86$ kJ/kg-K, and $h_{fg} = 2215$ kJ/kg.

(Hint: Here, T_w is not known. Therefore, $w_{v,w}$ and T_w must be iteratively determined by invoking equilibrium at the surface from Eqs. 10.11 and 10.37).

4. A 0.5 mm water droplet evaporates in hot dry air at 1000 K. Assume $T_w = T_{bp}$. Evaluate evaporation times t_{eva}, when p = 0.1, 1 and 10 bars. Comment on the results. (Hint: You will need to use Eq. 10.38 in which h_{fg}, T_{bp}, and properties are functions of p).

5. A 50 μm ethanol (C_2H_5OH) droplet evaporates in stationary N_2 environment at 800 K. Calculate evaporation time using (a) equilibrium assumption and $T_w = T_l$, and (b) $T_w = T_{bp}$.

6. Consider combustion of a 100 μm ethanol (C_2H_5OH) droplet in stagnant dry air at 1 atm and 300 K. Assuming $T_l = T_w = T_{bp}$, estimate (a) initial burning rate, (b) ratio of flame radius to droplet radius, (c) burning time, and (d) flame temperature.

7. In the previous problem, if the relative velocity between the droplet and air is 65 m/s, estimate the droplet burn time. Evaluate Nusselt number using

$$Nu = 2 + 0.555 \, Pr^{0.33} \left[\frac{Re}{1 + 1.232/(Re \, Pr^{1.33})} \right]^{0.5}.$$

Assume $PF_c = 1$ (Hint: Numerical integration is required).

8. Repeat the previous problem, if p = 10 atm and $T_\infty = 800$ K. Neglect flame radiation.

9. A 100 μm wood particle, initially at $T_{pi} = 300$ K contains 30% moisture on dry basis. The particle is introduced in a furnace at 1200 K with relative velocity 20 m/s. The dry particle density is 780 kg/m³. Estimate the drying time and the distance traveled while drying. Assume $\epsilon = 1$ and $cv_{dp} = 1000$ J/kg-K. Is drying dominated by convection or radiation?

10. In the previous problem, the dried particle first heats up from 0.5 $(T_{pi} + T_{bp})$ to $T_{vol} = 450 + 273 = 723$ K. The particle then devolatilizes at constant temperature T_{vol}. Estimate

   a. Heating up time (Hint: Numerical integration is required)
   b. Time required for 90% devolatilization using data from Table 10.4

Comment on the results.

11. Repeat Problem 10.8 in the text for $w_{O_2,\infty} = 0.05, 0.1, 0.15,$ and 0.2. Also, evaluate $\dot{m}_w$ and t_{burn}, using Nusselt's relation 10.86. Comment on the result.

12. Justify Eq. 10.83 for Model 1.

13. Show that in Model 2, Φ can also be taken as

$$\Phi = Cp_m \, T - \left(\frac{w_{CO_2}}{1 + r_{st4}} \right) \Delta H_4.$$

Hence, show that

$$Q_w = \dot{m}_w \, \Delta H_3 + \dot{m}_{CO} \, \Delta H_4 - 4\,\pi\, r_w^2 \,\epsilon\, \sigma\, (T_\infty^4 - T_w^4),$$

$$B_h = \frac{\Phi_\infty - \Phi_w}{\Phi_w - \Phi_T}$$

$$= \frac{(1 + r_{st4})\, cp_m\, (T_\infty - T_w) + \omega_{CO_2,w}\, \Delta H_4}{(1 + r_{st4})\, \{\Delta H_{eff} + (1 + r_{st3})\, \Delta H_4\} - (r_{st3} + \omega_{CO_2,w})\, \Delta H_4},$$

$$\dot{m}_w = 4\,\pi\, r_w\, \Gamma_m\, \ln\,(1 + B_h),$$

where ΔH_{eff} is given by Eq. 10.122.
14. It is of interest to plot variations of Φ, ω_{CO_2}, ω_{N_2}, and T as functions of radius r. Hence, using relations shown in Eqs. 10.97 to 10.100, and derivations of the previous problem, show that

$$\Phi = \frac{\omega_{O_2}}{r_{st4}} + \frac{\omega_{CO_2}}{1 + r_{st4}} = \left(\frac{\omega_{O_2,\infty}}{r_{st4}}\right) \exp\left(-\frac{A}{r}\right) \text{ for all r } \quad A = \frac{\dot{m}_w}{4\,\pi\,\Gamma}.$$

$$\omega_{N_2} = \omega_{N_2,\infty} \exp\left(-\frac{A}{r}\right) \text{ for all r }.$$

$$\omega_{CO_2} = \omega_{CO_2,w} + (1 + r_{st4})\,(\Phi - \Phi_w) \text{ for } r \le r_f,$$
$$= \text{From equation } 10.120 \text{ for } r > r_f.$$

$$T = T_w + \left[\frac{\omega_{CO_2} - \omega_{CO_2,w}}{cp_m\,(1 + r_{st4})}\right] \times F_1 \text{ for } r \le r_f,$$

$$F_1 = \Delta H_4 + \left(\frac{r_{st4}}{\omega_{O_2,\infty}}\right)\left[cp_m\,(T_\infty - T_w) + \Delta H_{eff} + \left(1 + \frac{r_{st3}\,r_{st4}}{1 + r_{st4}}\right)\Delta H_4\right],$$

$$T = \text{From equation } 10.121 \text{ for } r > r_f.$$

15. Using results of Model 2 for $\dot{m}_w$, $\omega_{CO_2,w}$, $\omega_{CO_2,f}$, and r_f in Problems 10.8 and 10.9 in the text, plot variations of Φ, ω_{O_2}, ω_{CO_2}, ω_{CO}, ω_{N_2}, and T as functions of radius ratio r/r_w. Extend calculations up to $(r/r_w) = 20$. Note values of T_f, $\omega_{CO,w}$, and $\omega_{N_2,w}$.
16. Using Model 1, determine char particle temperature T_w for $0.1 < r_w < 150\ \mu$m. Assume $T_\infty = 1600$ K. Allow for radiation with $\epsilon = 1$. Plot variations of T_w, $\dot{m}_w$, $\omega_{O_2,w}$, R_{phy}, and R_{kin} with r_w with appropriate scalings. Comment on the result.

Chapter 11
Combustion Applications

11.1 Introduction

As mentioned in Chap. 1, the main purpose of combustion science is to aid in performance prediction or design of practical equipment as well as to aid in prevention of fires and pollution. This is done by analyzing chemical reactions with increasing refinements, namely, stoichiometry, chemical equilibrium, chemical kinetics, and finally, fluid mechanics. Such analyses of practical equipment are accomplished by modeling. The mathematical models can be designed to varying degrees of complexity, starting from simple zero-dimensional thermodynamic stirred-reactor models to one-dimensional plug-flow models and, finally, to complete three-dimensional models. At different levels of complexity, the models result in a system of a few algebraic equations, a system of ordinary differential equations, a system of partial differential equations.

Practical combustion equipment typically has very complex geometry and flow paths. The governing equations of mass, momentum, and energy must be solved with considerable empirical inputs. The size of the mathematical problem is dictated by the reaction mechanism chosen and by the chosen dimensionality of the model. The choice of the model thus depends on the complexity of the combustion phenomenon to be analyzed, the objectives of the analysis, and the resources available to track the mathematical problem posed by the model.

In this chapter, a few relevant problems are formulated and solved as *case studies* to help in understanding of the *art of modeling* as applied to design and performance prediction of practical combustion equipment. In each case, material of previous chapters will be recalled, along with some new information specific to the case at hand.

© Springer Nature Singapore Pte Ltd. 2020
A. W. Date, *Analytic Combustion*,
https://doi.org/10.1007/978-981-15-1853-9_11

11.2   Wood-Burning Cookstove

In developing countries, a large portion of people living in rural areas carry out cooking in cookstoves that burn firewood, cow-dung cakes, or agricultural residues. It is estimated [29] that an average family of five typically burns one ton of fuelwood for cooking and 0.8 tons of fuelwood for space-heating, amounting to nearly 60% to 80% of total energy consumption in rural households. Hidden behind such consumption is the drudgery of women who typically spend two to three hours per day for collection of fuelwood. It is, therefore, important to redesign existing wood-burning stoves to obtain higher thermal and combustion efficiencies to reduce drudgery and indoor pollution. Such a redesigned stove must be simple and easy to fabricate and retrofit, and also must facilitate dissemination of improved stoves to large numbers of families in need.

Families adopt a large variety of stove designs, depending on family size and cooking practice [24, 31, 105]. Some stoves are simple three-stone open fires, whereas others are more sophisticated multi-pot stoves made of clay with enclosed combustion spaces and damping-control over flow of naturally aspirated air that is required for combustion. Most importantly, in almost all traditional stoves, there is no provision of a grate; the fuelwood is simply rested on the floor. Sometimes, chimney is provided to drive away the smoke.

Considerable research is directed toward evolving new stove designs that have much higher efficiencies, provide for clean combustion, and are easy to manufacture and disseminate. Comprehensive reviews of the literature pertaining to the above aspects can be found in [76, 117].

Because of the large variations in design , each stove type must be experimentally tested separately [52] or analyzed through a model [77]. However, to find general directions for improved design, tests can be carried out in a laboratory on an *experimental stove*. Such tests and modeling are reported in Ref. [12] with respect to the CTARA[1] experimental stove shown in Fig. 11.1.

11.2.1 CTARA Experimental Stove

Figure 11.2 shows a schematic of the experimental stove design. To carry out modeling, the combustion space of the cylindrical experimental-stove is divided into four zones, each of which is treated as a well-stirred reactor. Zone 1 comprises the space below the grate (diameter D_{gr}) placed at height H_{gr} inside the stove of total height H_t. There are $N_{pr} = 5$ semicircular primary ports (diameter D_{pr}) at the bottom of the stove. A conical clay enclosure, which is placed on the grate to height H_{th}, forms Zone 2. In this zone, wood is inserted through a semicircular opening (diameter D_{ent}). The throat (diameter D_{th}) of the conical clay enclosure is connected to a metal enclosure with a flared top, which forms Zone 3. This enclosure has $N_{hole} = 24$ tiny

---

[1]Center for Technology Alternatives for Rural Areas (CTARA) IIT Bombay.

Fig. 11.1 CTARA experimental stove: (left) assembly, (right) parts (outer body, grate, metal enclosure, and conical clay enclosure.)

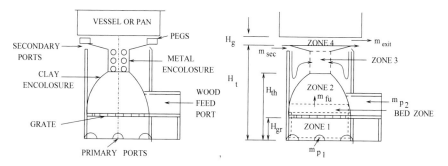

Fig. 11.2 CTARA experimental stove; (left) stove design, (right) zoning of combustion space

holes (diameter D_{hole}) through which secondary air entrains from $N_{sec} = 6$ semicircular ports (diameter D_{sec}) on stove outer body. This air is preheated to temperature T_{sec} through contact with the outer surfaces of the clay and metal enclosures. Finally, the space between the stovetop and pan bottom (distance H_g) is designated as Zone 4, through which the burned products of combustion exit. In addition, a wood-bed zone is demarcated just above the grate.

11.2.2 Modeling Considerations

In naturally aspirated wood-burning stoves, the airflow is driven by buoyancy forces that overcome the flow resistances inside the stove. This airflow, in turn, determines the wood-burning rate, as well as the overall thermal and combustion efficiencies. Large-diameter (typically, $D_{wo} = 0.5$–5 cm) wood is fed intermittently into a stove; the frequency depends on the firepower required by the user. In addition, wood-burning itself is a transient phenomenon accompanied by volatiles evolution and char burning layer by layer. Given the complex geometry of the combustion space, the flow and associated heat/mass transfer and combustion phenomena are thus three-dimensional and time-dependent.

Most experimental studies on stoves are concerned with the determination of thermal efficiency (η_{th}) [7, 33, 36] and/or emissions [13, 66] mainly for *water-boiling* tasks. A single experiment is typically run for a period of one hour [12, 33]. As such, the reported measured data on efficiency and emissions are *averaged* over the one hour period. This implies that the thermal mass effects of the stove body are also considered negligible. As such, a *steady-state* model of the stove is desirable. However, even after neglecting thermal mass effects, the main difficulty in modeling lies in finding plausible expressions for *steady-state* burning rate of wood. This matter is discussed below.

Kinetically Controlled Burning Rate

Let $\dot{m}_{fu}$ (kg/s) be the average wood-burning rate. To model this burning rate, we invoke the Stefan flow model introduced in Chap. 6. It is thus assumed that wood is being fed into the stove *exactly at its burning rate*, with constant burning surface area A_{wo}. Then,

$$\dot{m}_{fu} = \dot{m}_{char} + \dot{m}_{vol} \quad \dot{m}_{vol} = x_{vol} \, \dot{m}_{fu} \quad \dot{m}_{char} = (1 - x_{vol}) \, \dot{m}_{fu}, \qquad (11.1)$$

where x_{vol} is the volatile mass fraction in wood. Therefore, the char fraction is $1 - x_{vol}$.

In the literature, experimental data on instantaneous burning rates of large diameter wood are scarce. However, Tinney [120] and Blackshear and Murthy [17] reported such data. Tinney has reported experimental data on burning of dry wooden dowels 5 inches long (1/4, 3/8, and 1/2 in diameter) placed in a preheated air furnace (3 ft^3) whose temperature was held constant in the range 300 to 650 °C. Continuous weight-loss histories were recorded using a strain-gauge-cantilever-beam weighing system. In Ref. [17], temperature histories are recorded at different radii (including surface and centerline) of a cellulosic cylinder using thermocouples. For the furnace temperature of 600 °C, for example, the wood surface temperature was found to increase with time from about 300 °C to 550 °C. Temperature histories at all radii followed similar trends, but the temperatures decreased with reducing radii, with temperature levels at the centerline being lowest. The rate of increase of temperature at any radius, however, showed two distinct phases: in phase 1 at lower temperatures, the rate of temperature increase is rapid, whereas in phase 2 at higher temperatures, the rate is slower. Phase 1 typically occupied about 20% to 40%, whereas phase 2 occupied 80% to 60% of the total burning time. From these observations, we view the char burning and volatiles-evolution as two *sequential* processes and make the following postulates. Thus,

1. Volatiles evolution will dominate at lower temperatures and therefore is assumed to occupy 40% of the total time in phase 1.
2. Char burning will dominate at higher temperatures and therefore is assumed to occupy 60% of the total time in phase 2.

Table 11.1 Rate constants [120]

D_{wo}	θ_{vol}	θ_{char}	E_{vol}	E_{char}
(m)	(hr^{-1})	(hr^{-1})	(K)	(K)
0.0254	2.16×10^{11}	1.44×10^{12}	15000	21500
0.0126	1.26×10^{12}	4.30×10^{12}	15000	19900
0.0063	2.7×10^{12}	7.20×10^{12}	15000	18300

From these postulates, we adopt the following expressions for the *average steady-state burning flux* (kg/m²-s) as

$$\frac{\dot{m}_{fu}}{A_{wo}} = \frac{\dot{m}_{char} + \dot{m}_{vol}}{A_{wo}} \simeq \frac{\rho_{wo} D_{wo}}{4} \left[\frac{R_{char} + R_{vol}}{2} \right], \tag{11.2}$$

$$R_{vol} = 0.4 \, x_{vol} \, \theta_{vol} \, \exp\left\{ - \left(\frac{E_{vol}}{T_{wo,vol}} \right) \right\}, \tag{11.3}$$

$$R_{char} = 0.6 \, (1 - x_{vol}) \, \theta_{char} \, \exp\left\{ - \left(\frac{E_{char}}{T_{wo,char}} \right) \right\}, \tag{11.4}$$

$$\theta_{vol} = 16.944 \times 10^8 \times \exp(-133.0 \times D_{wo}) \ (s^{-1}), \tag{11.5}$$

$$\theta_{char} = 34.277 \times 10^8 \times \exp(-84.4 \times D_{wo}) \ (s^{-1}), \tag{11.6}$$

$$E_{vol} = 15000 \ (K), \quad \text{and} \tag{11.7}$$

$$E_{char} = 17244.5 + 1.6754 \times 10^5 \times D_{wo} \ (K), \tag{11.8}$$

where $T_{wo,vol}$ and $T_{wo,char}$ are representative *surface temperatures* during the volatiles evolution and char-burning phases, respectively. The rate constants θ and E are evaluated from curve fit to the discrete data for wooden dowels of 3 different diameters (see Table 11.1).

We now make a drastic assumption and postulate that representative wood surface temperatures in the two phases can be taken as

$$T_{wo,vol} = T_{wo} - 50 \quad \text{and} \quad T_{wo,char} = T_{wo} + 50, \tag{11.9}$$

where T_{wo} is the *average steady-state wood surface temperature of wood.*[2] Equation 11.2 can thus be used to determine T_{wo} when the burning mass flux ($\dot{m}_{fu}/A_{wo}$) is known.

Physically Controlled Burning Rate

To determine the steady-state burning flux, we carry out energy balance at the wood surface. Then, following the Reynolds flow model of Spalding [110] (see Chap. 6), it is assumed that the heat transfer by convection and radiation ($q_{rad+conv}$) to the wood

---

[2]The choice of 50 in Eq. 11.9 is arbitrary. However, computational experience gathered through sensitivity analysis of the model has suggested that this value will suffice.

surface will be balanced by the burning-rate multiplied by the enthalpy change of the transferred material at the surface. Thus,

$$q_{rad+conv} = \frac{\dot{m}_{fu}}{A_{wo}} \{h_{wo} - (h_T + \lambda_{wo})\} , \quad \text{or}$$

$$\frac{\dot{m}_{fu}}{A_{wo}} = \frac{q_{rad} + q_{conv}}{h_{wo} - h_T - \lambda_{wo}}, \tag{11.10}$$

$$h_T = Cp_{wood} (T_T - T_{ref}) \tag{11.11}$$

$$\lambda_{wo} = 10467.5 (4926 D_{wo} + 38) \quad (\text{J/kg}) \quad [103] \tag{11.12}$$

where $h_{wo} = cp_m (T_{wo} - T_{ref})$ is the enthalpy of the gas phase at the surface temperature T_{wo} and h_T is the enthalpy of the *transferred substance*, or that of the wood-substance deep inside the wood. Following [34, 120], we take $T_T \simeq (T_{wo} - 150)$ K. Finally, λ_{wo} represents the *latent heat* of wood. This latent heat is notional that mimics wood-burning as volatile liquid fuel burning. Equation 11.12 is a curve-fit to experimental data presented by Simmons and Lee [103]. The expressions for q_{rad} and q_{conv} will be developed during the course of further analysis.

Thus, equating the kinetically controlled Eq. 11.2 and the physically controlled Eq. 11.10 enables determination of *steady-state* wood surface temperature T_{wo} and the wood-burning flux $\dot{m}_{fu}/A_{wo}$. These two equations[3] are very important because they provide the freedom for thermo-fluid-dynamic modeling of the combustion space based on the laws governing transport of bulk and species mass, momentum (or force balance), and energy. In the next subsection, the laws are applied to the four selected zones of the combustion space.

11.2.3 Zonal Modeling

Zone 1:

To carry out modeling, we first postulate flow patterns in Zone 1 (Fig. 11.3) below the grate and in Zone 2 (Fig. 11.4) above the grate. Thus, it is assumed that the aspirated air enters through the semicircular primary ports with velocity V_{in}, expands and bends to an imaginary area $A_m = 0.5\times$ the grate area A_{gr}, and then again expands to occupy the total grate area. The flow then overcomes the resistance of the grate bars.

Thus, N_{pr} semicircular primary ports experience resistance owing to entry ($k_{in} = 0.6$), expansion ($k_{exp1} = (1 - A_{pr}/A_m)^2$), and merging and bending ($k_{bend} = 1.0$). The air then further expands ($k_{exp2} = (1 - A_m/A_{gr})^2$) from A_m to A_{gr} and experiences resistance ($k_{gr} = 0.75 \{(\delta + D_{rod})/\delta) - 1\}^{1.33}$) offered by the grate. Thus, invoking the momentum balance principle (that is, buoyancy force = flow resistance), we have

---

[3]Recall that in Chap. 10, kinetic and physically controlled rates were equated in this way with respect to coal particle burning.

Fig. 11.3 CTARA
experimental stove: Zone 1

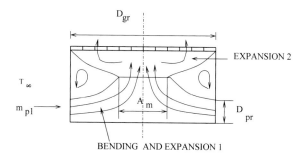

$$\Delta p_{o-gr} = g \, H_{gr} \, (\rho_\infty - \rho_{z2})$$

$$= \rho_\infty \left\{ N_{pr}(k_i + k_{bend}) + k_{exp1} \right\} \frac{V_{in}^2}{2}$$

$$+ \rho_m \, k_{exp2} \frac{V_m^2}{2} + \rho_{gr} \, (1 + k_{gr}) \frac{V_{gr}^2}{2}, \qquad (11.13)$$

where the velocities at different locations are evaluated from mass balance principle as

$$\dot{m}_{p1} = \rho_\infty \, A_{pr} \, V_{in} = \rho_m \, A_m \, V_m = \rho_{gr} \, A_{gr} \, V_{gr}, \qquad (11.14)$$

and $A_{pr} = N_{pr} \, (\pi/8) \, D_{pr}^2$. Substitution of Eq. (11.14) in Eq. (11.13) gives

$$g \, H_{gr} \, (\rho_\infty - \rho_{z2}) = \left[\frac{N_{pr} \, (k_{in} + k_{exp1}) + k_{bend}}{\rho_\infty \, A_{pr}^2} + \frac{k_{exp2}}{\rho_{z1} \, A_m^2} + \frac{(1 + k_{gr})}{\rho_{z2} \, A_{gr}^2} \right]$$

$$\times \frac{(\dot{m}_{p1})^2}{2}, \qquad (11.15)$$

where zonal densities $\rho_{z2,z1} = \rho_\infty \, (T_\infty/T_{z2,z1})$, T_{z2} is the average temperature of Zone 2, and T_{z1} is the average temperature of Zone 1. T_∞ is the ambient temperature. The k-values[4] are taken from Kazantsev [63]. Knowing T_{z2}, $\dot{m}_{p1}$ can be determined. In this zone, there are no chemical reactions, so species equations are not invoked.

Bed Zone: (See Figs. 11.2 and 11.4)

In this zone, the char burns and volatiles evolve at the rate given by Eqs. 11.2 and/or 11.10. The composition of gaseous products of char-burning and volatiles, however, must be determined with care. The procedure is as follows:

---

[4]From zero-dimensional modeling, $\Delta p = k \, V^2/2$ [126], where k captures the effects of friction and momentum change.

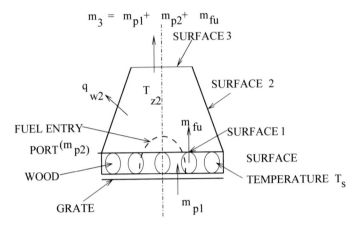

Fig. 11.4 CTARA experimental stove: bed zone + Zone 2

Composition of Volatiles

Although different wood species will have different chemical formulas, we take a generic formula [119]. Thus, the chemical formula for wood is taken as

$$C_{4.42}\ H_{6.1}\ O_{2.42}\ N_{0.01}\ (H_2O)_l \quad \text{and} \quad l = \frac{98}{18} \times \frac{f}{1-f},$$

where f is moisture fraction in wood on wet basis.
   Further,

$$m_{wo,dry} = 4.42 \times 12 + 6.1 \times 1 + 2.42 \times 16$$
$$+ 0.01 \times 14 = 98, \tag{11.16}$$

$$m_{char} = m_{wo,dry} \times (1 - x_{vol}) = 98 \times (1 - x_{vol}), \tag{11.17}$$

$$n_{char} = \frac{m_{char}}{M_{char}} = \frac{m_{char}}{12} = 8.167 \times (1 - x_{vol}). \tag{11.18}$$

It is now assumed that the volatiles will comprise five species $j = CO_2$, CO, H_2, H_2O, and $HC \equiv C_m H_{2m+2}$ where the value of m is to be chosen by the analyst. Thus, we write

$$C_{4.42-n_{char}}\ H_{6.1}\ O_{2.42} = n_{CO}\ CO + n_{CO_2}\ CO_2 + n_{H_2}\ H_2 + n_{H_2O}\ H_2O + n_{HC}\ HC. \tag{11.19}$$

Carrying out element balance, we have

$$\text{C-balance} \rightarrow 4.42 - n_{char} = n_{CO} + n_{CO_2} + m\ n_{HC}. \tag{11.20}$$

$$\text{H-balance} \rightarrow 6.1 = 2\ n_{H_2} + 2\ n_{H_2O} + (2m + 2)\ n_{HC}, \tag{11.21}$$

$$\text{O-balance} \rightarrow 2.42 = n_{CO} + 2\ n_{CO_2} + n_{H_2O}. \tag{11.22}$$

Table 11.2 Mass fraction ratios [14]

$T_{reactor}$	873 K	973 K	1073 K
$\omega_{CO}/\omega_{CO_2}$	1.515	2.25	2.7
$\omega_{H_2O}/\omega_{CO_2}$	1.78	1.491	1.15

Besides the above 3 constraints, we need 2 more constraints to determine 5 unknowns. For example, Shah and Date [100], using experimental data of Ragland [94] adopted following relations:

$$\frac{\omega_{CO}}{\omega_{CO_2}} = 1.591 \quad \text{and} \quad \frac{\omega_{H_2O}}{\omega_{CO_2}} = 2.174 \,. \tag{11.23}$$

The ratios are independent of the temperature. However, experimental data of Boroson et al. [14] suggest that the ratios are in fact functions of the reactor/furnace temperature as shown in Table 11.2.

In the steady-state analysis of wood-burning stoves, one deals with space- and time-averaged temperatures. The zonal averaged temperatures above the wood surface are found to be in the range $650\,K < T_{z2} < 750\,K$. Hence, to obtain realistic estimates, we adopt, following values for the ratios:

$$\frac{\omega_{CO}}{\omega_{CO_2}} = 1.3 \quad \text{and} \quad \frac{\omega_{H_2O}}{\omega_{CO_2}} = 1.9 \,. \tag{11.24}$$

Note that both values are smaller than those shown in Eq. 11.23 and respond to the tendencies exhibited in Table 11.2.

Thus, inter-substitutions between Eqs. 11.20, 11.21, 11.22, and 11.24 give the number of moles (n_j) of each species j. Note that n_j will be different for different values of m in the HC formula and for different values of volatile fraction x_{vol} in the wood. The value of m must be selected such that all predicted values of n_j are positive. Among the several feasible values of m, here we take $HC \equiv C_7H_{16}$ (that is, m = 7) to be a suitable hydrocarbon *to effectively represent both light and heavy (soot and tar) hydrocarbons*. The evolution rates of all species j are thus determined from

$$\dot{m}_{j,vol} = \left(n_j \frac{M_j}{M_{vol}} \right) \times \dot{m}_{vol} \quad \rightarrow \quad j = CO, CO_2, H_2, H_2O, HC, \tag{11.25}$$

where $\dot{m}_{vol}$ is given by Eq. 11.1.

Char Surface Reactions

The char surface reactions, in general, produce CO, CO_2, and H_2, which then burn in the gas phase. In wood-burning stoves, however, because the wood surface temperatures are very low ($T_{wo} < 1200\,K$), production of CO_2 must dominate overproduction of CO [30, 68]. As such, the char surface reaction

$$RO: \quad C^* + O_2 = CO_2 \quad \rightarrow \quad \Delta H_c = 14 \times 10^6 \ (J/kg)$$

is the only reaction[5] assumed so that the rate of CO_2 from the surface, and of O_2 to the surface will be given by

$$\dot{m}_{CO_2,add} = \dot{m}_{char} \ (M_{CO_2}/M_{char}) \quad \text{and} \quad \dot{m}_{O_2,add} = - \dot{m}_{char} \ (M_{O_2}/M_{char}) \ .$$

These flow rates are now added to the gas phase. Similarly, the unbound moisture $\dot{m}_{vap,add} = (18 \, l/M_{wo}) \times \dot{m}_{fu}$ is also added to the gas phase. For all other species, $\dot{m}_{j,add} = 0$. Finally, the mass fractions of species leaving the fuel bed $(\omega_{j,bed})$ are evaluated from the species balance

$$\omega_{j,bed} = \frac{\dot{m}_{p1} \, \omega_{j,\infty} + \dot{m}_{j,vol} + \dot{m}_{j,add}}{\dot{m}_{bed}} \tag{11.26}$$

$$\dot{m}_{bed} = \dot{m}_{p1} + \dot{m}_{vap} + \dot{m}_{fu} \tag{11.27}$$

where $j = CO_2$, CO, H_2, H_2O, C_7H_{16}, O_2, and N_2.

Zone 2 (See Fig. 11.4)

In this zone, the primary air leaving the grate overcomes the resistance of the bed zone and then mixes with the ambient air entering from the fuel entry port $(\dot{m}_{p2})$. The mixed gases then converge and leave through the throat area (A_{th} and surface s_3—see Fig. 11.4). Applying momentum balance to this zone, we have

$$g \left\{ H_{th} - (H_{gr} + N_{row} \times D_{wo}) \right\} (\rho_\infty - \rho_{s3}) = \frac{k_{bed}}{2 \, \rho_{bed}} \left(\frac{\dot{m}_{bed}}{A_{gr}} \right)^2 + \frac{k_{ent}}{2 \, \rho_\infty} \left(\frac{\dot{m}_{p2}}{A_{ent}} \right)^2$$
$$+ \frac{(1 + k_{cont})}{2 \, \rho_3} \left(\frac{\dot{m}_{bed} + \dot{m}_{p2}}{A_{th}} \right)^2$$
$$- \frac{1}{2 \, \rho_{gr}} \left(\frac{\dot{m}_{bed}}{A_{gr}} \right)^2 . \tag{11.28}$$

Here, $k_{ent} = 0.6$ and $k_{bed} \simeq 1.2 \, (0.667 \, N_{row} + 0.5)$, where inclusion of number of rows N_{row} of wood pieces is an acknowledgment that the bed resistance will increase with increase in number of rows. Further, $k_{cont} = 0.5 \sin (\Phi) \, (1 - A_{th}/A_{gr})$ [63] is the resistance coefficient for contraction from A_{gr} to A_{th}. Here, Φ is the half-angle of the conical clay enclosure. $\rho_{s3} = \rho_\infty \times (T_{s3}/T_\infty)$ represents the density of surface s3 at the throat. Knowing T_{s3}, the quadratic equation 11.28 determines $\dot{m}_{p2}$.

---

[5]The heat of combustion ΔH_{char} for this reaction is 32.6 MJ/kg when $C^* \equiv$ graphite [122]. But, for lighter wood char, it is taken as 14 MJ/kg.

To determine T_{s3}, the following energy equation is solved with reference to Fig. 11.4.

$$\dot{m}_{s3}\, c p_m\, (T_{s3} - T_{ref}) = \dot{Q}_{char} + \dot{Q}_{vol,z2} - q_{rads1}\, A_{gr} - q_{conv,bed}\, A_{wo}$$
$$+ \dot{m}_{bed}\, c p_m\, (T_{wo} - T_{ref}) + \dot{m}_{p2}\, c p_\infty\, (T_\infty - T_{ref})$$
$$- (q_{rads2} + \alpha_{s2}\, (T_{z2} - T_{s2i}))\, A_{s2}\,, \tag{11.29}$$

where $\dot{Q}_{char} = \dot{m}_{char}\, \Delta H_{char}$ and α_{s2} is the convection coefficient at the conical wall surface 2. $\dot{Q}_{vol,z2}$ represents heat released from volatiles combustion. The radiation heat transfers $q_{rad,s1,s2}$ (fluxes to surfaces 1 and 2) are determined from enclosure theory (see Sect. 11.2.4).

The volatiles combustion is modeled using Hautmann's [43] quasi-global reaction mechanism comprising four reaction steps whose rate constants are given in Table 5.7. Then, the net consumption/generation rates of each species $R_{j,net}\,(T_{z2}, \omega_{j,z2})$ are evaluated at temperature T_{z2} and $\omega_{j,z2}$, and mass fractions $\omega_{j,s3}$ leaving the zone are evaluated from

$$\omega_{j,s3} = \frac{\dot{m}_{bed}\, \omega_{j,bed} + \dot{m}_{j,p2}\, \omega_{j,\infty} + R_{j,net}\, \Delta V_{z2}}{\dot{m}_{s3}}\,, \tag{11.30}$$

$$\dot{m}_{s3} = \dot{m}_{bed} + \dot{m}_{p2}, \tag{11.31}$$

and $j = CO_2$, CO, H_2, H_2O, C_7H_{16}, C_2H_4, O_2, and N_2. Of course, $\omega_{j,\infty}$ refers only to ambient air species O_2, N_2 and H_2O. Finally, $Q_{vol,z2} = \sum R_{j,net}\,(T_{z2}, \omega_{j,z2})$ $\Delta V_{z2}\, \Delta H_j$ is evaluated.

Zone 3 (See Fig. 11.5)

In this zone, the secondary air flow rate $\dot{m}_{sec}$ is predicted from force-balance Eq. 11.32:

$$g\,(H_t + H_g - H_{th})\,(\rho_\infty - \rho_{ex}) = \frac{(1 + k_{exp,eff})}{2\,\rho_{ex}}\left(\frac{\dot{m}_{s3} + \dot{m}_{sec}}{A_{gap}}\right)^2$$
$$+ \frac{k_{sec,eff}}{2\,\rho_{sec}}\left(\frac{\dot{m}_{sec}}{A_{sec}}\right)^2$$
$$- \frac{1}{2\,\rho_{s3}}\left(\frac{\dot{m}_{s3}}{A_{th}}\right)^2\,, \tag{11.32}$$

where the subscript ex connotes exit, and resistances owing to expansion and turning from Zones 3 to 4 in the metal enclosure below the pan ($k_{exp,eff}$), and the entry of secondary air through semicircular holes on the stove's outer body and in the cylindrical metal enclosure ($k_{sec,eff}$) are taken as

Fig. 11.5 CTARA
experimental stove: Zone 3
and 4

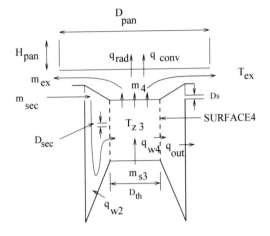

$$k_{exp,eff} = \left(1 - \frac{A_{th}}{A_{pan}}\right)^2 + \left(1 - \frac{D_{th}}{D_{pan}}\right)^2 \text{ and} \tag{11.33}$$

$$k_{sec,eff} = k_{sec}\left[1 + \left(\frac{k_{hole}}{k_{sec}}\right)\left(\frac{\rho_{sec}}{\rho_\infty}\right)\left(\frac{A_{sec}}{A_{hole}}\right)^2\right]. \tag{11.34}$$

Solution of quadratic Equation 11.32 requires temperatures T_{sec}, T_4, and T_{ex}. Thus, T_4 is determined from the energy balance

$$\dot{m}_4\, C_{pm}\, (T_4 - T_{ref}) = \dot{Q}_{vol,z3} + \dot{m}_{s3}\, cp_m\, (T_{s3} - T_{ref}) + \dot{m}_{sec}\, cp_{sec}\, (T_{sec} - T_{ref})$$
$$- (q_{rad,4} + \alpha_{s4i}\, (T_{z3} - T_{s4i}))\, A_{s4}\,, \tag{11.35}$$

where T_{sec} is evaluated from the energy balance for the space between the stove's outer body and the enclosures as derived below:

$$\dot{m}_{sec}\, cp_{sec}\, (T_{sec} - T_{ref}) = \dot{m}_{sec}\, cp_\infty\, (T_\infty - T_{ref}) + \alpha_{s4o}\, (T_{s4,o} - T_{sec})\, A_{w4}$$
$$+ \alpha_{s2o}\, (T_{s2o} - T_{sec})\, A_{s2}$$
$$- U_{out}\, (T_{sec} - T_\infty)\, A_{out}\,, \tag{11.36}$$

$$T_{s2,i} = \frac{\alpha_{s2i}\, T_{z2} + q_{rad2} + Z_2\, T_{sec}}{Z_2 + \alpha_{s2i}}\,, \tag{11.37}$$

$$T_{s4,i} = \frac{\alpha_{s4i}\, T_{z3} + q_{rad4} + Z_4\, T_{sec}}{Z_4 + \alpha_{s4i}}\,, \tag{11.38}$$

$$T_{s2o} = Z_2\left(\frac{T_{s2i}}{\alpha_{s2o}} + \frac{b_2}{K_2}\, T_{sec}\right)\,, \tag{11.39}$$

$$T_{s4o} = Z_4\left(\frac{T_{s4i}}{\alpha_{s4o}} + \frac{b_4}{K_4}\, T_{sec}\right)\,, \text{ and} \tag{11.40}$$

$$U_{out} = \left(\frac{1}{\alpha_{outi}} + \frac{1}{\alpha_{outo}} + \frac{b_{out}}{K_{out}} \right)^{-1} , \tag{11.41}$$

where

$$Z_2 = \left(\frac{b_2}{K_2} + \frac{1}{\alpha_{s2o}} \right)^{-1} \quad \text{and} \quad Z_4 = \left(\frac{b_4}{K_4} + \frac{1}{\alpha_{s4o}} \right)^{-1} .$$

In Eq. 11.41, b_2, b_4, and b_{out} represent wall thicknesses (having conductivity K), i and o connote the inner and outer sides of the respective surfaces, and subscript out refers to the stove's outer body.

To evaluate $Q_{vol,z3}$, we solve species Eq. 11.42:

$$\omega_{j,4} = \frac{\dot{m}_{s3}\, \omega_{j,s3} + \dot{m}_{j,sec}\, \omega_{j,\infty} + R_{j,net}\, (T_{z3}, \omega_{j,z3})\, \Delta V_{z3}}{\dot{m}_4} , \tag{11.42}$$

$$\dot{m}_4 = \dot{m}_{ex} = \dot{m}_{s3} + \dot{m}_{sec}, \tag{11.43}$$

where $Q_{vol,z3} = \sum R_{j,net}\, (T_{z3}, \omega_{j,z3})\, \Delta V_{z3}\, \Delta H_j$.

Zone 4 (see Fig. 11.5)

In this zone, again the following energy and species equations are solved:

$$\dot{m}_{ex}\, (T_{ex} - T_4) = Q_{vol,z4} + (q_{rad} + q_{conv})_{pan}\, A_{pan} , \tag{11.44}$$

$$\omega_{j,ex} = \frac{\dot{m}_{s4}\, \omega_{j,s4} + R_{j,net}\, (T_{z4}, \omega_{j,z4})\, \Delta V_{z4}}{\dot{m}_{ex}} , \tag{11.45}$$

$$Q_{vol,z4} = \sum R_{j,net}\, (T_{z4}, \omega_{j,z4})\, \Delta V_4\, \Delta H_j , \tag{11.46}$$

$$q_{conv,pan} = \alpha_{pan}\, (T_{z4} - T_{pan}) , \tag{11.47}$$

$$\frac{\alpha_{pan}\, k}{D_{th}} = 0.5\, (1.65\, Re_{D_{th}}^{0.5} + 2.733\, Re_{D_{th}}^{0.59}) \tag{11.48}$$

where pan temperature T_{pan} is specified according to the cooking task. For example, $T_{pan} \simeq 388$ K for boiling and $\simeq 473$ K for a roasting task. Correlation for α_{pan} is taken from experimental measurements of Bhandari et al. [12]. $Re_{D_{th}}$ is Reynolds number based on throat diameter D_{th}.

Finally, the zonal average temperatures and mass fractions are evaluated by upstream weighting from

$$T_{z1} = 0.8\, T_\infty + 0.2\, T_{wo} , \qquad \omega_{j,z1} = \omega_{j,\infty} , \tag{11.49}$$

$$T_{z2} = 0.6\, T_{wo} + 0.4\, T_{s3} , \qquad \omega_{j,z2} = 0.8\, \omega_{j,s2} + 0.2\, \omega_{j,s3} , \tag{11.50}$$

$$T_{z3} = 0.8\, T_{s3} + 0.2\, T_4 , \qquad \omega_{j,z3} = 0.8\, \omega_{j,s3} + 0.2\, \omega_{j,4} , \tag{11.51}$$

$$T_{z4} = 0.6\, T_4 + 0.4\, T_{ex} , \quad \text{and}\ \omega_{j,z4} = 0.8\, \omega_{j,4} + 0.2\, \omega_{j,ex} . \tag{11.52}$$

11.2.4   Radiation Model

To evaluate radiation fluxes to surfaces s1, s2, s4, and the pan bottom ($q_{rad,s1,s2,s4,pan}$), the entire combustion space is viewed as an enclosure. Then, applying the enclosure theory with a participating medium, we have

$$q_{rads1} = \sigma \left[F_{1,3}\, \tau_{1,3}\, T_{s3}^4 + F_{1,s2}\, \tau_{1,s2}\, T_{s2,i}^4 + \epsilon_{z2}\, T_{z2}^4 - T_{s1}^4 \right], \tag{11.53}$$

$$q_{rads2} = \sigma \left[F_{s2,3}\, \tau_{s2,3}\, T_{s3}^4 + F_{2,1}\, \tau_{w2,1}\, T_{s1}^4 + F_{w2,w2}\, \tau_{w2,w2}\, T_{s2,i}^4 + \epsilon_{z2}\, T_{z2}^4 \right]$$
$$- \sigma\, T_{s2,i}^4, \tag{11.54}$$

$$q_{rads4} = \sigma \left[F_{s4,3}\, \tau_{s4,3}\, T_{s3}^4 + F_{s4-pan}\, \tau_{s4-pan}\, T_{pan}^4 + F_{w4,w4}\, \tau_{s4,s4}\, T_{s4,i}^4 + \epsilon_{z3}\, T_{z3}^4 \right]$$
$$- \sigma\, T_{s4,i}^4, \tag{11.55}$$

$$q_{rad,pan} = \sigma \left[F_{pan-3}\, \tau_{pan-3}\, T_{s3}^4 + F_{pan-s4}\, \tau_{pan-s4}\, T_{s4,i}^4 + \epsilon_{z3}\, T_{z3}^4 \right]$$
$$- \sigma\, T_{pan}^4, \tag{11.56}$$

where $T_{s1} = T_{wo}$, F's are shape factors, and transmitivities (τ) and emissivities (ϵ) are evaluated as functions of mean-beam-length L_b and temperature [102] according to

$$\tau (T, L_b) = 1 - \epsilon (T, L_b), \tag{11.57}$$
$$\epsilon (T, L_b) = \exp \{ A + B \ln (C \times L_b) \}, \tag{11.58}$$
$$A = 0.848 + 9.02 \times 10^{-4} \times T, \tag{11.59}$$
$$B = 0.9589 + 4.8 \times 10^{-6} \times T, \quad \text{and } C = 0.2. \tag{11.60}$$

Values of L_b are derived by evaluating mean distances between surfaces. For evaluating zonal gas emissivity, $L_b = 3.6\times$ volume/surface area. Constants in expressions A, B, and C are evaluated at 700 K [102]. This temperature is representative of zonal temperatures $T_{z2,z3}$. Shape factors are functions of the stove/enclosure geometry.[6]

11.2.5   Output Parameters

The preceding equations are solved iteratively with under relaxation factor 0.5. Convergence is measured by noting fractional change in temperatures T_{wo}, T_{z2}, T_{z3}, T_{sec}, and T_{exit} between successive iterations. When the maximum absolute change $<10^{-4}$,

---

[6]Radiation shape factors are

$$\text{Zone 2}: F_{1,2} = 0.953,\ F_{1,3} = 0.0468$$
$$F_{w2,1} = 0.477,\ F_{w2,w2} = 0.48,\ F_{w2,3} = 0.043.$$
$$\text{Zones 3 and 4}: F_{3,w4} = 0.855,\ F_{3,pan} = 0.144$$
$$F_{w4,3} = 0.19,\ F_{w4,w4} = 0.62,\ F_{w4,pan} = 0.199$$
$$F_{pan,w4} = 0.85,\ F_{pan,3} = 0.145.$$

convergence is declared. Fewer than twenty iterations are typically required for convergence. The computed data are found to be independent of the initially guessed values assigned to different variables. Finally, the following output parameters of the stove performance are evaluated:

$$X_{air} = \frac{\dot{m}_{p1} + \dot{m}_{p2} + \dot{m}_{sec}}{6.51 \times \dot{m}_{fu}} , \tag{11.61}$$

$$\text{Power } P \ (kW) = \dot{m}_{fu} \times CV_{wo} , \tag{11.62}$$

$$\dot{Q}_{abs} = (q_{rad} + q_{conv})_{pan} \, A_{pan}$$
$$- \alpha_{side} \, A_{side} \, (T_{pan,side} - T_{\infty}) \quad (W) , \tag{11.63}$$

$$\alpha_{side} = 1.42 \left[\frac{(T_{pan,side} - T_{\infty})}{H_{pan}} \right]^{0.25} \frac{W}{m^2\text{-K}} [51] ,$$

$$\eta_{th} = \frac{\dot{Q}_{abs}}{P \times 1000} , \quad \text{and} \tag{11.64}$$

$$\eta_{comb} = \frac{Q_{vol,2} + Q_{vol,3} + Q_{vol,4} + Q_{char}}{P \times 1000} . \tag{11.65}$$

In Eq. 11.61, the excess air factor X_{air} is evaluated noting the stoichiometric A/F = 6.51 for wood. The calorific value of wood CV_{wo} (kJ/kg) accounting for wood-moisture content f is given by

$$CV_{wo} = 4.18 \, [4500 \, (1 - f) - 5.832 \, \{100 \, f + 54.3 \, (1 - f)\}] . \tag{11.66}$$

The net heat absorbed by the pan is evaluated as heat absorbed from combustion products at pan bottom minus heat lost by natural convection from the vertical pan side (with $A_{side} = \pi \, D_{pan} \, H_{pan}$ and $T_{pan,side} = T_{pan} - 50$). η_{th} and η_{comb} represent thermal and combustion efficiencies, respectively. In all computations, the various heat transfer coefficients (W/m²-K) are taken as $\alpha_{s2i} = \alpha_{s2o} = 6$, $\alpha_{s4i} = \alpha_{s4o} = \alpha_{outi} = \alpha_{outo} = 5$. Also, $T_{ref} = 300 \, K$ is specified.

11.2.6 Reference Stove Specifications

Table 11.3 shows the specifications of the reference stove. For these specifications, the computed results are shown in Table 11.4. The principal outputs of efficiency, excess air factor, and power are shown in the last column of this table. The first column shows that the wood surface temperature is 677.8 K whereas the exit temperature of the burned products is 395.9 K. $T_{sec} = 345.5$K shows that there is an increase of nearly 45 °C in the secondary air temperature because of heat gained from the clay and metal enclosures. Radiation (q_{rads1}) and convection ($q_{conv,bed}$) are negative,

Table 11.3 Specifications for the reference stove

Heights cm	Diameters cm	Areas cm^2	Volumes cm^3	Others
$H_t = 31$	$D_{gr} = 22$	$A_{gr} = 380$	$\Delta V_{bed} = 479$	
$H_{gr} = 7.5$	$D_{th} = 8$	$A_{th} = 50$	$\Delta V_{z2} = 2269$	
$H_{th} = 22$	$D_{pr} = 4.5$	$A_{pr} = 39.7$	$\Delta V_{z3} = 452.4$	$N_{pr} = 5$
$H_g = 2$	$D_s = 3$	$A_s = 42.4$	$\Delta V_{z4} = 1509$	$N_s = 6$
	$D_{sec} = 1$	$A_{sec} = 18.85$		$N_{sec} = 24$
		$A_{pan} = 754.5$		
	$D_{ent} = 10$	$A_{ent} = 39.26$		
		$A_{s2} = 758.7$		$b_2 = 5$ mm
		$A_{s4} = 226.2$		$b_4 = 1$ mm
		$A_{out} = 2142$		$b_{out} = 1$ mm
		$A_{gap} = 194.8$		

Fuel properties: $\rho_{wo} = 800 \frac{\text{kg}}{\text{m}^3}$, $x_{vol} = 0.75$, $f = 0.15$
Wood Dimensions: $D_{wo} = 1.26$ cm, $A_{wo} = 400$ cm^2
Grate Dimensions: $D_{rod} = 2$ cm, rod spacing $\delta = 2$ cm
Pan specification: $T_{pan} = 388$ K, $D_{pan} = 31$ cm, $H_{pan} = 15$ cm
Ambient: $T_\infty = 300$ K, $\omega_{H_2O,\infty} = 0$

Table 11.4 Predicted parameters—reference stove

Temperatures (K)	Heat fluxes (W/m^2)	Heat release (W)	Mass flow Rates (kg/s)	Outputs
$T_{wo} = 677.8$	$q_{rads1} = -2629.9$	$Q_{char} = 794.2$	$\dot{m}_{fu} = 2.27e-4$	$\eta_{th} = 29.4\%$
$T_{z2} = 676.2$	$q_{rads2} = 1450.8$	$Q_{vol,z2} = 0.68e-3$	$\dot{m}_{p1} = 1.58e-3$	$\eta_{comb} = 24.15\%$
$T_{s3} = 673.8$	$q_{rads4} = 536.7$	$Q_{vol,z3} = 0.26e-4$	$\dot{m}_{p2} = 2.16e-3$	$X_{air} = 4.42$
$T_{sec} = 345.5$	$q_{rad,pan} = 4885.4$	$Q_{vol,z4} = 0.16e-6$	$\dot{m}_{sec} = 2.78e-3$	$\epsilon_{z2} = 0.0158$
$T_{z3} = 646.0$	$q_{conv,bed} = -2463.7$		$\dot{m}_{char} = 5.673e-5$	$\epsilon_{z3} = 0.0091$
$T_{s4} = 549.6$			$\dot{m}_{vol} = 1.702e-4$	
$T_4 = 535.05$				$\alpha_{pan} = 56.73$
$T_{ex} = 395.9$				$\alpha_{bed} = 13.04$
$T_{s2,i} = 635.3$				
$T_{s4,i} = 549.6$				$P = 3.29$ (kW)

indicating heat transfer *from the wood surface*.[7] All other radiation fluxes are *to the respective surfaces*, and are thus positive. Column 3 shows that char combustion provides the major heat release ($\dot{Q}_{char}$), and there is hardly any heat generation ($\dot{Q}_{vol}$) from volatiles combustion. This is attributed to very low temperatures in Zone 2 and Zone 3 leading to very poor combustion efficiency.

The next column shows primary flow rates $\dot{m}_{p1} = 5.688$ kg/hr and $\dot{m}_{p2} = 7.77$ kg/hr, whereas the secondary air flow rate $\dot{m}_{sec} = 9.94$ kg/hr. The mixing of

---

[7]This may appear anomalous. However, these fluxes are evaluated from *average surface and zonal temperatures*. As such, they represent only global balance, and not local balance.

Table 11.5 Variations of ω_j with surface locations—reference stove

Location	ω_{O_2}	ω_{CO_2}	ω_{CO}	ω_{H_2}	ω_{H_2O}	ω_{HC}	ω_{C_2H4}	ω_{N_2}
∞	0.232	0.0	0.0	0.0	0.0	0.0	0.0	0.768
(z2)	0.117	0.128	0.020	1.52×10^{-5}	0.0480	0.0274	0.001	0.659
(z3)	0.179	0.059	0.0092	7.1×10^{-6}	0.0221	0.0126	0.00046	0.718
(z4)	0.201	0.0348	0.00543	4.14×10^{-6}	0.0130	0.00745	0.000271	0.738
(Exit)	0.201	0.0348	0.00543	4.14×10^{-6}	0.0130	0.00745	0.000271	0.738

this preheated secondary air with burned gases yields $T_4 = 535$ K. Heat absorbed at the pan-bottom then reduces temperature, yielding exit temperature $T_{ex} = 395.9$ K. In the last column, emissivities of Zones 2 and 3 are predicted at 0.0158 and 0.0091, respectively, and the power $= 3.29$ kW.

These trends of temperature are also corroborated by the mass fraction variations along the flow passage in Table 11.5. Thus, following char combustion and volatiles evolutions in Bed Zone, ω_{O_2} and ω_{N_2} decline, owing to the evolution of other product species. However, as additional air ingresses $\dot{m}_{p2}$ (in Zone 2) and $\dot{m}_{sec}$ (in Zone 3) take place, the mass fractions of oxygen and nitrogen rise again. There is hardly any change in the pan-bottom region because of very low temperature and mass fractions there. The exit mass fractions of CO and HC show that combustion is incomplete leading to low combustion efficiency. Also, at exit, $\omega_{CO}/\omega_{CO_2} = 0.00543/0.0348 = 0.156$, which is typically measured in the range 0.12–0.16 in different stoves [66].

11.2.7 Comparison with Measurements

Table 11.6 shows the comparison of predictions for the above-referenced stove with previous experimental data. The experimentally measured efficiency in a water boiling test was found to be $\eta_{th} = 28.3\%$ [12], which approximately matches with the predicted value $\eta_{th} = 29.4\%$. The excess air factor $X_{air} = 4.42$, and stove power is 3.29 kW. Both these are nearly corroborated in the experiments. Finally, the measured exit $\omega_{CO}/\omega_{CO_2}$ ratios are reported to be 0.12–0.16 [13, 66] in different stoves. These are in reasonable accord with the present prediction of 0.156. There are no reported measured values of η_{comb} found in the literature. Thus, all parameters are predicted

Table 11.6 Comparison of predictions with measurements

Item	Experiment	Prediction
η_{th}	28.3% [12]	29.4%
η_{comb}	–	24.15%
X_{air}	4 to 6 [12, 66]	4.42
$\omega_{CO}/\omega_{CO_2}$	0.12–0.16 [13, 66]	0.156
P (kW)	3.62 [12]	3.29

within the plausible range, which provides confidence in the various assumptions made in the model. As such, the model is run with different geometric and operating variables to test the main tendencies of the stove performance. In all runs, the wood-burning surface area is retained at $A_{wo} = 400 \, cm^2$.

11.2.8 Effect of Parametric Variations

Different parameters are varied one at a time, keeping all other parameters constant to derive directions for improved stove design. Only the principal outputs of η_{th}, η_{comb}, X_{air}, Power P, CO/CO$_2$ ratio, and unburned fractions ω_{CO} and ω_{HC} are considered.

Effect of Geometric Parameters

Table 11.7 shows the effect of geometric parameters. The table shows that as the total height H_t of the stove is decreased, the thermal efficiency η_{th} increases without loss of power. While wood surface temperature T_{wo} remains unaffected by stove height, the flue gas temperature at exit T_{exit} decreases with decrease in H_t; thus, suggesting enhanced heat transfer to the pan-bottom and enhanced thermal efficiency.

The effect of throat height H_{th} shows a similar inverse trend for η_{th}, but stove power decreases with a reduction in H_{th}. An increase in grate height H_{gr} increases η_{th} and excess air, but with a reduction in power. Finally, reducing H_g to 1 cm gives a dramatic increase in η_{th} without loss of power. These tendencies suggest that stove height $H_t \simeq 23$ cm and $H_g = 1$ cm help in obtaining higher η_{th} without loss of power. The effect of changes in H_{th} is marginal. In all cases, $\omega_{CO}/\omega_{CO_2} \simeq 0.156$ is maintained.[8]

Further, the table shows that grate diameter D_{gr} and secondary port diameter D_s have little effect on η_{th}. However, a reduction in throat diameter D_{th} and an increase in the primary port diameter D_{pr} improve η_{th} significantly, along with yielding a significant reduction in stove power. As expected, a reduction in the pan diameter D_{pan} reduces η_{th} because of decreased heat transfer surface area. Again, in all cases, $\omega_{CO}/\omega_{CO_2} \simeq 0.156$ is maintained.

Stove designers would like to change the reference stove design in such a way that η_{th} is enhanced while unburned ω_{CO} and ω_{HC} at exit decrease. It is seen from the table that this simultaneous energy and environmental improvement is obtained in respect of H_{th}, H_{gr}, D_{pr}, D_{th}, and D_{pan}. For the remaining parameters, improved energy performance is obtained at the expense of the environment. Finally, in all cases, the excess air factors X_{air} are found to vary between 4 and 5 in all cases. This observation accords with measurements reported in Refs. [12, 66].

---

[8] In stove literature, sometimes, the *emission factor* (EF) is quoted in g/kg of fuel. It can be deduced from the predicted values of species mass fractions at stove exit as

$$EF_j = \frac{\dot{m}_j}{\dot{m}_{fu}} = \omega_{j,ex} \, (1 + 6.51 \times X_{air}) \times 1000 \quad \left(\frac{g}{kg}\right).$$

Table 11.7 Effect of geometric parameters

Length cm	T_{wo} K	η_{th}	T_{ex} K	X_{air}	P kW	$\omega_{CO,ex}$	$\omega_{HC,ex}$	$\frac{\omega_{CO}}{\omega_{CO_2}}\vert_{ex}$
H_t								
31*	677.83	0.294	395.94	4.42	3.29	0.543E-02	0.745E-02	0.156
26	677.79	0.346	376.05	4.23	3.28	0.566E-02	0.777E-02	0.156
23	677.78	0.409	347.29	4.12	3.28	0.581E-02	0.797E-02	0.156
H_{th}								
20	675.60	0.319	389.41	4.53	3.02	0.529E-02	0.727E-02	0.156
22*	677.83	0.294	395.94	4.42	3.29	0.543E-02	0.745E-02	0.156
25	680.49	0.273	400.55	4.30	3.64	0.558E-02	0.765E-02	0.156
H_{gr}								
10	673.80	0.389	377.05	4.68	2.82	0.513E-02	0.704E-02	0.156
7.5*	677.83	0.294	395.94	4.42	3.29	0.543E-02	0.745E-02	0.156
5.0	681.05	0.228	407.10	4.20	3.71	0.569E-02	0.782E-02	0.156
H_g								
1.0	677.33	0.447	330.10	3.80	3.23	0.626E-02	0.860E-02	0.156
2.0*	677.83	0.294	395.94	4.42	3.29	0.543E-02	0.745E-02	0.156
3.0	677.92	0.247	413.29	4.59	3.30	0.523E-02	0.717E-02	0.156
D_{pr}								
3.0	681.32	0.186	401.33	3.75	3.75	0.635E-02	0.872E-02	0.156
4.5*	677.83	0.294	395.94	4.42	3.29	0.543E-02	0.745E-02	0.156
5.5	673.86	0.443	385.17	5.22	2.83	0.462E-02	0.634E-02	0.156
D_{gr}								
18	675.13	0.281	401.57	5.06	2.97	0.477E-02	0.654E-02	0.156
20	676.54	0.287	398.89	4.71	3.13	0.510E-02	0.700E-02	0.156
22*	677.83	0.294	395.94	4.42	3.29	0.543E-02	0.745E-02	0.156
D_{th}								
7.0	673.34	0.426	350.17	4.47	2.77	0.537E-02	0.737E-02	0.156
8.0*	677.83	0.294	395.94	4.42	3.29	0.543E-02	0.745E-02	0.156
9.0	680.87	0.226	417.86	4.41	3.69	0.543E-02	0.745E-02	0.156
D_s								
2.0	677.59	0.316	393.00	4.08	3.26	0.586E-02	0.805E-02	0.156
3.0*	677.83	0.294	395.94	4.42	3.29	0.543E-02	0.745E-02	0.156
4.0	677.87	0.289	396.41	4.51	3.29	0.532E-02	0.731E-02	0.156
D_{pan}								
22	677.80	0.168	456.30	4.37	3.29	0.548E-02	0.752E-02	0.156
25	677.82	0.206	437.25	4.41	3.29	0.543E-02	0.746E-02	0.156
31*	677.83	0.294	395.94	4.42	3.29	0.543E-02	0.745E-02	0.156

*Refers to the reference stove

Table 11.8 Effect of fuel and ambient parameters

	T_{wo} K	η_{th}	T_{ex} K	X_{air}	P kW	$\omega_{CO,ex}$	$\omega_{HC,ex}$	$\frac{\omega_{CO}}{\omega_{CO_2}}\vert_{ex}$
x_{vol}								
0.7	677.61	0.348	391.19	4.61	3.14	0.521E-02	0.525E-02	0.133
0.75*	677.83	0.294	395.94	4.42	3.29	0.543E-02	0.745E-02	0.156
0.8	678.12	0.240	400.78	4.23	3.45	0.565E-02	0.983E-02	0.189
D_{wo} (cm)								
1.26*	677.83	0.294	395.94	4.42	3.29	0.543E-02	0.745E-02	0.156
2.00	686.06	0.302	401.22	5.93	2.44	0.408E-02	0.561E-02	0.156
3.00	705.08	0.352	400.97	7.58	1.87	0.322E-02	0.441E-02	0.156
ρ_{wo} $(\frac{kg}{m^3})$								
700	681.67	0.299	395.05	4.37	3.33	0.549E-02	0.753E-02	0.156
800*	677.83	0.294	395.94	4.42	3.29	0.543E-02	0.745E-02	0.156
900	674.47	0.290	396.66	4.46	3.26	0.537E-02	0.737E-02	0.156
T_∞ (C)								
10	679.23	0.268	399.39	4.53	3.47	0.530E-02	0.728E-02	0.156
27*	677.83	0.294	395.94	4.42	3.29	0.543E-02	0.745E-02	0.156
45	676.24	0.324	391.00	4.32	3.10	0.555E-02	0.761E-02	0.156
$\omega_{H_2O,\infty}$								
0.0*	677.83	0.294	395.94	4.42	3.29	0.543E-02	0.745E-02	0.156
0.01	677.96	0.292	397.12	4.40	3.31	0.545E-02	0.748E-02	0.156
0.05	678.50	0.283	401.56	4.32	3.37	0.555E-02	0.761E-02	0.156

*Refers to the reference stove

Effect of Fuel and Ambient Parameters

Different types of wood have different volatiles fractions and densities. Also, stove-users use wood of different diameters and moisture content. Table 11.8 shows the effect of such wood parameters. With increase in x_{vol}, η_{th} decreases sharply with a dramatic increase in unburned hydrocarbon fraction $\omega_{HC,ex}$. Similarly, ratio $\omega_{CO}/\omega_{CO_2}$ at exit also increases with increase in x_{vol}. Similarly, increasing the wood diameter D_{wo} increases η_{th} but the power decreases, indicating a reduced rate of burning, but with a considerable reduction in unburned fractions. Notice that T_{wo} and X_{air} increase with increase in D_{wo}. This suggests that the stove runs "hotter" with larger diameter wood. The effect of wood density ρ_{wo} is seen to be marginal.

The effect of ambient parameters is assessed through changes in T_∞ $\omega_{H_2O,\infty}$. The table shows that with an increase in ambient temperature, η_{th} improves but with reduced stove power. The unburned mass fractions, however, increase with an increase in T_∞. With an increase in $\omega_{H_2O,\infty}$, the η_{th} decreases as expected, and there is a marginal increase in unburned mass fractions.

Table 11.9 Effect of cooking tasks

T_{pan} C	T_{wo} K	η_{th}	T_{ex} K	X_{air}	P kW	$\omega_{CO,ex}$	$\omega_{HC,ex}$	$\frac{\omega_{CO}}{\omega_{CO_2}}\big\vert_{ex}$	Remarks
115*	677.83	0.294	395.94	4.42	3.29	0.543E-02	0.745E-02	0.156	Boiling
150	677.85	0.224	421.08	4.47	3.29	0.537E-02	0.737E-02	0.156	Frying
150	677.85	0.236	421.08	4.47	3.29	0.537E-02	0.737E-02	0.156	$H_{pan} = 5\,cm$
200	677.87	0.118	458.31	4.53	3.29	0.530E-02	0.727E-02	0.156	Roasting
200	677.87	0.156	458.31	4.53	3.29	0.530E-02	0.727E-02	0.156	$H_{pan} = 2\,mm$
200	677.79	0.234	416.54	4.15	3.28	0.576E-02	0.791E-02	0.156	$H_t = 23\,cm$

*Refers to the reference stove

Table 11.10 Effect of wood surface area

A_{wo} cm²	T_{wo} K	η_{th}	T_{ex} K	X_{air}	P kW	$\omega_{CO,ex}$	$\omega_{HC,ex}$	$\frac{\omega_{CO}}{\omega_{CO_2}}\big\vert_{ex}$
400*	677.83	0.294	395.94	4.42	3.29	0.543E-02	0.745E-02	0.156
600	670.32	0.312	387.38	3.87	3.70	0.616E-02	0.846E-02	0.156
800	664.69	0.363	362.71	3.37	3.96	0.703E-02	0.965E-02	0.156
1000	661.30	0.390	342.66	3.00	4.32	0.784E-02	0.108E-01	0.156

*Refers to the reference stove

Effect of Cooking Tasks

The effect of cooking tasks is assessed through changes in pan temperature T_{pan} (see Table 11.9). It is seen that as T_{pan} increases, η_{th} dramatically drops. This effect, to the best of our knowledge, has not been measured in previous work. It must be remembered that the frying and roasting tasks (which require higher pan-bottom temperature) are usually accomplished in pans in which $H_{pan} << 15\,cm$. Accounting for this observation reduces the heat loss from the pan side, and somewhat higher thermal efficiencies can be restored as seen in the table. Lowest efficiency is obtained for roasting tasks. However, decreasing stove height H_t from 31 cm (as in the reference case) to 23 cm will restore η_{th}, as shown in the last entry of the table.

Effect of Wood Surface Area

All previous results are obtained for $A_{wo} = 400\,cm^2$. In Table 11.10, therefore, the effect of the increasing[9] the wood-burning surface area is evaluated. It is seen that the stove power increases with an increase in A_{wo}, as expected. η_{th} and unburned mass fractions also increase with increase in A_{wo}, but the excess air factor X_{air} decreases. De Iepelerie et al. [31] have given rules of thumb for estimating the maximum power obtainable from a stove of given dimensions. They recommend the combustion volume to be 0.6 Lit/kW and, the grate power to be 10–15 W/cm². The

---

[9]With wood diameter $D_{wo} = 1.26\,cm$ and length 20 cm, for $A_{wo} = 400\,cm^2$, the grate ($D_{gr} = 22\,cm$) will accommodate five pieces of wood. With an increase in surface area, the number of wood pieces (or the number of layers N_{row}, see Eq. 11.28) and hence, resistance k_{bed} will increase proportionally.

Fig. 11.6 Top view of twisted tape swirler placed in the metal enclosure [12]

reference stove has grate area $A_{gr} = 380$ cm^2 and combustion volume of 2.8 Lit. As such, we can expect maximum power $P_{max} \simeq 5$ kW. The last entry in Table 11.10 corroborates this estimate.

Effect of Swirl Inducement

Inducement of swirl in the flow of burned products is known to improve mixing and, hence, combustion. Swirl inducement can also enhance impingement heat transfer coefficient α_{pan} to the pan-bottom at the expense of enhanced resistance to flow through the stove body. Swirl can be imparted to the burned products by ensuring that the secondary air flows into the metal enclosure tangentially. Alternatively, swirl can be induced more positively by inserting a twisted-tape (see Fig. 11.6) or a bladed swirler (for example) in the metal enclosure [12]. Here, the effect of swirl is notionally accounted for by enhancing α_{pan} (Eq. 11.47), loss coefficients k_{cont} (Eq. 11.28), $k_{sec,eff}$, and $k_{exp,eff}$ (Eq. 11.32) by a factor $K_{swirl} > 1$. The results of the computations are shown in Table 11.11. With an increase in swirl, η_{th} indeed improves, as has been experimentally observed ($\eta_{th} \simeq 46\%$) in Ref. [12]. The excess air decreases, owing to enhanced resistance, but the power level is relatively unaffected. Again, improvement in η_{th} is accompanied by higher unburned fractions.

More recently, Nuntadusit et al. [86] have reported use of multiple twisted-tapes for enhancing flow impingement heat transfer to a flat surface through laboratory

Table 11.11 Effect of swirl inducement

K_{swirl}	T_{wo} K	η_{th}	T_{ex} K	X_{air}	P kW	$\omega_{CO,ex}$	$\omega_{HC,ex}$	$\frac{\omega_{CO}}{\omega_{CO_2}}\vert_{ex}$
1.0*	677.83	0.294	395.94	4.42	3.29	0.543E-02	0.745E-02	0.156
1.25	677.26	0.373	364.92	4.15	3.22	0.576E-02	0.791E-02	0.156
1.5	676.67	0.465	322.80	3.88	3.15	0.615E-02	0.845E-02	0.156

*Refers to the reference stove

experimentation. Honkalaskar et al. [50], on the other hand, have reported over 25% reduction in rural household-level fuelwood consumption through use of multiple twisted tapes placed below the pan in a mud cookstove.

11.2.9 Overall Conclusions

1. The results presented in the previous subsection show how a simple thermochemical model of a wood-burning stove can help in the study of design, operating, and fuel property parameters of the stove.
2. The principal ingredients of the steady-state model are the kinetically and physically controlled wood-burning flux-rate expressions, Eqs. 11.2 and 11.10. These are derived from curve-fits to the experimental data. Other empirical parameters are taken from the literature [63, 102]. None of the parameters is *tuned* to the specificities of the CTARA stove. To that extent, the model can be said to be general enough from which to draw inferences for improved stove design and operating practices. The overall model framework can also be extended to other stove designs for a quick evaluation of their performance [7, 77].
3. For the range of feasible parameters considered, it is found that the combustion in a stove is incomplete, mainly owing to lack of volatiles burning as a result of low temperatures. Unburned fractions increase in general with an increase in thermal efficiency, with some exceptions. Compared with the reference stove design, setting $H_t = 23$ cm, $H_g = 1$ cm, $D_{pr} = 5.5$ cm, and $D_{th} = 7.0$ cm will improve thermal efficiency. These findings are corroborated in several improved stove designs advocated for dissemination (see, for example, [3]). For all parametric variations, the CO/CO_2 ratio at stove-exit is found to be about 0.156 in most cases[10] for $x_{vol} = 0.75$. Lower x_{vol} results in a lower CO/CO_2 ratio, and vice versa.
4. Imparting swirl to the burned products improves stove efficiency. This has been demonstrated through laboratory experiment, Household-level kitchen-performance tests as well as through modeling. The excess air factor X_{air} is influenced mainly by dimensions D_{gr}, D_{pr}, and H_g and by the wood diameter D_{wo}, whereas, for a given wood surface area, stove power is influenced mainly by wood properties f, x_{vol}, and D_{wo}, by stove dimensions H_t, H_{th}, D_{pr}, and D_{th}, and by ambient temperature T_∞. Finally, the maximum stove power estimated from rules of thumb provided in Ref. [31] is verified through the present stove model.

---

[10]Of course, the exact magnitude of the ratio at exit is governed by the assumed values shown in Eq. 11.24. Similar assumptions have been made in respect of (a) Time split (see Eqs. 11.3 and 11.4) for volatile (0.4) and char (0.6) burning, (b) Effective temperatures during volatile and char evolutions (see Eq. 11.9) and (c) Definition of transferred substance state (see Eq. 11.11). All these assumptions must be subjected to sensitivity analysis as reported in Shah and Date [100].

11.3   Vertical Shaft Brick Kiln

Production of construction bricks consumes significant amounts of energy through coal burning. For example, in a country such as India, nearly 140 billion bricks are produced per year in approximately 100,000 enterprises consuming 8% of total coal consumption or, nearly 25 million tons of coal per year. Energy costs amount to nearly 35% to 40% of the brick production cost.

Much of the total brick production is carried out in *scove clamps* (65%) and in *Bull's trench kilns* (35%). Figure 11.7 shows construction of both types. In the scove clamp, the green bricks (moisture content <10%) are stacked in layers along with coal powder or even with wooden logs. After stacking, the fuel is set on fire. The bricks are heated and baked over a period of forty to forty five days. The clamp is insulated with layers of ash put on the exposed surfaces.

Unlike the scove clamp, in which bricks are produced in *batch mode*, the Bull's trench kiln (BTK) produces bricks is a semicontinuous manner. The BTK is also called a *moving fire* kiln because in this oval-shaped kiln, bricks are fired over only a small zone, as shown in Fig. 11.7. In this firing zone, coal powder is intermittently fed from the top of the kiln by a fireman by observing the state of combustion. The air for combustion is drawn through the already-fired bricks upstream of the direction of the flame-travel, so the combustion air is preheated. Similarly, the hot burned products of combustion are passed through the green bricklayers ahead of the firing zone before venting through a *fixed chimney* as shown in the figure. Both air preheating and green-brick drying by burned products result in energy savings.

The energy consumption in scove-clamp is typically measured at 2.5–3.0 MJ/kg of brick, whereas that in the BTK is measured at 1.2–1.5 MJ/kg of brick [79]. In spite of this relative advantage of the BTK, the scove clamp has remained popular because it can be constructed anywhere near the availability of soil, whereas for the BTK,

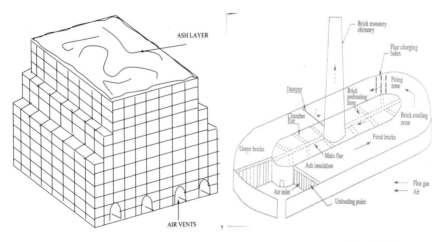

Fig. 11.7 Conventional brick kilns: (left) scove clamp, (right) Bull's trench kiln (BTK)

Fig. 11.8 Two-shaft vertical shaft brick kiln at Varanasi, India

the kiln is fixed in a suitable location and the soil must be carted to the location. In both types of kilns, heat loss to the ground (estimated at 25% on BTK and 40% for the scove clamp) cannot be avoided [79].

It is to reduce energy consumption further that the *vertical shaft brick kiln* (VSBK) of original Chinese design, was introduced in India under the United Nations Development Program (UNDP) in 1998. Figure 11.8 shows a photo of a typical kiln. By 2004, there were 180 such kilns operating in India; the first 6 kilns constructed in different parts of India were monitored for their energy and environment performance by Maithel [79] through field-level experiments, and the measurements were verified through modeling.

11.3.1 VSBK Construction and Operation

Figure 11.9 shows the main features of the VSBK. The brick shaft (1 × 1 or 2 m in cross section and 4–6 m tall) is enclosed in a brickwork (80 cm thick), leaving a peripheral air-gap (1–4 cm) between the shaft and the brickwork. Green bricks with powdered (2–5 mm diameter) coal in the interbrick spacings are fed from the top of the kiln. The shaft rests on cross-steel bars at the bottom (about 1.5 m from the ground) of the kiln. The coal firing zone is at the mid-height of the shaft. Typically, four fired-brick layers are removed from the bottom of the shaft every two hours by transferring the shaft load on a screw jack located on the ground. The space thus vacated is filled with fresh green brick layers at the top of the kiln. As a result of coal combustion, air induction in the kiln is by natural draft. The air flows through the interbrick spacings in the shaft, as well as through the peripheral airspace between the shaft and the brick-work. Below the firing zone, the fresh draft air is preheated. The burned gases, on the other hand, dry the green bricks above the firing zone before leaving through the chimney.

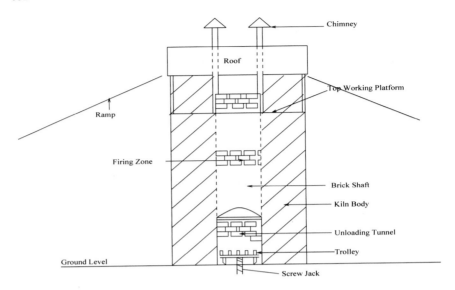

Fig. 11.9 Main features of the VSBK

Initially, the entire shaft is stacked with green bricks with coal powder in the interbrick spaces. A fire is lit from the bottom of the kiln. It takes typically three hours for the fire to travel up to the mid-height of the shaft. At this stage, the first layer of burned bricks is withdrawn from the bottom. The space vacated is filled with fresh green bricks at the top, and the firing of a new layer entering the firing zone is carried out for a further 2–2.5 h. This process of discontinuous feeding and withdrawal of bricks is now continued as long as desired. Typically, a single shaft kiln, yields about 3000 bricks in twenty four hrs. However, a two-shaft kiln shown in Fig. 11.8 with production rate of 6000–7000 bricks per day, is preferred on the grounds of economics.

Maithel [79] carried out measurements of vertical temperature profile in the kiln by designing a *traveling thermocouple* arrangement, as shown in Fig. 11.10. The chromel–alumel thermocouples were specially manufactured in which the wires (0.8 mm diameter) were embedded in mineral insulation. This assembly was then inserted inside a steel sheath tube (6 mm dia). The long coiled thermocouples were then inserted in a hole (80 mm deep, 6 mm diameter) in one green brick and fixed there. This brick was then located at the desired cross-sectional location at the top of the shaft during the loading operation. After positioning, the thermocouple was suspended from a hook-pulley hung from the roof of the kiln. The thermocouple then traveled down with the brick and the temperatures were recorded at regular time intervals until the brick was unloaded at the shaft bottom. In addition, the shaft-side temperatures were recorded by means of fixed thermocouples inserted through spyholes along the brickwork surrounding the shaft.

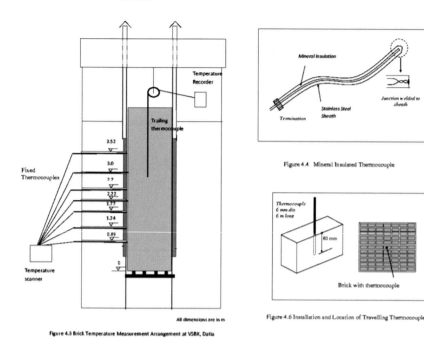

Fig. 11.10 Temperature measurements in a VSBK

Maithel [79] also measured mass fractions of CO_2 and O_2 as well as the air induction rates for given ambient conditions and green-brick throughput rates. The air temperatures at ambient and exit conditions were also measured using thermocouples. The coal feed rates were also recorded. From these measurements, energy audits of the kilns were built-up. Table 11.12 shows the energy balance statement of one such kiln located at Amaravati in the state of Maharashtra, India. The statement shows that for a production rate of approximately 5600 kg of bricks per day, the specific energy consumption is 0.84 MJ/kg of bricks (assumed coal calorific value 28.36 MJ/kg). This is a typical figure for most other VSBKs operating in India.

11.3.2 Modeling Assumptions

Because the brick movement is intermittent, coal combustion, heat transfer, and fluid flow in a VSBK are three-dimensional time-dependent phenomena. The complexities involved in modeling can be appreciated from the recognition that the naturally aspirated air flows tortuously through the spaces between bricks, coal combustion is both external (in the interspaces) and internal (mixed with green bricks), chemical reactions occur among the constituents of the brick material, and, finally, brick drying takes place owing to passing of hot gases near the exit. To simplify matters, we carry out one-dimensional plug-flow reactor modeling (see Fig. 11.11), in which

Table 11.12 Typical Heat balance statement for VSBK. Production rate = 5600 kg/day

Item	MJ/day	%
Heat input	4708	100
Loss from outside surface	308	6.5
Loss from exposed bricks at bottom	344	7.5
Loss in water vapor in bricks	1240	26.3
Loss in vapor due to moisture and H_2 in coal	286	6.0
Heat required for chemical reactions in brick	975	20.7
Sensible heat in exit gases	733	15.5
Sensible heat in fired bricks withdrawn	356	8.0
Miscellaneous	466	10.0

Fig. 11.11 Domain of computation—one-dimensional model

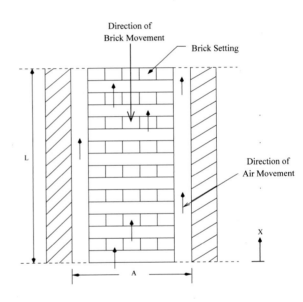

1. Steady state is assumed and the downward brick flow rate $\dot{m}_{br}$ is evaluated from the average brick throughput rate for the day.
2. The airflow is assumed to be vertically upward. Its rate is unknown and therefore guessed[11] as $\dot{m}_a = \lambda \dot{m}_{brick}$. The value of λ was derived from observed experimental conditions.
3. It is assumed that 20% air flows through the gap between the brick shaft and the enclosing brickwork and remaining 80% air flows through the brick shaft. These percentages were determined from numerical experiments to have the best fit for the Datia-I kiln [59]. The same proportions are used for all kilns.
4. Variations in brick-setting density with axial (upward) distance are ignored.

---

[11]This guessing obviates the need to solve the bulk mass conservation and momentum equations, simplifying modeling further.

5. Coal particles in the interbrick spaces are assumed to move at the brick velocity and have the same temperature as the bricks, except in one chemical reaction involving CO production, for which it is assumed that the coal particle temperature T_p exceeds the brick temperature T_{br}, as $T_p = T_{br} + 100$.
6. Air properties are assumed to be functions of temperature only; effects of species fractions are ignored.
7. Radiation between the brick shaft and the brickwork inner wall is accounted for with the emissivity of both surfaces 0.9.
8. Radiation and convection loss from the brickwork sides (vertical) and the exposed bricks at the kiln bottom (horizontal) are evaluated from textbook [51] correlations.
9. The effect of coal ash on the change in void fraction of the interspaces is ignored.
10. Different specific heats are used for green and fired bricks.
11. Both volatilization and char combustion are considered in the coal combustion model.
12. The coal combustion heat generated is accounted for with bricks, whereas the heat generated in the gas phase is accounted for with air.

11.3.3 Coal Burning

Coal burning is assumed to take place via (1) devolatilization and (2) char oxidation at the surface. Following Kobayashi et al. [65], the devolatilization rate (kg/s) is evaluated from

$$\dot{m}_{vol} = \rho_{rc} \, [0.39 \, k_1 + 0.8 \, k_2] \,, \tag{11.67}$$

where $k_1 = 370 \times exp \left[-8858/T_p\right], k_2 = 1.46 \times 10^{11} exp \left[-30196/T_p\right] m^3/s, \rho_{rc}$ is raw coal density, and T_p is coal particle temperature. Its value is taken to be same as T_{br}. The composition of the volatiles is determined from the ultimate and the proximate analysis of the coal used (ssh 25.5%, char 44.8%, and volatiles 29.4%). The element-balance method for this determination is given in Ref. [79]. The evolved species are: CO, H_2, H_2O, NO, N_2, S, and hydrocarbons HC $\equiv C_3H_8$. HC is taken to represent both heavy (soot) and light hydrocarbons.

The **heterogeneous char oxidation** reaction is modeled as

$$R_1 \; 2C + O_2 \rightarrow 2CO \tag{11.68}$$
$$\text{with } k_1 = 20.4 \times 10^{-4} \, R_u \, T_a/M_{O_2} \, exp \left[-9658/T_p\right] \,. \tag{11.69}$$
$$R_2 \; C + CO_2 \rightarrow 2CO \tag{11.70}$$
$$\text{with } k_2 = 20 \, R_u \, T_a/M_{CO_2} exp \left[-25136.5/T_p\right] \,. \tag{11.71}$$

Thus, the char burning rate of a particle of radius r_{wi} is given by

$$\dot{m}_{c1,i} = 2 \, k_1 \, (4 \, \pi \, r_{wi}^2) \, \rho_m \, \frac{M_C}{M_{O_2}} \, \omega_{O_2} \,, \tag{11.72}$$

$$\dot{m}_{c2,i} = 2 \, k_2 \, (4 \, \pi \, r_{wi}^2) \, \rho_m \, \frac{M_C}{M_{CO_2}} \, \omega_{CO_2} \, , \tag{11.73}$$

where ρ_m is the gas mixture density. With these evaluations, the total char burn rate in a volume $A \, \Delta x$ of the kiln is given by

$$\dot{m}_{char} = \sum_{i=1}^{NI} (\dot{m}_{c1,i} + \dot{m}_{c2,i}) \, n_{p,i} \, A \, \Delta x \, , \tag{11.74}$$

where $n_{p,i}$ is the number of coal particles of group i per m^3 volume, and NI is the total number of groups. $n_{p,i}$ are determined from sieve analysis.

In the gas phase, products of surface reactions and devolatilization burn by combining with oxygen in **homogeneous reactions**. It is assumed that NO and N_2 do not react. As such, their compositions change owing to volatilization only. For the other species, chemical reactions are assumed to follow the Hautmann [43] quasi-global reaction mechanism (see Chap. 5).

11.3.4 Model Equations

Air Side

On the air side, the air mass flow rate $\dot{m}_a = \rho_a \, A \, u_a$ is specified from the experimental data. $\dot{m}_a$ remains constant with x, the distance from the shaft bottom. Knowing $\rho_a = \rho_{a,\infty} \, T_\infty / T_a$, u_a is known. This u_a is used in Nusselt number correlations. Thus,

$$\dot{m}_a \, \frac{d \, \Phi}{dx} = \frac{d}{dx} \left(\Gamma \, \frac{d \, \Phi}{dx} \right) + S_\Phi \, , \tag{11.75}$$

where $\Phi = T_a$ and ω_j where j = O_2, CO, CO_2, H_2, C_2H_4, C_3H_8, NO, and N_2. $\Gamma = k_a/cp_a$ for $\Phi = T_a$ and $\Gamma = \rho_a \, D_a$ for $\Gamma = \omega_j$. The mass diffusivity for all species is assumed to be the same, and equal to $D_{CO_2-N_2} = 1.65 \times 10^{-5} \, (T/393)^{1.5}$ m^2/s.

The source term S_{T_a} is evaluated as

$$S_{T_a} = (Q_{comb} + Q_{br-a} - Q_{a-w} - Q_{evap})/cp_a \, , \tag{11.76}$$

where combustion heat is evaluated as $Q_{comb} = \dot{m}_{vol} \times cv_{vol} + \dot{m}_{char} \times cv_{char}$ and cv_{vol} is evaluated from volatiles composition ($cv_{vol} = 32.52$ MJ/kg and $cv_{char} = 33.1$ MJ/kg). The heat gain from hot shaft bricks Q_{br-a} is evaluated by assuming heat transfer coefficient (a function of brick-setting density) between closely packed bricks and air [67], and the convection–radiation loss from the air to the brick work inner wall Q_{a-w} is evaluated from the heat transfer coefficient for flow between parallel plates. The emissivity of the brick surface is taken as 0.9. Finally, the heat used to evaporate water from the green bricks is evaluated as

$$Q_{evap} = \Delta Y (1 - Y_0) \, \dot{m}_{br} \, h_{fg},$$ (11.77)

$$\frac{Y}{Y_0} = exp \left[-\frac{(\pi/B)^2 \, D_w \, (L - x)}{u_{br}} \right],$$ (11.78)

where Y is the mass fraction of water in the brick and Y_0 is the initial moisture in the brick, D_w is diffusivity of water in the brick (a function of T_{br} and Y), and B = width of the brick as stacked in x-direction.

Similarly, the source terms for ω_j are

$$S_{O_2} = -\left[\frac{1}{2} \times R_{CO} \frac{M_{O_2}}{M_{CO}} + R_{C_2H_4} \frac{M_{O_2}}{M_{C_2H_4}} \right.$$

$$\left. + R_{H_2} \frac{M_{O_2}}{M_{H_2}} + \sum_{i=1}^{NI} \frac{n_{p,i}}{2} \dot{m}_{cl,i} \frac{M_{O_2}}{M_C} \right],$$ (11.79)

$$S_{CO_2} = R_{CO} \frac{M_{CO_2}}{M_{CO}} - \sum_{i=1}^{NI} \frac{n_{p,i}}{2} \dot{m}_{c2,i} \frac{M_{CO_2}}{M_C},$$ (11.80)

$$S_{CO} = \beta_{CO} \dot{m}_{vol} + 2 R_{C_2H_4} \frac{M_{CO}}{M_{C_2H_4}} - R_{CO}$$

$$+ \left(\sum_{i=1}^{NI} n_{p,i} \dot{m}_{cl,i} + 2 \dot{m}_{c2,i} \right) \frac{M_{CO}}{M_C},$$ (11.81)

$$S_{H_2O} = \beta_{H_2O} \dot{m}_{vol} - R_{H_2} \frac{M_{H_2O}}{M_{H_2}}$$

$$+ \Delta Y (1 - Y_0) \dot{m}_{br},$$ (11.82)

$$S_{H_2} = \beta_{H_2} \dot{m}_{vol} + 2 R_{C_2H_4} \frac{M_{H_2}}{M_{C_2H_4}} - R_{H_2},$$ (11.83)

$$S_{C_3H_8} = \beta_{C_3H_8} \dot{m}_{vol} - R_{C_3H_8},$$ (11.84)

$$S_{NO} = \beta_{NO} \dot{m}_{vol}, \quad \text{and}$$ (11.85)

$$S_{N_2} = \beta_{N_2} \dot{m}_{vol}.$$ (11.86)

where βs are species fractions in the devolatilized products.

Brick Side

The brick side energy equation reads as

$$\dot{m}_{br} \, cp_{br} \frac{d \, T_{br}}{dx} = \frac{d}{dx} \left(k_{br} \frac{d \, T_{br}}{dx} \right) + S_{T_{br}},$$ (11.87)

where the brick flow rate $\dot{m}_{br} = \rho_{br} (A - A_{gap}) u_{br}$ is known. Thus, knowing ρ_{br} and the airgap, the brick velocity u_{br} is calculated. The brick specific heat $cp_{br} = 840$ J/kg-K is assumed [104], and brick conductivity $k_{br} = 0.41 + 3.5 \times 10^{-4} (T_{br} - 273)$ (W/m-K). The source term is evaluated as

$$S_{T_{br}} = - \left(Q_{br-a} + Q_{br-w} \right),$$ (11.88)

where the heat gain from coal combustion Q_{br-a} is convective heat transfer to air and Q_{br-w} is convection + radiation heat transfer to the inner wall of the brickwork.

Coal Particle Radius

The mass conservation equation for coal determines the coal particle radius $r_{w,i}$. It is assumed that the particles are carried downwards at the same velocity as the bricks. Thus,

$$\frac{4}{3} \rho_c \, \pi \, u_{br} \, \frac{d \, r_{w,i}^3}{dx} = \dot{m}_{c1,i} + \dot{m}_{c2,i} + \frac{\dot{m}_{vol}}{n_{p,i}}.$$ (11.89)

11.3.5   Inlet and Exit Conditions

At the bottom Inlet ($x = 0$), the mass fractions specified are $\omega_{O_2} = 0.233/(1 + \Psi)$, where Ψ is ambient specific humidity, $\omega_{H_2O} = \Psi$; and $\omega_{N_2} = 1 - \omega_{O_2} - \omega_{H_2O}$. All other species mass fractions are set to zero. The inlet air temperature $T_a = T_\infty$, and the inlet brick temperature is evaluated from convection + radiation heat loss conditions prevailing at the kiln bottom. The particle radius, though expected to be zero at $x = 0$, is evaluated by linear extrapolation from $d^2 \, r_{w,i}/dx^2 = 0$.

At Top Exit ($x = L$), T_a and all ω_j are evaluated from $d^2 \, \Phi/d \, x^2 = 0$. The value of brick temperature $T_{br} = T_\infty$ and $r_{w,i}$ = initial unburned particle radius. The initial coal particle radii are divided into four groups (NI = 4).

Together with the boundary conditions, the equations for T_a, ω_j, T_{br}, and $r_{w,i}$ are solved by finite difference using an upwind difference scheme. The main input parameters are: $\dot{m}_{br}$, ρ_{br}, L, A, brickwork thickness, initial $r_{w,i}$, $\dot{m}_{coal}$, T_∞, Ψ, and Y_0. The value of $\dot{m}_a$ is assumed. Computations were carried out using 1000 grid points. Convergence was measured using the maximum fractional change between iterations. Its value was set at 10^{-4}. Eleven hundred iterations were required for convergence.

11.3.6   Results for the Reference Case

For the specifications of the Amaravati kiln, computations were performed with following input parameters: $\dot{m}_{br} = 0.109$ kg/s, $\rho_{br} = 2000$ kg/m^3, $\dot{m}_{coal} = 3.2$ gm/s, L = 4.5 m, A = 1.8 m^2, brickwork thickness = 1.6 m, initial $r_{w,i}$ = 2 (11%), 3 (28%), 4 (55%), and 10 (6%) mm, $T_\infty = 297$ K, $\Psi = 0.016$, initial moisture in green bricks $Y_0 = 0.04$, and $X_{clay} = 0.35$. The void fraction of the brick setting is 0.36. It is assumed that $\dot{m}_a = \dot{m}_{br}$ or $\lambda = 1$.

Figure 11.12 (left) shows the comparison between the predicted brick and air temperatures and those measured [79]. The maximum predicted $T_{br} = 1154$ K, whereas

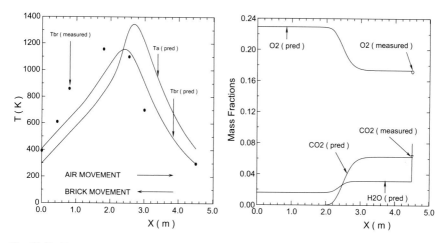

Fig. 11.12 Temperature and mass fraction variations—VSBK, Amaravati

the maximum measured value is 1153 K, although the locations of the maxima do not match. In fact, the measured temperatures near the bottom end of the kiln are consistently higher than the predicted ones. One reason for this mismatch is that in the actual kiln, part of the coal was mixed with bricks that burns somewhat later in the brick travel. In the computations, all coal is assumed to be placed in the spaces between the bricks (that is, external coal). The predicted T_{br} at the kiln bottom (385 K) matches well with the measured value of 398 K. The measured T_a at the kiln top is 398 K and the predicted one is 411 K. The calculated specific energy consumption is 0.716 MJ/kg against the experimentally estimated value of 0.84 MJ/kg (see Table 11.12). This discrepancy is attributed to somewhat under evaluation of the calorific value of coal in the model. The value evaluated from volatiles composition and corresponding heating values of reactions is 24.39 MJ/kg. In the field experiment, the specific consumption is evaluated from assumed calorific value of coal (28.61 MJ/kg).

Figure 11.12 (right) shows the variation of mass fractions for O_2, CO_2, and H_2O. As expected, oxygen is not consumed until the combustion zone is approached when its mass fraction decreases. The carbon dioxide shows the opposite trend as expected. The water vapor increases in concentration first via devolatilization of coal and then, suddenly, near the exit, owing to evaporation of water from the green bricks. Evaporation is permitted only when $T_{br} > 100°$ C and therefore takes place over a very short distance of a few centimeters near the exit. The exit mass fractions for O_2 and CO_2 match very well with the experimental values.

Figure 11.13 shows the variation of particle radii. It is seen that in the direction of the brick movement, the larger particles begin to burn earlier than the smaller ones, a consequence of the char burn rates (see Eqs. 11.72 and 11.73) which are proportional to r_w^2. As explained earlier, particles of all sizes are assumed to have the same temperature as that of the bricks. It is conceivable, of course, that particles

Fig. 11.13 Particle radius
variations—VSBK,
Amaravati

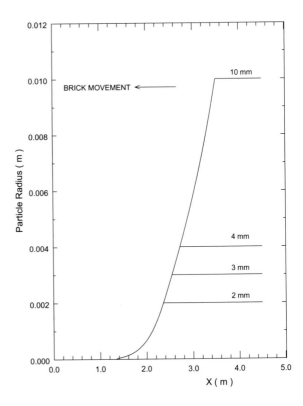

of different diameters may *heat up* at different rates, and the particle radius histories
will then deviate from the predicted ones in a real kiln. Such an experimental inquiry
has not been made to enable comparison. All particles burn out in the combustion
zone at $x \simeq 1.35$ m from the kiln bottom.

11.3.7 Parametric Studies

Effect of Production Rate

There is a tendency among kiln operators to throughput bricks at a rate larger than
that used in the reference case, and the experience suggests a reduction in specific
energy consumption (SEC—MJ/kg). To verify this practical experience, a factor
$\beta = \dot{m}_{br}/\dot{m}_{br,ref}$ is varied from 0.7 to 1.4. Table 11.13 shows that the SEC reduces
with increase in β, but with increased CO_2 at exit and a reduced maximum brick
temperature. The latter may affect the quality of bricks, whereas the former incurs
enhanced environmental cost. Thus, it is important to choose the production rate
judiciously. For all values of β, $\lambda = 1$ is assumed in this case study.

Table 11.13 Effect of production rate

β	SEC	$T_{br,max}$ (K)	$\omega_{CO_2,exit}$
0.7	0.97	1305	0.158
0.8	0.895	1245	0.169
1.0	0.716	1154	0.174
1.2	0.596	1094	0.183
1.4	0.514	1056	0.190

Effect of Kiln Height

Practical experience with kilns in different parts of India shows that the SEC varies inversely with the kiln height. For example, the kiln at Varanasi [79] is 5.8 m tall (against 4.5 m in the reference case). Other parameters of the kiln are nearly the same as the reference case ($\dot{m}_{coal} = 0.02$ gm/s is smaller). For the kiln at Varanasi, the present model predicted (details not shown here) SEC $= 0.528$ (<0.716), thus confirming the trend observed in practical kilns. Of course $T_{br,max}$ decreased to 1100 K (<1153 K), whereas the CO_2 mass fraction increased to 0.19 (>0.17). Keeping in mind the mechanical strength of the stacked bricks in the shaft, it is not possible to increase the kiln height indefinitely.

11.3.8 Overall Conclusions

In spite of the several empirical inputs, the thermochemical model has succeeded in predicting the main trends observed through experimentation on full-scale practical kilns. To overcome discrepancies between model and out-of-pile experimental values of the heating value of fuel, it may be desirable to use a higher order reaction mechanism involving the presence of radical species. In the experiments, the brick temperatures measured through spy holes were somewhat lower than those measured through the central traveling thermocouple; thus suggesting cross-sectional variations in brick temperature. This suggests that a three-dimensional CFD model must be deployed. Further refinements with respect to combustion of in-brick fuel are likely to predict axial variations of brick temperature more accurately. Finally, the model has shown that construction of even taller kilns will bring down the specific fuel consumption still further.

11.4 Gas Turbine Combustion Chamber

We now consider a case study involving gaseous fuel combustion.

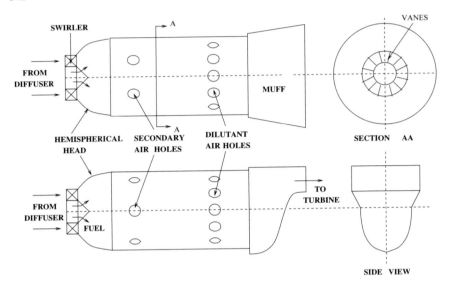

Fig. 11.14 Can-type gas turbine combustor [15]

11.4.1 Combustor Designs

For aircraft propulsion, as well as for stationary power generation, the gas turbine (GT) engine has emerged as an important device of high efficiency and low weight-to-power ratio (see Fig. 1.4). It is an external combustion air-breathing engine in which the combustion chamber must be carefully designed to heat the compressed air from the compressor before it enters the turbine.

Figure 11.14 shows the main features of the *inner shell* of a can-type *diffusion flame* (non-premixed) annular combustor. The outer shell (not shown) is called the *casing*. The compressed air first enters the *diffuser*. A portion of this air is captured by the inner shell through a vane swirler. The remaining air is led through the annular space between the inner and the outer shells. The hemispherical upstream end of the inner shell is called the *dome*. The body of the inner shell is called the *flame tube*, which has discrete peripheral holes/slots to admit the annular air, which is called the *secondary air*. The dome houses the *kerosene fuel nozzle/atomizer* at its center. The nozzle is surrounded by the *annular swirler*, which imparts swirl to the compressed air before it enters the *primary zone* of the flame tube. Thus, the fuel and air streams are separated but mix vigorously in the primary zone. As combustion proceeds through the primary, intermediate, and dilution zones, annular air is admitted to provide additional combustion air, which also provides necessary cooling of the flame tube. The downstream end of the flame tube is called the *muff*. The muff is of rectangular cross section, whereas the flame tube is of circular cross section. Sometimes, a portion of the cooling annular air is admitted through tiny holes in the muff to provide cooling to the *root zones* of the turbine blades, over which the hot combustion products flow.

The shaping and sizing of the combustor, the locations of annular holes, the shaping of the diffuser, dome, and the muff and the like are governed by several considerations. Thus, the design is dictated by the following:

1. The fuel must encounter fresh air in the primary zone to provide continuous combustion through proper mixing. Also, the diffusion flame must be contained within the flame tube.

2. The portion of annular air and its distribution through the holes must be such as to provide *film cooling* of the inner surface of the flame tube, so metallurgical limits are not exceeded.

3. The temperature profile of the gases exiting the flame tube must be as uniform as possible. This uniformity is of importance to thermally protect the first row of the *nozzle guide vanes* placed before the turbine rotor. The temperature uniformity of the exiting air is measured by the *temperature traverse quality factor* TTQF. It is defined as

$$\text{TTQF} = \frac{T_{max,exit} - \overline{T}_{exit}}{\overline{T}_{exit} - \overline{T}_{inlet}}. \qquad (11.90)$$

The smaller the TTQF, the more uniform is the exit gas temperature profile.

4. The pressure drop in the combustor must not exceed 4% to 6% of the elevated inlet pressure so the thrust/power produced by the engine is not sacrificed.

5. The *combustion efficiency* must be near 100% while making sure that the levels of CO and NO in the combustor exhaust are less than 25 ppm.

Based on these considerations, several patented designs have emerged. For example, to accommodate the combustor in the limited space available, in modern jet aircraft engines, the combustor is *kidney shaped.*

11.4.2 Idealization

Flow in a GT combustion chamber represents a challenging situation in CFD. This is because the flow is three-dimensional, elliptic, and turbulent, and involves chemical reactions and effects of radiation. Besides, the fluid properties are functions of temperature and composition of products of combustion, and the true geometry of the chamber (a compromise amongst several factors) is always very complex.

Here (see Fig. 11.15), we consider an *idealized* chamber geometry. Unlike the real chamber, the chamber here is taken to be axisymmetric with exit radius $R = 0.0625$ m and length $L = 0.25$ m. In actual combustion, aviation fuel (kerosene) is used, but we assume that fuel is vaporized and enters the chamber with air in *stoichiometric* proportions—that is, 1 kg of fuel is *premixed* with 17.16 kg of air. Thus, the stoichiometric air/fuel ratio is $R_{stoic} = 17.16$. The fuel–air mixture enters radially through a circumferential slot (width $= 3.75$ mm, located at $0.105 L$) with a

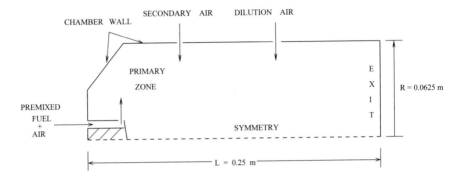

Fig. 11.15 Idealized gas turbine combustion chamber

velocity 111.0 m/s and temperature 500 °C (773 K). Additional air is injected radially through the cylindrical portion (called casing) of the chamber through two circumferential slots.[12] The first slot (width = 2.25 mm, located at 0.335 L) injects air (called *secondary air*) to sustain chemical reaction in the primary zone and the second slot (width = 2.25 mm, located at 0.665 L) provides additional air (called *dilution air*) to dilute the hot combustion products before they leave the chamber. The secondary air is injected with a velocity of 48 m/s and temperature 500 °C. The dilution air is injected at 42.7 m/s and 500 °C. The mean pressure in the chamber is 8 bar and the molecular weights of fuel, air, and combustion products are taken as 16.0, 29.045, and 28.0, respectively. The heat of combustion ΔH_c of fuel is 49 MJ/kg.

With these specifications, we have a domain that captures the main features of a typical GT combustion chamber. The top of Fig. 11.16 shows the grid (in *curvilinear coordinates*) generated to fit the domain. A 50 (axial) × 32 (radial) grid is used. Equations for $\Phi = u_1, u_2, k, \epsilon$, and T must be solved in axisymmetric mode (see Chap. 6). In addition, equations for scalar variables ω_{fu} and a composite variable $\Psi = \omega_{fu} - \omega_{air}/R_{stoic}$ must also be solved. The latter variable is admissible because a simple one-step chemical reaction

$$(1) \quad \text{kg of fuel} + (R_{st}) \text{ kg of air} \rightarrow (1 + R_{st}) \text{ kg of products}$$

is assumed to take place. Thus, there are eight variables to be solved simultaneously. The source terms of flow variables remain unaltered from those introduced in Chap. 6 but, those of T, ω_{fu} and conserved scalar Ψ are as follows:

$$S_{\omega_{fu}} = - R_{fu} \quad S_{\Psi} = 0 \quad S_T = \frac{R_{fu} \, H_c}{Cp} , \quad \text{and}$$

$$R_{fu} = C \, \rho \, \omega_{fu} \, \frac{\epsilon}{e} , \tag{11.91}$$

---

[12]In actual practice, radial injection is carried out through discrete holes. However, accounting for this type of injection will make the flow three-dimensional—hence, the idealization of a circumferential slot.

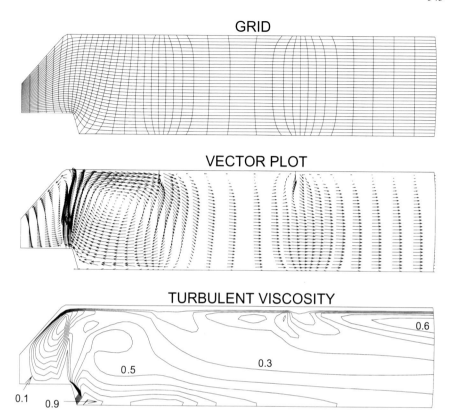

Fig. 11.16 G-T combustion chamber—grid and flow variables

where the volumetric fuel burn rate R_{fu} kg/m^3-s is specified following Spalding [111], with C = 1. This eddy break-up model is chosen because it is assumed that the fuel burning rate is completely controlled by the turbulent timescale ϵ/e. For further variations on Spalding's model, see Ref. [78]. In the specification of S_T, radiation contribution is ignored.

The combustion chamber walls are assumed adiabatic. The inflow boundary specifications, however, require explanation. At the primary slot, $u_1 = 0$, $u_2 = 100$ m/s, k = $(0.005 \times u_2)^2$, $\epsilon = C_\mu\, \rho\, k^2 / (\mu\, R_\mu)$ where viscosity ratio $R_\mu = \mu_t / \mu = 10$, T = 773 K, $\omega_{fu} = (1 + R_{stoic})^{-1}$, and $\Psi = 0$. At the secondary and dilution slots, $u_2 = -48$ m/s and -42.5 m/s, respectively, and k = $(0.0085 \times u_2)^2$, $R_\mu = 29$, T = 773 K, $\omega_{fu} = 0$, and $\Psi = -1/R_{stoic}$ are specified. Finally, the fluid viscosity is taken as $\mu = 3.6 \times 10^{-4}$ N-s/m^2 and specific heats of all species are assumed constant at cp = 1500 J/kg-K. The density is calculated from $\rho = 8 \times 10^5\, M_{mix} / (R_u\, T)$, where R_u is the universal gas constant and $M_{mix}^{-1} = \omega_{fu}/M_{fu} + \omega_{air}/M_{air} + \omega_{pr}/M_{pr}$, and product mass fraction $\omega_{pr} = 1 - \omega_{fu} - \omega_{air}$.

11.4.3   *Computed Results*

This case study is taken from the author's book [30]. In the middle of Fig. 11.16, the vector plot is shown. The plot clearly shows the strong circulation in the primary zone, with a reverse flow near the axis necessary to sustain combustion. All scalar variables are now plotted as $(\Phi - \Phi_{min}) / (\Phi_{max} - \Phi_{min})$ in the range $(0, (0.1), 1.0)$. For turbulent viscosity, $\mu_{t,min} = 0$ and $\mu_{t,max} = 0.029$; for temperature, $T_{min} = 773$ K and $T_{max} = 2456$ K (adiabatic temperature $= 2572$ K); for fuel mass fraction, $\omega_{fu,min} = 0$ and $\omega_{fu,max} = 0.055066$ and, for the composite variable, $\Psi_{min} = -0.058275$ and $\Psi_{max} = 0$. The bottom of Fig. 11.16 shows that high turbulent viscosity levels occur immediately downstream of the fuel injection slot and secondary and dilution air slots because of high levels of mixing.

The top of Fig. 11.17 shows that the fuel is completely consumed in the primary zone. Sometimes, it is of interest to know the values of the mixture fraction $f = f_{stoic} + \Psi (1 - f_{stoic})$, where $f_{stoic} = (1 + R_{stoic})^{-1}$. From the contours of Ψ

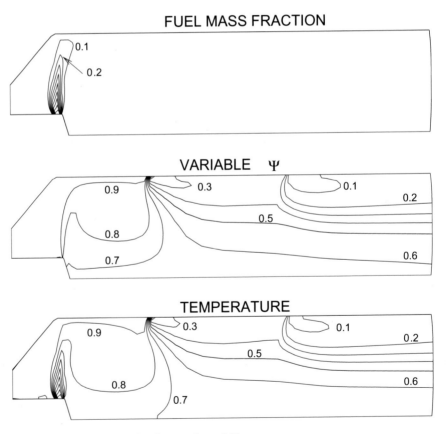

Fig. 11.17 GT combustion chamber—scalar variables

shown in the middle of Fig. 11.17, therefore the values of f and concentrations of air and products can be deciphered. The temperature contours shown at the bottom of Fig. 11.17 are similar to those of Ψ. This is not surprising because although T is not a conserved property, enthalpy $h = Cp\,T + \omega_{fu}\,\Delta H_c$, like Ψ, is a conserved property, and $\omega_{fu} \simeq 0$ over the greater part of the domain. The temperatures, as expected, are high in the primary zone and in the region behind the fuel injection slot. However, the temperature profile is not at all uniform in the exit section. Combustion chamber designers desire a high uniformity of temperature in the exit section to safeguard the operation of the turbine downstream. Such a uniformity is often achieved by nonaxisymmetric narrowing of the exit section (see Fig. 11.14). However, accounting for this feature will make the flow three-dimensional and is hence not considered here.

Combustion chamber flows are extensively investigated through CFD for achieving better profiling of the casing, for determining geometry of injection holes to achieve high levels of mixing, for determining the exact location of injection ports to minimize NOx formation, to achieve uniformity of exit temperatures, to take account of liquid-fuel injection from burners and consequent fuel break up into droplets.

Appendix A
Thermochemistry Data

A.1 Curve-Fit Coefficients

The curve-fit coefficients for gas-phase thermochemistry data of C–H–O–N systems[1] are given in Tables A.1, A.2, and A.3. The values of properties $\overline{C}_p$ (kJ/kmol-K), $\Delta \overline{h}_s = \overline{h} - \overline{h}^0{}_{f,ref}$ (kJ/kmol), and $\overline{s}$ (kJ/kmol-K) are evaluated as follows:

$$\overline{C}_p = A + B\,\theta + C\,\theta^2 + D\,\theta^3 + \frac{E}{\theta^2}, \tag{A.1}$$

$$\frac{\Delta \overline{h}_s}{1000} = A\,\theta + B\,\frac{\theta^2}{2} + C\,\frac{\theta^3}{3} + D\,\frac{\theta^4}{4} - \frac{E}{\theta} + F - \frac{\overline{h}^0{}_{f\,ref}}{1000}, \tag{A.2}$$

$$\overline{s}_{P_{ref}} = A\ln\theta + B\,\theta + C\,\frac{\theta^2}{2} + D\,\frac{\theta^3}{3} + E\,\frac{\theta^2}{2} + G, \tag{A.3}$$

where $\theta = T/1000$. The subscript *ref* denotes $T_{ref} = 298$ K and $p_{ref} = 1$ bar.

Graphite C^* Data

The specific heat of solid graphite[2] is given by

$$\begin{aligned} \overline{C}p = {}& 24.3 + 9.446 \times 10^{-4}\,T - 5.1252 \times 10^6\,T^{-2} \\ & + 1.5858 \times 10^9\,T^{-3} - 1.44 \times 10^{11}\,T^{-4} \end{aligned} \tag{A.4}$$

with $\overline{h}^0{}_{f\,ref} = 0$ and $\overline{s}_{P_{ref}} = 5.69$. The values of $\Delta \overline{h}_s\,(T)$ and $\overline{s}_{P_{ref}}\,(T)$ are evaluated using Eqs. 3.56 and 3.59, respectively (Tables A.4, A.5, A.6, A.7, A.8, A.9, A.10, A.11, A.12, A.13, A.14, A.15, A.16 and A.17).

---

[1] Please see the website of the National Institute of Standards and Technology, http://webbook.nist.gov/chemistry/form-ser.html.

[2] Please see the website of the International Nuclear Safety Society, http://www.insc.anl.gov/matprop/graphite/ent-hc/assessment.php.

© Springer Nature Singapore Pte Ltd. 2020
A. W. Date, *Analytic Combustion*,
https://doi.org/10.1007/978-981-15-1853-9

Table A.1 Curve-fit coefficients for $\overline{C}p$, $\Delta \overline{h}_s$, $\overline{s}^0$ for C, CO, CO$_2$, O, and O$_2$

Species	CO	CO$_2$	O	O$_2$
T (K)	298–1300 (1300–6000)	298–1200 (1200–6000)	298–6000	298–6000
A	25.56759 (35.15070)	24.99735 (58.16639)	21.1861	29.659
B	6.09613 (1.300095)	55.18696 (2.720074)	−0.502314	6.137261
C	4.054656 (−0.205921)	−33.69137 (−0.492289)	0.168694	−1.186521
D	−2.671301 (0.01355)	7.948387 (0.038844)	−0.008962	0.09578
E	0.131021 (−3.28278)	−0.136638 (−6.447293)	0.075664	−0.219663
F	−118.0089 (−127.8375)	−403.6075 (−425.9186)	243.1306	−9.861391
G	227.3665 (231.712)	228.2431 (263.6125)	187.2591	237.948
$h^0_{f_{ref}}$	−110530	−393520	249170	0.0
$\overline{s}_{p_{ref}}$	197.66	213.79	161.05	205.07

Table A.2 Curve-fit coefficients for $\overline{C}p$, $\Delta \overline{h}_s$, $\overline{s}^0$ for N, NO, NO$_2$, N$_2$, and N$_2$O

Species	N	NO	NO$_2$	N$_2$	N$_2$O
T (K)	298–6000	298–1200 (1200–6000)	298–1200 (1200–6000)	298–6000	298–1400 (1400–6000)
A	21.13581	23.83491 (35.99169)	16.10857 (56.82541)	26.092	27.67988 (60.30274)
B	−0.388842	12.58878 (0.95717)	75.89525 (0.738053)	8.218801	51.14898 (1.034566)
C	0.043545	−1.139011 (−0.148032)	−54.3874 (−0.144721)	−1.976141	−30.64454 (−0.192997)
D	0.024685	−1.497459 (0.009974)	14.30777 (0.00977)	0.159274	6.847911 (0.012540)
E	−0.025678	0.214194 (−3.004088)	0.239423 (−5.459911)	0.044434	−0.157906 (−6.860254)
F	466.311	83.35783 (73.10787)	26.17464 (2.846456)	−7.98923	71.24934 (48.61390)
G	178.8263	237.1219 (246.1619)	240.5386 (290.5056)	221.02	238.6164 (272.5002)
$h^0_{f_{ref}}$	472683	90290	33100	0.0	82048.24
$\overline{s}_{p_{ref}}$	153.28	210.76	240.04	191.56	219.96

Table A.3 Curve-fit coefficients for $\overline{C}p$, $\Delta\overline{h}_s$, $\overline{s}^0$ for H, OH, H_2, and H_2O

Species	H	OH	H_2	H_2O
T (K)	298–6000	298–1300 (1300–6000)	298–1000 (1000–2500) (2500–6000)	500–1700 (1700–6000)
A	20.78603	32.27768 (28.74701)	33.066178 (18.563083) (43.413560)	30.092 (41.96426)
B	4.850638×10^{-10}	−11.36291 (4.714489)	−11.363417 (12.257357) (−4.293079)	6.832514 (8.622053)
C	$-1.582916 \times 10^{-10}$	13.60545 (−0.814725)	11.432816 (−2.859786) (1.272428)	6.793435 (−1.49978)
D	1.525102×10^{-11}	−3.846486 (0.054748)	−2.772874 (0.268238) (−0.096876)	−2.53448 (0.098119)
E	3.196347×10^{-11}	−0.001335 (−2.747829)	−0.158558 (1.97799) (−20.533862)	0.082139 (−11.15764)
F	211.802	29.75113 (26.41439)	−9.980797 (−1.147438) (−38.515158)	−250.881 (−272.1797)
G	139.8711	225.5783 (214.1166)	172.707974 (156.288133) (162.081354)	223.3967 (219.7809)
$\overline{h}^0_{f_{ref}}$	21800	38990	0.0	−241830
$\overline{s}_{p_{ref}}$	114.72	183.71	130.68	188.84

Table A.4 Properties of CO_2; reaction: $C^* + O_2 = CO_2$

T K	$\overline{C}p$ kJ/kmol-K	$\Delta \overline{h}_s$ kJ/kmol	$\overline{h}_f^0$ kJ/kmol	$\overline{s}^0$ kJ/kmol-K	$\overline{g}^0$ kJ/kmol
200	31.33	−3387.8	−393511.3	200.1	−394091.3
298	37.12	0.0	−393520.0	213.8	−394414.0
300	37.22	63.4	−393507.9	214.0	−394437.9
400	41.34	4000.1	−393573.3	225.3	−394693.3
500	44.61	8303.2	−393707.2	234.9	−394957.2
600	47.32	12904.0	−393883.1	243.3	−395203.1
700	49.57	17751.7	−394073.5	250.8	−395473.5
800	51.44	22804.9	−394263.2	257.5	−395623.2
900	53.00	28029.4	−394446.4	263.6	−395706.4
1000	54.30	33396.6	−394623.1	269.3	−395923.1
1100	55.40	38883.5	−394795.8	274.5	−396005.8
1200	56.31	44468.8	−394969.8	279.4	−396169.8
1300	57.14	50142.8	−395143.1	283.9	−396183.1
1400	57.83	55892.2	−395319.5	288.2	−396299.5
1500	58.40	61704.6	−395503.9	292.2	−396403.9
1600	58.90	67570.4	−395699.7	296.0	−396339.7
1700	59.33	73482.2	−395909.1	299.6	−396419.1
1800	59.70	79434.2	−396133.7	303.0	−396493.7
1900	60.04	85421.6	−396374.4	306.2	−396374.4
2000	60.34	91440.6	−396631.8	309.3	−396431.8
2100	60.61	97487.9	−396906.4	312.2	−396486.4
2200	60.85	103560.8	−397198.3	315.1	−396538.3
2300	61.07	109657.0	−397507.5	317.8	−396587.5
2400	61.28	115774.6	−397834.0	320.4	−396394.0
2500	61.47	121911.8	−398177.7	322.9	−396177.7
2600	61.64	128067.2	−398538.3	325.3	−396198.3
2800	61.95	140427.2	−399309.8	329.9	−396229.8
3000	62.23	152846.0	−400147.0	334.2	−395647.0
3200	62.47	165316.5	−401048.0	338.2	−395288.0
3400	62.69	177833.5	−402011.2	342.0	−395211.2
3600	62.89	190392.4	−403035.5	345.6	−394755.6
3800	63.08	202989.8	−404120.0	349.0	−393860.0
4000	63.25	215623.2	−405264.0	352.2	−393264.0
4200	63.42	228290.5	−406467.7	355.3	−392607.6
4400	63.58	240990.5	−407731.3	358.3	−392331.2
4600	63.74	253722.3	−409055.8	361.1	−391575.8
4800	63.90	266485.8	−410442.5	363.8	−390762.5
5000	64.06	279281.0	−411893.5	366.4	−389893.5

Table A.5 Properties of CO; reaction: $C^* + \frac{1}{2} O_2 = CO$

T K	$\overline{C}p$ kJ/kmol-K	$\Delta \overline{h}_s$ kJ/kmol	$\overline{h}_f^0$ kJ/kmol	$\overline{s}^0$ kJ/kmol-K	$\overline{g}^0$ kJ/kmol
200	30.20	−2888.8	−111378.1	185.9	−128538.1
298	29.15	0.0	−110530.0	197.6	−137156.3
300	29.15	60.0	−110504.2	197.8	−137354.2
400	29.30	2977.7	−110096.9	206.2	−146336.9
500	29.82	5932.1	−110025.0	212.8	−155425.0
600	30.47	8946.0	−110189.9	218.3	−164489.8
700	31.17	12028.0	−110513.5	223.1	−173583.5
800	31.88	15180.6	−110941.7	227.3	−182541.7
900	32.55	18402.4	−111439.0	231.1	−191449.0
1000	33.18	21689.5	−111983.0	234.5	−200283.0
1100	33.73	25035.6	−112561.2	237.7	−209086.2
1200	34.20	28432.9	−113167.0	240.7	−217927.0
1300	38.47	26820.6	−118850.5	241.5	−229090.5
1400	38.28	30657.4	−119144.6	244.3	−237514.7
1500	38.14	34478.1	−119508.2	247.0	−246108.2
1600	38.04	38287.1	−119931.5	249.4	−254331.5
1700	37.97	42087.3	−120407.4	251.7	−262697.3
1800	37.92	45881.4	−120929.5	253.9	−271139.5
1900	37.88	49671.1	−121493.2	255.9	−279383.2
2000	37.86	53457.8	−122094.6	257.9	−287794.6
2100	37.84	57242.6	−122730.4	259.7	−296085.4
2200	37.84	61026.6	−123397.5	261.5	−304347.5
2300	37.84	64810.2	−124093.9	263.2	−312693.9
2400	37.84	68594.1	−124817.5	264.8	−320777.5
2500	37.85	72378.8	−125566.5	266.3	−328691.4
2600	37.86	76164.4	−126339.5	267.8	−336939.5
2800	37.89	83739.8	−127952.2	270.6	−353212.2
3000	37.93	91321.9	−129647.4	273.2	−368897.4
3200	37.97	98911.4	−131418.2	275.7	−385018.3
3400	38.01	106508.8	−133259.9	278.0	−401009.9
3600	38.05	114114.2	−135168.4	280.2	−416688.5
3800	38.09	121727.9	−137140.8	282.2	−431830.8
4000	38.13	129349.6	−139175.0	284.2	−447575.1
4200	38.17	136979.3	−141269.9	286.0	−462569.9
4400	38.21	144617.0	−143424.4	287.8	−478044.3
4600	38.25	152262.6	−145638.4	289.5	−493398.4
4800	38.29	159916.2	−147912.2	291.1	−508392.2
5000	38.33	167577.8	−150246.5	292.7	−523496.5

Table A.6 Properties of C*

T	$\overline{C}p$	$\Delta\overline{h}_s$	$\overline{h}_f^0$	$\overline{s}^0$	$\overline{g}^0$
K	kJ/kmol-K	kJ/kmol	kJ/kmol	kJ/kmol-K	kJ/kmol
200	4.58	−684.8	0.0	3.0	0.0
298	8.53	0.0	0.0	5.7	0.0
300	8.59	17.1	0.0	5.7	0.0
400	11.80	1035.8	0.0	8.7	0.0
500	14.65	2363.7	0.0	11.6	0.0
600	16.86	3944.6	0.0	14.5	0.0
700	18.53	5717.8	0.0	17.2	0.0
800	19.79	7636.5	0.0	19.8	0.0
900	20.78	9667.0	0.0	22.2	0.0
1000	21.56	11785.4	0.0	24.4	0.0
1100	22.20	13974.4	0.0	26.5	0.0
1200	22.72	16221.1	0.0	28.4	0.0
1300	23.17	18516.2	0.0	30.3	0.0
1400	23.55	20852.4	0.0	32.0	0.0
1500	23.88	23224.1	0.0	33.6	0.0
1600	24.17	25627.2	0.0	35.2	0.0
1700	24.44	28058.0	0.0	36.7	0.0
1800	24.68	30513.9	0.0	38.1	0.0
1900	24.90	32992.7	0.0	39.4	0.0
2000	25.10	35492.4	0.0	40.7	0.0
2100	25.29	38011.6	0.0	41.9	0.0
2200	25.46	40549.1	0.0	43.1	0.0
2300	25.63	43103.7	0.0	44.2	0.0
2400	25.79	45674.6	0.0	45.3	0.0
2500	25.94	48261.0	0.0	46.4	0.0
2600	26.08	50862.3	0.0	47.4	0.0
2800	26.36	56107.1	0.0	49.3	0.0
3000	26.62	61405.6	0.0	51.2	0.0
3200	26.87	66754.8	0.0	52.9	0.0
3400	27.11	72152.7	0.0	54.5	0.0
3600	27.34	77597.4	0.0	56.1	0.0
3800	27.56	83087.6	0.0	57.6	0.0
4000	27.78	88622.1	0.0	59.0	0.0
4200	28.00	94200.2	0.0	60.4	0.0
4400	28.21	99821.0	0.0	61.7	0.0
4600	28.42	105483.9	0.0	62.9	0.0
4800	28.63	111188.4	0.0	64.1	0.0
5000	28.83	116934.0	0.0	65.3	0.0

Table A.7 Properties of H_2O; reaction: $H_2 + \frac{1}{2} O_2 = H_2O$

T K	$\overline{C}p$ kJ/kmol-K	$\Delta\overline{h}_s$ kJ/kmol	$\overline{h}_f^0$ kJ/kmol	$\overline{s}^0$ kJ/kmol-K	$\overline{g}^0$ kJ/kmol
200	33.76	−3289.5	−240991.0	175.4	−232771.0
298	33.59	0.0	−241830.0	188.8	−228583.9
300	33.60	66.3	−241834.3	189.0	−228484.3
400	34.26	3455.8	−242842.1	198.8	−223922.1
500	35.22	6928.2	−243847.1	206.5	−219047.1
600	36.32	10504.2	−244798.2	213.1	−214018.2
700	37.50	14194.6	−245677.8	218.7	−208787.8
800	38.74	18006.2	−246480.9	223.8	−203520.9
900	40.00	21942.8	−247207.7	228.5	−198067.7
1000	41.27	26006.0	−247861.2	232.7	−192561.2
1100	42.52	30195.5	−248445.9	236.7	−187010.9
1200	43.75	34509.5	−248966.3	240.5	−181406.3
1300	44.94	38944.3	−249428.2	244.0	−175718.2
1400	46.06	43494.6	−249837.2	247.4	−170107.1
1500	47.11	48153.7	−250198.8	250.6	−164398.8
1600	48.07	52913.3	−250519.8	253.7	−158679.8
1700	48.91	57760.5	−250811.1	256.6	−152891.1
1800	49.75	62694.3	−251071.4	259.4	−147121.4
1900	50.51	67708.3	−251304.8	262.2	−141294.8
2000	51.20	72794.8	−251516.5	264.8	−135616.5
2100	51.84	77947.3	−251710.7	267.3	−129805.6
2200	52.41	83160.1	−251890.5	269.7	−123960.5
2300	52.95	88428.4	−252058.6	272.0	−117968.6
2400	53.44	93747.9	−252217.1	274.3	−112177.0
2500	53.89	99114.8	−252367.1	276.5	−106492.1
2600	54.32	104525.6	−252511.1	278.6	−100411.1
2800	55.08	115466.9	−252787.7	282.7	−89127.7
3000	55.74	126550.4	−253054.7	286.5	−77104.7
3200	56.32	137758.2	−253319.9	290.1	−65319.9
3400	56.83	149074.8	−253591.6	293.6	−53841.6
3600	57.28	160487.4	−253877.1	296.8	−41837.1
3800	57.68	171984.8	−254183.8	299.9	−29793.8
4000	58.04	183557.5	−254518.0	302.9	−18518.0
4200	58.36	195197.6	−254884.9	305.7	−6244.9
4400	58.65	206898.6	−255288.8	308.4	5411.3
4600	58.91	218655.0	−255732.8	311.1	17047.1
4800	59.16	230462.9	−256219.0	313.6	28901.0
5000	59.40	242319.1	−256748.1	316.0	41002.0

Table A.8 Properties of H_2

T K	$\overline{C}p$ kJ/kmol-K	$\Delta\overline{h}_s$ kJ/kmol	$\overline{h}_f^0$ kJ/kmol	$\overline{s}^0$ kJ/kmol-K	$\overline{g}^0$ kJ/kmol
200	27.26	−2772.7	0.0	119.4	0.0
298	28.84	0.0	0.0	130.7	0.0
300	28.85	53.5	0.0	130.9	0.0
400	29.18	2959.1	0.0	139.2	0.0
500	29.26	5882.0	0.0	145.7	0.0
600	29.32	8811.1	0.0	151.1	0.0
700	29.44	11748.7	0.0	155.6	0.0
800	29.62	14701.3	0.0	159.5	0.0
900	29.88	17676.1	0.0	163.1	0.0
1000	30.21	20680.1	0.0	166.2	0.0
1100	30.58	23718.9	0.0	169.1	0.0
1200	30.99	26797.1	0.0	171.8	0.0
1300	31.42	29917.7	0.0	174.3	0.0
1400	31.86	33082.1	0.0	176.6	0.0
1500	32.30	36290.3	0.0	178.8	0.0
1600	32.73	39541.6	0.0	180.9	0.0
1700	33.14	42834.9	0.0	182.9	0.0
1800	33.54	46168.7	0.0	184.8	0.0
1900	33.92	49541.4	0.0	186.7	0.0
2000	34.28	52951.3	0.0	188.4	0.0
2100	34.62	56396.6	0.0	190.1	0.0
2200	34.95	59875.6	0.0	191.7	0.0
2300	35.26	63386.6	0.0	193.3	0.0
2400	35.56	66928.0	0.0	194.8	0.0
2500	35.83	70497.6	0.0	196.2	0.0
2600	36.11	74095.1	0.0	197.7	0.0
2800	36.62	81369.6	0.0	200.3	0.0
3000	37.09	88741.4	0.0	202.9	0.0
3200	37.53	96203.3	0.0	205.3	0.0
3400	37.94	103750.4	0.0	207.6	0.0
3600	38.34	111379.3	0.0	209.8	0.0
3800	38.74	119087.5	0.0	211.9	0.0
4000	39.12	126873.0	0.0	213.8	0.0
4200	39.49	134733.5	0.0	215.8	0.0
4400	39.85	142667.0	0.0	217.6	0.0
4600	40.19	150670.7	0.0	219.4	0.0
4800	40.52	158741.9	0.0	221.1	0.0
5000	40.83	166876.9	0.0	222.8	0.0

Table A.9 Properties of H; reaction: $\frac{1}{2}$ H$_2$ = H

T	$\overline{C}p$	$\Delta \overline{h}_s$	$\overline{h}_f^0$	$\overline{s}^0$	$\overline{g}^0$
K	kJ/kmol-K	kJ/kmol	kJ/kmol	kJ/kmol-K	kJ/kmol
200	20.79	−2040.8	217345.5	106.4	208005.5
298	20.79	0.0	218000.0	114.7	203293.7
300	20.79	37.8	218011.0	114.8	203206.0
400	20.79	2116.4	218636.8	120.8	198156.8
500	20.79	4195.0	219254.0	125.5	192929.0
600	20.79	6273.6	219868.0	129.3	187618.0
700	20.79	8352.2	220477.8	132.5	182187.8
800	20.79	10430.8	221080.2	135.2	176720.2
900	20.79	12509.4	221671.3	137.7	171136.3
1000	20.79	14588.0	222248.0	139.9	165448.0
1100	20.79	16666.6	222807.2	141.9	159722.2
1200	20.79	18745.2	223346.7	143.7	153986.7
1300	20.79	20823.8	223865.0	145.3	148270.0
1400	20.79	22902.4	224361.3	146.9	142321.4
1500	20.79	24981.0	224835.8	148.3	136485.8
1600	20.79	27059.7	225288.9	149.6	130648.9
1700	20.79	29138.3	225720.8	150.9	124655.9
1800	20.79	31216.9	226132.5	152.1	118672.5
1900	20.79	33295.5	226524.8	153.2	112809.8
2000	20.79	35374.1	226898.5	154.3	106698.4
2100	20.79	37452.7	227254.4	155.3	100729.4
2200	20.79	39531.3	227593.5	156.3	94603.5
2300	20.79	41609.9	227916.6	157.2	88651.6
2400	20.79	43688.5	228224.5	158.1	82544.5
2500	20.79	45767.1	228518.3	158.9	76518.3
2600	20.79	47845.7	228798.2	159.7	70588.2
2800	20.79	52002.9	229318.1	161.3	58098.1
3000	20.79	56160.1	229789.4	162.7	46039.4
3200	20.79	60317.3	230215.7	164.0	33895.7
3400	20.79	64474.5	230599.3	165.3	21499.3
3600	20.79	68631.7	230942.1	166.5	9182.1
3800	20.79	72788.9	231245.2	167.6	−3024.9
4000	20.79	76946.1	231509.6	168.7	−15690.4
4200	20.79	81103.3	231736.5	169.7	−27823.4
4400	20.79	85260.5	231927.0	170.7	−40433.0
4600	20.79	89417.7	232082.3	171.6	−52657.7
4800	20.79	93575.0	232204.0	172.5	−65155.9
5000	20.79	97732.2	232293.8	173.3	−77206.3

Table A.10 Properties of OH; reaction: $\frac{1}{2}$ H$_2$ + $\frac{1}{2}$ O$_2$ = OH

T K	$\overline{C}p$ kJ/kmol-K	$\Delta \overline{h}_s$ kJ/kmol	$\overline{h}_f^0$ kJ/kmol	$\overline{s}^0$ kJ/kmol-K	$\overline{g}^0$ kJ/kmol
200	30.49	−2969.2	38763.0	171.6	35803.0
298	29.98	0.0	38990.0	183.7	34281.6
300	29.97	52.2	38998.3	183.9	34243.3
400	29.65	3032.1	39033.8	192.5	32633.8
500	29.51	5989.1	38974.8	199.1	31049.7
600	29.52	8939.6	38862.8	204.4	29532.8
700	29.67	11898.2	38720.2	209.0	27940.2
800	29.92	14876.9	38560.4	213.0	26360.4
900	30.27	17885.7	38393.2	216.5	24938.3
1000	30.67	20932.2	38225.0	219.7	23425.0
1100	31.12	24021.6	38059.7	222.7	21889.7
1200	31.59	27156.9	37899.6	225.4	20499.6
1300	31.99	30335.4	37741.7	228.0	18956.7
1400	32.50	33560.5	37589.8	230.3	17639.8
1500	32.95	36833.3	37445.9	232.6	16145.9
1600	33.36	40148.9	37306.6	234.7	14826.6
1700	33.73	43503.2	37169.1	236.8	13284.1
1800	34.06	46892.9	37031.5	238.7	11921.6
1900	34.38	50315.2	36892.9	240.6	10577.8
2000	34.67	53767.7	36752.0	242.3	9252.0
2100	34.94	57248.2	36608.5	244.0	7838.6
2200	35.19	60754.8	36462.0	245.7	6322.0
2300	35.43	64285.8	36312.1	247.2	5147.1
2400	35.65	67839.7	36158.7	248.7	3878.7
2500	35.86	71415.1	36002.0	250.2	2377.0
2600	36.05	75010.7	35841.6	251.6	1131.5
2800	36.41	82257.8	35508.1	254.3	−1731.9
3000	36.73	89572.7	35158.3	256.8	−4141.7
3200	37.02	96947.9	34791.4	259.2	−6808.6
3400	37.27	104377.2	34406.0	261.5	−9624.0
3600	37.50	111855.1	34000.2	263.6	−12079.8
3800	37.71	119376.8	33571.9	265.6	−14308.1
4000	37.90	126938.4	33119.4	267.6	−17280.6
4200	38.08	134536.4	32640.7	269.4	−19439.3
4400	38.24	142168.2	32134.3	271.2	−22205.7
4600	38.39	149831.5	31599.0	272.9	−24520.9
4800	38.54	157525.0	31034.1	274.5	−26805.9
5000	38.68	165247.6	30438.9	276.1	−29311.1

Table A.11 Properties of O_2

T K	$\overline{C}p$ kJ/kmol-K	$\Delta \overline{h}_s$ kJ/kmol	$\overline{h}_f^0$ kJ/kmol	$\overline{s}^0$ kJ/kmol-K	$\overline{g}^0$ kJ/kmol
200	25.35	−2711.7	0.0	194.2	0.0
298	28.91	0.0	0.0	205.1	0.0
300	28.96	34.2	0.0	205.2	0.0
400	30.56	3017.6	0.0	213.8	0.0
500	31.56	6126.7	0.0	220.8	0.0
600	32.32	9322.5	0.0	226.6	0.0
700	32.96	12587.4	0.0	231.6	0.0
800	33.52	15911.6	0.0	236.0	0.0
900	34.02	19288.8	0.0	240.0	0.0
1000	34.49	22714.3	0.0	243.6	0.0
1100	34.92	26184.9	0.0	246.9	0.0
1200	35.33	29697.5	0.0	250.0	0.0
1300	35.71	33249.7	0.0	252.8	0.0
1400	36.08	36839.3	0.0	255.5	0.0
1500	36.42	40464.4	0.0	258.0	0.0
1600	36.75	44122.9	0.0	260.4	0.0
1700	37.06	47813.3	0.0	262.6	0.0
1800	37.35	51534.0	0.0	264.7	0.0
1900	37.63	55283.3	0.0	266.8	0.0
2000	37.90	59060.0	0.0	268.7	0.0
2100	38.15	62862.7	0.0	270.5	0.0
2200	38.39	66690.0	0.0	272.3	0.0
2300	38.62	70540.8	0.0	274.0	0.0
2400	38.84	74414.0	0.0	275.7	0.0
2500	39.05	78308.5	0.0	277.3	0.0
2600	39.25	82223.2	0.0	278.8	0.0
2800	39.62	90109.9	0.0	281.7	0.0
3000	39.95	98067.4	0.0	284.5	0.0
3200	40.27	106089.7	0.0	287.1	0.0
3400	40.56	114172.0	0.0	289.5	0.0
3600	40.83	122310.5	0.0	291.8	0.0
3800	41.09	130502.2	0.0	294.1	0.0
4000	41.34	138745.1	0.0	296.2	0.0
4200	41.59	147038.0	0.0	298.2	0.0
4400	41.84	155380.8	0.0	300.1	0.0
4600	42.10	163774.2	0.0	302.0	0.0
4800	42.36	172219.9	0.0	303.8	0.0
5000	42.65	180720.5	0.0	305.5	0.0

Table A.12 Properties of O; reaction: $\frac{1}{2} O_2 = O$

T K	$\overline{C}_p$ kJ/kmol-K	$\Delta \overline{h}_s$ kJ/kmol	$\overline{h}_f^0$ kJ/kmol	$\overline{s}^0$ kJ/kmol-K	$\overline{g}^0$ kJ/kmol
200	22.98	−2190.1	248335.8	152.1	237335.8
298	21.90	0.0	249170.0	161.0	231751.9
300	21.89	43.1	249196.0	161.2	231616.0
400	21.48	2209.2	249870.4	167.4	225670.4
500	21.28	4346.4	250453.0	172.2	219553.0
600	21.15	6467.6	250976.3	176.1	213296.3
700	21.07	8578.5	251454.8	179.3	207004.8
800	21.01	10682.0	251896.2	182.1	200616.2
900	20.96	12780.1	252305.7	184.6	194165.7
1000	20.92	14873.9	252686.8	186.8	187686.8
1100	20.89	16964.2	253041.8	188.8	181156.8
1200	20.86	19051.7	253373.0	190.6	174653.0
1300	20.84	21137.0	253682.2	192.3	168012.2
1400	20.83	23220.5	253970.8	193.8	161500.8
1500	20.82	25302.6	254240.4	195.3	154790.4
1600	20.81	27383.8	254492.3	196.6	148252.3
1700	20.80	29464.2	254727.5	197.9	141507.6
1800	20.80	31544.2	254947.2	199.1	134797.2
1900	20.80	33624.2	255152.5	200.2	128232.5
2000	20.80	35704.3	255344.3	201.2	121644.3
2100	20.81	37785.0	255523.7	202.3	114718.6
2200	20.82	39866.3	255691.3	203.2	108181.3
2300	20.83	41948.6	255848.2	204.2	101288.2
2400	20.84	44032.1	255995.1	205.0	94835.1
2500	20.86	46116.9	256132.7	205.9	88007.7
2600	20.87	48203.5	256261.9	206.7	81281.9
2800	20.92	52382.3	256497.3	208.3	67637.4
3000	20.96	56570.0	256706.3	209.7	54356.3
3200	21.02	60768.3	256893.5	211.1	40733.4
3400	21.08	64978.4	257062.4	212.3	27392.4
3600	21.15	69201.8	257216.5	213.5	13856.5
3800	21.23	73439.5	257358.4	214.7	288.4
4000	21.31	77692.8	257490.2	215.8	−13309.7
4200	21.39	81962.7	257613.7	216.8	−26726.3
4400	21.48	86250.1	257729.7	217.8	−40370.3
4600	21.58	90555.9	257838.8	218.8	−54041.2
4800	21.67	94880.9	257940.9	219.7	−67499.1
5000	21.77	99225.6	258035.3	220.6	−81214.7

Table A.13 Properties of N_2

T K	$\overline{C}p$ kJ/kmol-K	$\Delta\overline{h}_s$ kJ/kmol	$\overline{h}_f^0$ kJ/kmol	$\overline{s}^0$ kJ/kmol-K	$\overline{g}^0$ kJ/kmol
200	28.77	−2833.8	0.0	180.1	0.0
298	28.87	0.0	0.0	191.5	0.0
300	28.88	42.6	0.0	191.7	0.0
400	29.35	2952.9	0.0	200.1	0.0
500	29.91	5915.4	0.0	206.7	0.0
600	30.47	8934.2	0.0	212.2	0.0
700	31.02	12008.9	0.0	217.0	0.0
800	31.55	15137.9	0.0	221.1	0.0
900	32.06	18318.7	0.0	224.9	0.0
1000	32.54	21548.8	0.0	228.3	0.0
1100	32.99	24825.5	0.0	231.4	0.0
1200	33.42	28146.0	0.0	234.3	0.0
1300	33.81	31507.6	0.0	237.0	0.0
1400	34.18	34907.7	0.0	239.5	0.0
1500	34.53	38343.7	0.0	241.9	0.0
1600	34.85	41813.1	0.0	244.1	0.0
1700	35.15	45313.5	0.0	246.2	0.0
1800	35.43	48842.5	0.0	248.3	0.0
1900	35.68	52397.9	0.0	250.2	0.0
2000	35.91	55977.5	0.0	252.0	0.0
2100	36.12	59579.3	0.0	253.8	0.0
2200	36.31	63201.3	0.0	255.5	0.0
2300	36.49	66841.5	0.0	257.1	0.0
2400	36.64	70498.2	0.0	258.6	0.0
2500	36.78	74169.8	0.0	260.1	0.0
2600	36.91	77854.5	0.0	261.6	0.0
2800	37.11	85257.6	0.0	264.3	0.0
3000	37.27	92696.6	0.0	266.9	0.0
3200	37.38	100162.1	0.0	269.3	0.0
3400	37.46	107646.2	0.0	271.6	0.0
3600	37.50	115142.5	0.0	273.7	0.0
3800	37.53	122646.2	0.0	275.7	0.0
4000	37.55	130153.9	0.0	277.7	0.0
4200	37.55	137663.9	0.0	279.5	0.0
4400	37.57	145176.0	0.0	281.2	0.0
4600	37.59	152691.2	0.0	282.9	0.0
4800	37.63	160212.6	0.0	284.5	0.0
5000	37.69	167744.2	0.0	286.0	0.0

Table A.14 Properties of NO; reaction: $\frac{1}{2} N_2 + \frac{1}{2} O_2 = NO$

T K	$\overline{C}p$ kJ/kmol-K	$\Delta \overline{h}_s$ kJ/kmol	$\overline{h}_f^0$ kJ/kmol	$\overline{s}^0$ kJ/kmol-K	$\overline{g}^0$ kJ/kmol
200	31.65	−2989.2	90074.6	198.6	87784.6
298	29.86	0.0	90291.0	210.7	86595.8
300	29.85	56.4	90309.0	210.9	86574.0
400	29.93	3038.4	90344.2	219.5	85324.2
500	30.51	6058.5	90328.4	226.3	84053.4
600	31.25	9146.1	90308.8	231.9	82808.8
700	32.01	12309.3	90302.1	236.8	81552.1
800	32.74	15547.6	90313.9	241.1	80273.9
900	33.42	18856.2	90343.5	245.0	79048.5
1000	34.00	22227.8	90387.2	248.5	77837.3
1100	34.49	25653.1	90438.9	251.8	76523.9
1200	34.86	29119.2	90488.5	254.8	75308.5
1300	35.23	32624.3	90536.6	257.6	74026.6
1400	35.54	36163.1	90580.6	260.2	72800.6
1500	35.79	39729.9	90616.8	262.7	71491.8
1600	36.01	43320.4	90643.4	265.0	70243.4
1700	36.20	46931.2	90658.8	267.2	68898.8
1800	36.37	50559.7	90662.5	269.3	67622.5
1900	36.51	54203.8	90654.2	271.3	66334.2
2000	36.64	57861.6	90633.9	273.1	65133.9
2100	36.76	61531.9	90601.9	274.9	63826.9
2200	36.87	65213.3	90558.6	276.6	62618.6
2300	36.96	68904.9	90504.8	278.3	61179.8
2400	37.05	72605.8	90440.7	279.8	60080.7
2500	37.13	76315.2	90367.1	281.4	58617.1
2600	37.21	80032.5	90284.6	282.8	57524.7
2800	37.35	87488.5	90095.8	285.6	54815.7
3000	37.47	94970.1	89879.1	288.2	52379.0
3200	37.57	102474.2	89639.3	290.6	49959.3
3400	37.67	109998.3	89380.2	292.9	47390.2
3600	37.75	117540.4	89104.9	295.0	45004.9
3800	37.83	125098.8	88815.6	297.1	42455.6
4000	37.90	132672.2	88513.7	299.0	40313.8
4200	37.97	140259.4	88199.5	300.9	37589.5
4400	38.03	147859.6	87872.2	302.6	35292.2
4600	38.09	155471.9	87530.2	304.3	33020.3
4800	38.15	163095.9	87170.7	305.9	30770.7
5000	38.20	170731.1	86789.8	307.5	28039.8

Table A.15 Properties of NO_2; reaction: $\frac{1}{2} N_2 + O_2 = NO_2$

T K	$\overline{C}p$ kJ/kmol-K	$\Delta\overline{h}_s$ kJ/kmol	$\overline{h}_f^0$ kJ/kmol	$\overline{s}^0$ kJ/kmol-K	$\overline{g}^0$ kJ/kmol
200	35.21	−3522.2	33706.4	225.7	45416.4
298	36.97	0.0	33100.0	240.0	51233.3
300	37.03	63.9	33108.4	240.3	51333.4
400	40.18	3922.4	32528.3	251.3	57548.4
500	43.21	8094.4	32110.0	260.6	63885.0
600	45.82	12549.6	31860.0	268.8	70200.0
700	47.98	17243.5	31751.7	276.0	76621.7
800	49.72	22131.7	31751.2	282.5	82991.2
900	51.09	27174.6	31826.4	288.4	89471.5
1000	52.16	32339.2	31950.5	293.9	95800.5
1100	53.03	37600.2	32102.5	298.9	102172.5
1200	53.73	42940.0	32269.5	303.5	108649.5
1300	54.33	48344.1	32440.6	307.9	114860.6
1400	54.82	53802.3	32609.2	311.9	121299.2
1500	55.21	59304.4	32768.1	315.7	127643.1
1600	55.54	64842.7	32913.2	319.3	133953.3
1700	55.82	70411.2	33041.2	322.7	140141.1
1800	56.06	76005.4	33150.1	325.9	146460.2
1900	56.26	81621.5	33239.2	328.9	152939.2
2000	56.44	87256.5	33307.8	331.8	159107.8
2100	56.59	92907.9	33355.6	334.5	165445.5
2200	56.72	98573.8	33383.1	337.2	171653.1
2300	56.84	104252.3	33390.8	339.7	177945.7
2400	56.95	109942.2	33379.1	342.1	184339.1
2500	57.05	115642.0	33348.6	344.5	190473.6
2600	57.13	121350.8	33300.3	346.7	196840.3
2800	57.28	132791.9	33153.2	350.9	209413.2
3000	57.39	144259.2	32943.5	354.9	222093.5
3200	57.49	155748.2	32677.5	358.6	234757.4
3400	57.57	167255.0	32359.9	362.1	247239.9
3600	57.64	178776.7	31995.0	365.4	259695.0
3800	57.70	190310.8	31585.5	368.5	272695.5
4000	57.75	201855.4	31133.4	371.5	285333.4
4200	57.79	213408.8	30638.8	374.3	297968.9
4400	57.82	224969.7	30100.9	377.0	310381.0
4600	57.85	236537.0	29517.2	379.5	323687.2
4800	57.88	248109.9	28883.7	382.0	336323.7
5000	57.90	259687.7	28195.1	384.4	348695.1

Table A.16 Properties of N; reaction: $\frac{1}{2} N_2 = N$

T K	$\overline{C}p$ kJ/kmol-K	$\Delta \overline{h}_s$ kJ/kmol	$\overline{h}_f^0$ kJ/kmol	$\overline{s}^0$ kJ/kmol-K	$\overline{g}^0$ kJ/kmol
200	20.42	−2024.1	472075.8	145.1	461065.8
298	20.74	0.0	472683.0	153.3	455533.1
300	20.74	37.3	472699.0	153.4	455434.0
400	20.83	2116.5	473323.1	159.4	449583.1
500	20.85	4200.8	473926.1	164.0	443601.1
600	20.85	6286.2	474502.1	167.8	437482.1
700	20.84	8370.9	475049.4	171.1	431229.4
800	20.83	10454.3	475568.3	173.8	424968.3
900	20.81	12535.9	476059.6	176.3	418594.6
1000	20.79	14615.7	476524.3	178.5	412174.3
1100	20.77	16693.8	476964.1	180.5	405684.1
1200	20.76	18770.3	477380.3	182.3	399200.3
1300	20.74	20845.2	477774.4	183.9	392754.4
1400	20.73	22918.9	478148.1	185.5	386098.1
1500	20.72	24991.6	478502.8	186.9	379577.8
1600	20.72	27063.5	478839.9	188.2	372999.9
1700	20.71	29135.0	479161.2	189.5	366281.2
1800	20.71	31206.2	479467.9	190.7	359677.9
1900	20.72	33277.7	479761.8	191.8	353031.8
2000	20.72	35349.6	480043.8	192.9	346243.8
2100	20.73	37422.5	480315.8	193.9	339615.9
2200	20.75	39496.6	480578.9	194.8	333068.9
2300	20.77	41572.3	480834.6	195.8	326159.6
2400	20.79	43650.2	481084.1	196.6	319564.1
2500	20.82	45730.5	481328.6	197.5	312703.6
2600	20.85	47813.8	481569.6	198.3	306069.6
2800	20.93	51991.1	482045.3	199.9	292345.3
3000	21.02	56186.0	482520.7	201.3	278970.7
3200	21.14	60402.5	483004.4	202.7	265244.4
3400	21.29	64645.0	483504.9	203.9	251964.9
3600	21.45	68918.1	484029.8	205.2	237969.9
3800	21.64	73226.6	484586.5	206.3	224476.5
4000	21.86	77575.7	485181.8	207.4	210981.8
4200	22.10	81970.6	485821.7	208.5	197071.7
4400	22.37	86416.9	486511.9	209.6	182911.9
4600	22.67	90920.3	487257.7	210.6	169167.6
4800	23.00	95487.0	488063.7	211.5	155663.7
5000	23.36	100123.1	488934.0	212.5	141434.0

Table A.17 Properties of N_2O; reaction: $N_2 + O = N_2O$

T K	$\overline{C}p$ kJ/kmol-K	$\Delta\overline{h}_s$ kJ/kmol	$\overline{h}_f^0$ kJ/kmol	$\overline{s}^0$ kJ/kmol-K	$\overline{g}^0$ kJ/kmol
200	32.79	−3529.4	83542.7	205.7	108842.7
298	38.60	0.0	82048.2	219.9	121563.0
300	38.70	71.2	82033.7	220.2	121843.7
400	42.69	4149.8	81035.9	231.9	135275.9
500	45.82	8580.6	80367.0	241.8	148917.0
600	48.38	13294.5	79940.9	250.4	162680.9
700	50.49	18241.4	79702.2	258.0	176512.2
800	52.25	23381.3	79609.6	264.8	190329.7
900	53.69	28680.4	79629.8	271.1	204189.8
1000	54.87	34110.5	79736.0	276.8	218036.1
1100	55.85	39648.2	79906.7	282.1	231816.7
1200	56.65	45274.5	80125.0	287.0	245605.1
1300	57.34	50974.9	80378.5	291.5	259518.5
1400	57.91	56739.1	80659.1	295.8	273159.2
1500	58.41	62555.9	80957.8	299.8	287057.8
1600	58.84	68419.0	81270.3	303.6	300630.3
1700	59.19	74320.8	81591.3	307.2	314321.3
1800	59.50	80255.6	81917.1	310.6	328157.2
1900	59.76	86218.5	82244.6	313.8	341784.7
2000	59.99	92205.9	82572.3	316.9	355172.3
2100	60.18	98214.6	82898.5	319.8	369128.6
2200	60.36	104242.1	83222.8	322.6	382642.7
2300	60.52	110286.1	83544.2	325.3	396344.3
2400	60.66	116344.9	83862.8	327.9	409542.9
2500	60.78	122416.9	84178.4	330.4	423178.5
2600	60.89	128500.7	84490.9	332.7	437050.9
2800	61.09	140699.4	85107.8	337.3	463947.8
3000	61.25	152933.1	85714.7	341.5	491014.7
3200	61.38	165195.9	86313.8	345.4	518313.8
3400	61.49	177482.9	86906.5	349.2	544886.5
3600	61.58	189790.2	87494.1	352.7	571694.1
3800	61.66	202114.6	88077.1	356.0	598797.2
4000	61.73	214453.5	88655.0	359.2	625855.1
4200	61.78	226804.7	89226.3	362.2	652446.3
4400	61.83	239166.4	89788.6	365.1	678948.6
4600	61.87	251537.1	90338.2	367.8	706278.2
4800	61.91	263915.8	90870.6	370.4	733110.6
5000	61.94	276301.3	91379.7	373.0	759379.8

Appendix B
Curve-Fit Coefficients for ΔH_c, T_{ad}, K_p, Cp, h, and s

B.1 Heat of Combustion

The **heat of combustion** is correlated with Φ according to (see Table B.1)

$$\Delta H_c = a \qquad\qquad 0.4 \le \Phi \le 1. \tag{B.1}$$

$$\Delta H_c = a - b \ln \Phi \qquad 1.0 \le \Phi \le 1.5 \tag{B.2}$$

B.2 Adiabatic Flame Temperature ($0.4 \le \Phi \le 1.4$)

The **adiabatic flame temperature** is correlated as (see Table B.2)

$$\frac{T_{ad}}{1000} = a_1 + a_2 \Phi + a_3 \Phi^2 + a_4 \Phi^3 + a_5 \Phi^4$$
$$+ a_5 \Phi^4 + a_6 \Phi^5 + a_7 \Phi^6 . \tag{B.3}$$

B.3 Equilibrium Constant

The **equilibrium constant** is correlated as (see Table B.3)

$$log_{10} K_p = a_1 + a_2\,\theta + a_3\,\theta^2 + a_4\theta^3$$
$$+ a_5 \ln(\theta) + \frac{a_6}{\theta} \quad \text{where} \quad \theta = \frac{T}{1000} . \tag{B.4}$$

© Springer Nature Singapore Pte Ltd. 2020
A. W. Date, *Analytic Combustion*,
https://doi.org/10.1007/978-981-15-1853-9

B.4   Specific Heat, Enthalpy, Entropy

The **fuel specific heat (kJ/kmol-K), enthalpy (kJ/kmol), and entropy (kJ/kmol-K)** are correlated as (see Table B.4 and [20, 80])

$$\frac{C_p(T)}{R_u} = a_1 + a_2\,T + a_3\,T^2 + a_4\,T^3 + a_5\,T^4\,, \tag{B.5}$$

$$\frac{h(T)}{R_u} = a_1\,T + \frac{a_2}{2}\,T^2 + \frac{a_3}{3}\,T^3 + \frac{a_4}{4}\,T^4 + \frac{a_5}{5}\,T^5 + b1\,, \tag{B.6}$$

$$\frac{s(T)}{R_u} = a_1\,\ln T + a_2\,T + \frac{a_3}{2}\,T^2 + \frac{a_4}{3}\,T^3 + \frac{a_5}{4}\,T^4 + b2\,, \tag{B.7}$$

where $R_u = 8.314$ kJ/kmol-K.

Table B.1 ΔH_c at $T_{ref} = 298$ K and $p_{ref} = 1$ atm, Eqs. B.1, B.2

Fuel	a	b
CH_4	50.150	56.69
C_2H_4	47.2520	50.05
C_2H_6	47.599	53.49
C_3H_4	45.871	49.79
C_3H_8	46.458	52.32
C_4H_8	45.406	49.79
C_4H_{10}	45.839	51.71
C_5H_{10}	45.086	47.79
C_5H_{12}	45.449	51.34
C_6H_{12}	44.889	49.79
C_6H_{14}	45.198	51.09
C_7H_{14}	44.751	49.79
C_7H_{16}	45.018	50.19
C_8H_{16}	44.645	49.79
C_8H_{18}	44.882	50.77
C_9H_{18}	44.564	49.79
C_9H_{20}	44.776	50.66
$C_{10}H_{20}$	44.498	49.79
$C_{10}H_{22}$	44.691	50.58
$C_{11}H_{22}$	44.445	49.79
$C_{11}H_{24}$	44.621	50.51
$C_{12}H_{24}$	44.401	49.79
$C_{12}H_{26}$	44.555	50.45
$C_{7.76}H_{13.1}$	42.836	48.51
$C_{8.26}H_{15.5}$	43.575	49.30
$C_{10.8}H_{18.7}$	42.563	48.70
CH_3OH	21.030	21.41
C_2H_5OH	27.675	30.02

Table B.2 T_{ad} at $T_{ref} = 298$ K and $p_{ref} = 1$ atm, Eq. B.3

Fuel	CH_4	C_2H_4	C_2H_6	C_3H_6	C_3H_8
a1	−33.418620	50.340393	1.792035	3.391223	3.806911
a2	236.628209	−304.109243	8.302140	−1.830710	−1.423670
a3	−659.574620	760.621781	−57.435329	−30.708069	−40.465048
a4	960.231124	−979.191365	134.335015	98.035453	122.553574
a5	−761.824011	691.855692	−140.048628	−112.881861	−139.304938
a6	311.690171	−256.033794	67.805770	57.158644	70.457515
a7	−51.492821	38.900331	−12.498176	−10.790320	−13.352935

Fuel	C_4H_8	C_4H_{10}	C_5H_{10}	C_5H_{12}	C_6H_{12}
a1	3.449660	2.507647	3.364161	−1.445999	6.161763
a2	−2.220453	3.793982	−1.652343	27.670525	−19.510888
a3	−29.777019	−45.843903	−31.406251	−104.535326	14.612781
a4	96.812570	118.854126	99.168164	194.203944	37.760315
a5	−111.993484	−128.631246	−113.844894	−182.007618	−69.008947
a6	56.820382	63.385842	57.572729	83.198499	40.543393
a7	−10.737564	−11.796263	−10.861462	−14.811287	−8.225861

Fuel	C_6H_{14}	C_7H_{14}	C_7H_{16}	C_8H_{16}	C_8H_{18}
a1	4.327417	4.943773	3.877103	3.280516	2.912977
a2	−7.369111	−11.798567	−4.586560	−1.120968	1.167628
a3	−18.089831	−4.984894	−25.060672	−32.899013	−38.903915
a4	83.026048	63.288223	92.136428	101.269963	109.324404
a5	−103.217859	−87.067658	−109.797701	−115.462179	−121.413995
a6	53.968811	47.149472	56.468312	58.218970	60.523153
a7	−10.368917	−9.205929	−10.760026	−10.966423	−11.331699

Fuel	C_9H_{18}	C_9H_{20}	$C_{10}H_{20}$	$C_{10}H_{22}$	$C_{11}H_{22}$
a1	3.293601	2.958922	3.283484	2.994063	3.280371
a2	−1.205943	0.876199	−1.148853	0.661454	−1.138694
a3	−32.694484	−38.149769	−32.840706	−37.615730	−32.854315
a4	100.999323	108.310837	101.183730	107.622567	101.174702
a5	−115.261999	−120.661971	−115.387921	−120.172246	−115.359486
a6	58.140557	60.230441	58.184650	60.047292	58.164954
a7	−10.953729	−11.284997	−10.959900	−11.256836	−10.955399

(continued)

Table B.2 (continued)

Fuel	$C_{11}H_{24}$	$C_{12}H_{24}$	$C_{12}H_{26}$	$C_{7.76}H_{13.1}$	$C_{8.26}H_{15.5}$
a1	3.099466	3.283357	3.045509	3.839534	3.489864
a2	−0.140128	−1.158872	0.327521	−4.698130	−2.482144
a3	−35.190300	−32.810545	−36.733142	−23.696797	−29.437054
a4	103.880595	101.119482	106.410389	88.879398	96.598277
a5	−117.054984	−115.320145	−119.253866	−106.233405	−111.965389
a6	58.713141	58.149922	59.682798	54.607403	56.840826
a7	−11.026845	−10.952993	−11.197644	−10.386486	−10.743207
Fuel	$C_{10.8}H_{18.7}$	CH_3OH	C_2H_5OH		
a1	3.784620	−9.343315	−23.835810		
a2	−4.365100	82.041550	173.587997		
a3	−24.643015	−254.998922	−490.410990		
a4	90.186152	409.612090	724.365693		
a5	−107.221607	−351.092769	−581.697681		
a6	54.997086	152.332525	240.194863		
a7	−10.449157	−26.323396	−39.950311		

Table B.3 $log_{10}Kp$ versus $\theta = T(K)/1000$, Eq. B.4

Coefficient	$H_2 \rightleftharpoons 2H$	$O_2 \rightleftharpoons 2O$	$N_2 \rightleftharpoons 2N$	$H_2O \rightleftharpoons H_2 + \frac{1}{2}O_2$
a_1	5.364	6.434	6.404	2.696
a_2	−0.1766	−0.2755	−0.5302	−0.4523
a_3	0.01009	0.02396	0.06518	0.05166
a_4	−0.4135E-3	−0.1119E-2	−0.3807E-2	−0.2894E-2
a_5	0.8876	0.8258	1.199	0.9550
a_6	−22.49	−25.80	−48.99	−12.350
Coefficient	$H_2O \rightleftharpoons OH + \frac{1}{2}H_2$	$CO_2 \rightleftharpoons CO + \frac{1}{2}O_2$	$\frac{1}{2}N_2 + \frac{1}{2}O_2 \rightleftharpoons NO$	$CO_2 + H_2 \rightleftharpoons CO + H_2O$
a_1	3.555	4.804	0.6075	2.108
a_2	−0.4576	−0.3322	0.07028	0.1199
a_3	0.05404	0.0421	−0.01327	−0.9506E-2
a_4	−0.3214E-2	−0.2492E-2	0.8752E-3	0.3976E-3
a_5	0.8762	0.2787	−0.05411	−0.6761
a_6	−14.43	−14.73	−4.727	−2.378

Table B.4 C_p versus T (K), Eqs. B.5–B.7

	CH$_4$		C$_2$H$_4$	
	$T = 1000 \leq$ 6000 K	$T = 200 \leq 1000$ K	$T = 1000 \leq$ 6000 K	$T = 200 \leq 1000$ K
a1	1.6532623E+00	5.1491147E+00	3.9918272E+00	3.9592006E+00
a2	1.0026310E-02	-1.3662201E-02	1.0483391E-02	-7.5705137E-03
a3	-3.3166124E-06	4.9145392E-05	-3.7172134E-06	5.7098999E-05
a4	5.3648314E-10	-4.8424677E-08	5.9462837E-10	-6.9158835E-08
a5	-3.1469676E-14	1.6660344E-11	-3.5363039E-14	2.6988419E-11
b1	-1.0009594E+04	-1.0246598E+04	4.2686585E+03	5.0897760E+03
b2	9.9050628E+00	-4.6384884E+00	-2.6908176E-01	4.0973021E+00

	C$_2$H$_6$		C$_3$H$_6$	
	$T = 1000 \leq$ 6000 K	$T = 200 \leq 1000$ K	$T = 1000 \leq$ 6000 K	$T = 200 \leq 1000$ K
a1	4.0466641E+00	4.2914257E+00	6.0387023E+00	3.8346447E+00
a2	1.5353880E-02	-5.5015490E-03	1.6296393E-02	3.2907895E-03
a3	-5.4703949E-06	5.9943846E-05	-5.8213080E-06	5.0522800E-05
a4	8.7782654E-10	-7.0846647E-08	9.3593683E-10	-6.6625118E-08
a5	-5.2316753E-14	2.6868584E-11	-5.5860314E-14	2.6370747E-11
b1	-1.2447350E+04	-1.1522206E+04	-7.4171506E+02	7.8871712E+02
b2	-9.6869831E-01	2.6667899E+00	-8.4382599E+00	7.5340801E+00

	C$_3$H$_8$		C$_4$H$_8$	
	$T = 1000 \leq$ 6000 K	$T = 200 \leq 1000$ K	$T = 1000 \leq$ 5000 K	$T = 300 \leq 1000$ K
a1	6.6691976E+00	4.2109301E+00	2.0535841E+00	1.1811380E+00
a2	2.0610875E-02	1.7088650E-03	3.4350507E-02	3.0853380E-02
a3	-7.3651235E-06	7.0653016E-05	-1.5883197E-05	5.0865247E-06
a4	1.1843426E-09	-9.2006057E-08	3.3089662E-09	-2.4654888E-08
a5	-7.0691463E-14	3.6461845E-11	-2.5361045E-13	1.1110193E-11
b1	-1.6275407E+04	-1.4381088E+04	-2.1397231E+03	-1.7904004E+03
b2	-1.3194338E+01	5.6100445E+00	1.5556360E+01	2.1075639E+01

	C$_4$H$_{10}$		C$_5$H$_{10}$	
	$T = 1000 \leq$ 6000 K	$T = 200 \leq 1000$ K	$T = 1000 \leq$ 6000 K	$T = 200 \leq 1000$ K
a1	9.4454784E+00	6.1447401E+00	1.1950162E+01	5.8835915E+00
a2	2.5785662E-02	1.6450024E-04	2.5216000E-02	5.1040359E-03
a3	-9.2361319E-06	9.6784879E-05	-8.8568526E-06	9.7828663E-05
a4	1.4863176E-09	-1.2548621E-07	1.4260218E-09	-1.3238983E-07
a5	-8.8789121E-14	4.9784626E-11	-8.5479494E-14	5.3223394E-11
b1	-2.0138377E+04	-1.7598947E+04	-8.5211573E+03	-5.1682543E+03
b2	-2.6347759E+01	-1.0805888E+00	-3.6533772E+01	3.4198859E+00

(continued)

Table B.4 (continued)

	C_5H_{12}		C_6H_{12}	
	$T = 1000 \le$ 6000 K	$T = 200 \le 1000$ K	$T = 1000 \le$ 6000 K	$T = 300 \le 1000$ K
a1	1.6737270E+01	8.5485166E+00	1.6061609E+01	7.3150905E+00
a2	2.2392203E-02	−8.8817049E-03	2.7565056E-02	3.7115033E-03
a3	−6.1770554E-06	1.4308389E-04	−9.3297337E-06	1.2725032E-04
a4	1.0214492E-09	−1.7859233E-07	1.4934901E-09	−1.7155696E-07
a5	−6.6518312E-14	6.9748976E-11	−8.9881027E-14	6.8980594E-11
b1	−2.5761666E+04	−2.0749261E+04	−1.2804295E+04	−8.2091651E+03
b2	−6.4561909E+01	−8.9351826E+00	−5.6992559E+01	−5.9435437E-01
	C_6H_{14}		C_7H_{14}	
	$T = 1000 \le$ 6000 K	$T = 200 \le 1000$ K	$T = 1000 \le$ 6000 K	$T = 200 \le 1000$ K
a1	1.9515809E+01	9.8712117E+00	2.0032934E+01	8.7053986E+00
a2	2.6775394E-02	−9.3669900E-03	3.0187558E-02	2.8007449E-03
a3	−7.4978374E-06	1.6988787E-04	−9.9691290E-06	1.5520600E-04
a4	1.1951065E-09	−2.1501952E-07	1.5937646E-09	−2.0901403E-07
a5	−7.5195747E-14	8.4540709E-11	−9.6431403E-14	8.4050578E-11
b1	−2.9436247E+04	−2.3718550E+04	−1.7051261E+04	−1.1266149E+04
b2	−7.7489550E+01	−1.2499935E+01	−7.6677873E+01	−4.4649355E+00
	C_7H_{16}		C_8H_{16}	
	$T = 1000 \le$ 6000 K	$T = 200 \le 1000$ K	$T = 1000 \le$ 6000 K	$T = 200 \le 1000$ K
a1	2.0456520E+01	1.1153299E+01	2.4337877E+01	1.0148373E+01
a2	3.4857536E-02	−9.4941977E-03	3.2257457E-02	1.2543807E-03
a3	−1.0922685E-05	1.9557208E-04	−1.0338974E-05	1.8524552E-04
a4	1.6720178E-09	−2.4975366E-07	1.6535977E-09	−2.4908715E-07
a5	−9.8102485E-14	9.8487772E-11	−1.0090902E-13	1.0024793E-10
b1	−3.2555637E+04	−2.6768890E+04	−2.1438395E+04	−1.4326764E+04
b2	−8.0440502E+01	−1.5909684E+01	−9.8240513E+01	−8.5190178E+00
	C_8H_{18}		C_9H_{18}	
	$T = 1000 \le$ 6000 K	$T = 200 \le 1000$ K	$T = 1000 \le$ 5000 K	$T = 298.15 \le 1000$ K
a1	2.0943071E+01	1.2524548E+01	2.1815489E+01	2.6242942E+00
a2	4.4169102E-02	−1.0101883E-02	4.8037012E-02	7.1830270E-02
a3	−1.5326163E-05	2.2199261E-04	−1.7939230E-05	1.9249451E-05
a4	2.3054480E-09	−2.8486372E-07	3.1426572E-09	−7.2931057E-08
a5	−1.2976573E-13	1.1241014E-10	−2.0938779E-13	3.3915076E-11
b1	−3.5575509E+04	−2.9843440E+04	−6.3091173E+04	−5.5999299E+04
b2	−8.1063773E+01	−1.9710999E+01	−8.2916701E+01	2.4090565E+01

(continued)

Table B.4 (continued)

	C_9H_{20}		$C_{10}H_{20}$	
	$T = 1000 \leq$ 6000 K	$T = 200 \leq 1000$ K	$T = 1000 \leq$ 5000 K	$T = 298.15 \leq$ 1000 K
a1	2.5587752E+01	1.3984023E+01	2.4378494E+01	3.4760543E+00
a2	4.6077065E-02	-1.1722498E-02	5.3387966E-02	7.6512095E-02
a3	-1.6086063E-05	2.5231647E-04	-1.9961348E-05	2.9369696E-05
a4	2.5827441E-09	-3.2568036E-07	3.5008195E-09	-8.8800868E-08
a5	-1.5473469E-13	1.2910914E-10	-2.3345394E-13	4.0337257E-11
b1	-4.0074845E+04	-3.2825841E+04	-2.7299777E+04	-1.9441804E+04
b2	-1.0472247E+02	-2.3863375E+01	-9.4953711E+01	2.2270128E+01

	$C_{10}H_{22}$		$C_{11}H_{24}$	
	$T = 1000 \leq$ 6000 K	$T = 200 \leq 1000$ K	$T = 1000 \leq$ 6000 K	$T = 200 \leq 1000$ K
a1	2.9478296E+01	1.5432817E+01	3.4107065E+01	1.6758906E+01
a2	4.9051894E-02	-1.3297923E-02	5.0786599E-02	-1.3577182E-02
a3	-1.7031718E-05	2.8248058E-04	-1.7379740E-05	3.0821687E-04
a4	2.7291930E-09	-3.6592330E-07	2.7604815E-09	-4.0056266E-07
a5	-1.6337077E-13	1.4537212E-10	-1.6424873E-13	1.5927423E-10
b1	-4.4302262E+04	-3.5863283E+04	-4.8848625E+04	-3.8907687E+04
b2	-1.2406212E+02	-2.7945434E+01	-1.4754660E+02	-3.1562852E+01

	$C_{12}H_{26}$		CH_3OH	
	$T = 1000 \leq$ 6000 K	$T = 200 \leq 1000$ K	$T = 1000 \leq$ 6000 K	$T = 200 \leq 1000$ K
a1	3.7018793E+01	2.1326448E+01	3.5272680E+00	5.6585105E+00
a2	5.5472149E-02	-3.8639400E-02	1.0317878E-02	-1.6298342E-02
a3	-1.9207955E-05	3.9947611E-04	-3.6289294E-06	6.9193816E-05
a4	3.0817557E-09	-5.0668110E-07	5.7744802E-10	-7.5837293E-08
a5	-1.8480062E-13	2.0069788E-10	-3.4218263E-14	2.8042755E-11
b1	-5.2698446E+04	-4.2247505E+04	-2.6002883E+04	-2.5611974E+04
b2	-1.6145350E+02	-4.8584830E+01	5.1675869E+00	-8.9733051E-01

	C_2H_5OH		–	
	$T = 1000 \leq$ 6000 K	$T = 200 \leq 1000$ K		
a1	6.5624365E+00	4.8586957E+00		
a2	1.5204222E-02	-3.7401726E-03		
a3	-5.3896795E-06	6.9555378E-05		
a4	8.6225011E-10	-8.8654796E-08		
a5	-5.1289787E-14	3.5168835E-11		
b1	-3.1525621E+04	-2.9996132E+04		
b2	-9.4730202E+00	4.8018545E+00		

Appendix C
Properties of Fuels

Table C.1 gives the properties of liquid and gaseous fuels evaluated at $T_{ref} = 298\,\text{K}$ and $p_{ref} = 1\,\text{atm}$. The HHV/LHV data are obtained from [64] and the rest of the data are from Ref. [116].

Curve-fit coefficients [122] for thermal conductivity k_{fu} (W/m-K) are given in Table C.2 and for viscosity μ_{fu} (N-s/m^2) are given in Table C.3 for specified range ($T_{min} < T < T_{max}$) of temperatures. The properties are evaluated as under

$$k_{fu} = a_1 + a_2\,T + a_3\,T^2 + a_4\,T^3 + a_5\,T^4$$
$$+ a_6\,T^6 + a_7\,T^7 , \tag{C.1}$$
$$\mu_{fu} \times 10^6 = b_1 + b_2\,T + b_3\,T^2 + b_4\,T^3 + b_5\,T^4$$
$$+ b_6\,T^6 + b_7\,T^7 . \tag{C.2}$$

For temperatures beyond the maximum (T_{max}) of the specified range, $k_{fu,T} = k_{fu,T_{max}} \times (T/T_{max})^{0.5}$ and $\mu_{fu,T} = \mu_{fu,T_{max}} \times (T/T_{max})^{0.5}$.

© Springer Nature Singapore Pte Ltd. 2020
A. W. Date, *Analytic Combustion*,
https://doi.org/10.1007/978-981-15-1853-9

Table C.1 Properties at 298.15 K and 1 atm

	CH$_4$	C$_2$H$_4$	C$_2$H$_6$
Mol.wt(kg/kmol)	16.04	28.05	30.07
$\overline{h}_f^o$ (kJ/kmol)	−74600	52500	−84000
$\overline{g}_f^o$ (kJ/kmol)	−50500	68400	−32000
$\overline{s}^o$ (kJ/kmol-K)	186.3	219.3	229.1
HHV (kJ/kg) liq	−	−	51592.78
LHV (kJ/kg) liq	−	−	47198.59
HHV (kJ/kg) gas	55526.39	50287.79	51906.82
LHV (kJ/kg) gas	50036.56	47149.74	47512.63
T_{bp} (°C)	−162	−104	−89
h_{fg} (kJ/kg)	511.22	481.28	488.85
T_{ad} (K)	2242	2383	2251
ρ_{liq} (kg/m^3)	0.7168	0.1147	0.1356
$\overline{C}_p$ (kJ/kmol-K)	35.7	42.9	52.5
	C$_3$H$_6$	C$_3$H$_8$	C$_4$H$_8$
Mol.wt (kg/kmol)	42.08	44.1	56.11
$\overline{h}_f^o$ (kJ/kmol)	20000	−103800	100
$\overline{g}_f^o$ (kJ/kmol)	62800	−23400	71300
$\overline{s}^o$ (kJ/kmol-K)	266.6	270.2	305.6
HHV (kJ/kg) liq	−	49952.81	48056.96
LHV (kJ/kg) liq	−	45961.05	44918.92
HHV (kJ/kg) gas	48922.31	50329.66	48429.15
LHV (kJ/kg) gas	45784.26	46337.9	45291.11
T_{bp} (°C)	−47	−42	−6
h_{fg}(kJ/kg)	437.26	430.83	393.86
T_{ad}(K)	2374	2264	2353
ρ_{liq}(kg/m^3)	610.0	584.0	625.5
$\overline{C}_p$(kJ/kmol-K)	64.3	73.6	85.7

(continued)

Table C.1 (continued)

	C_4H_{10}	C_5H_{10}	C_5H_{12}
Mol.wt(kg/kmol)	58.12	70.14	72.15
$\overline{h}_f^o$ (kJ/kmol)	−125600	−46000(l), −21200(g)	−173500(l), −146900(g)
$\overline{g}_f^o$ (kJ/kmol)	−17200	79100	−9300(l), −8400(g)
$\overline{s}^o$ (kJ/kmol-K)	310.1	262.6(l), 345.8(g)	262.7(l), 349(g)
HHV (kJ/kg) liq	49136.32	47761.53	48638.51
LHV (kJ/kg) liq	45349.26	44623.49	44977.07
HHV (kJ/kg) gas	49513.16	48129.07	49006.05
LHV (kJ/kg) gas	45726.11	44991.03	45349.26
T_{bp} (°C)	−1	30	36
h_{fg}(kJ/kg)	385.40	359.28	357.58
T_{ad}(K)	2270	2340	2272
ρ_{liq}(kg/m^3)	578.8	642.9	626.2
$\overline{C}_p$(kJ/kmol-K)	97.5	154.0(l), 109.6(g)	167.2(l), 120.2(g)
	C_6H_{12}	C_6H_{14}	C_7H_{14}
Mol.wt(kg/kmol)	84.16	86.18	98.9
$\overline{h}_f^o$ (kJ/kmol)	−74100(l), −43500(g)	−198800(l), −167100(g)	−97900(l), −62300(g)
$\overline{g}_f^o$ (kJ/kmol)	83600(l), 84450(g)	−3800(l), −250(g)	95800
$\overline{s}^o$ (kJ/kmol-K)	295.1(l), 384.6(g)	296.1(l), 388.4(g)	423.6
HHV (kJ/kg) liq	47575.44	48317.5	47438.19
LHV (kJ/kg) liq	44437.39	44739.8	44300.15
HHV (kJ/kg) gas	47940.65	48680.38	47801.08
LHV (kJ/kg) gas	44802.61	45107.34	44663.04
T_{bp} (°C)	63	69	94
h_{fg}(kJ/kg)	336.26	335.34	314.45
T_{ad}(K)	2331	2277	2325
ρ_{liq}(kg/m^3)	673.2	659.4	697.0
$\overline{C}_p$(kJ/kmol-K)	183.3(l), 132.3(g)	195.6(l), 143.1(g)	211.8(l), 155.2(g)

Table C.1 (continued)

	C_7H_{16}	C_8H_{16}	C_8H_{18}
Mol.wt(kg/kmol)	100.21	112.22	114.23
$\overline{h}_f^o$ (kJ/kmol)	−224200 (l), −187700(g)	−121800(l), −81400(g)	−250100(l), −208600(g)
$\overline{g}_f^o$ (kJ/kmol)	8000	104200	16400
$\overline{s}^o$ (kJ/kmol-K)	427.9	462.5	466.7
HHV (kJ/kg) liq	48077.9	47335.84	47889.47
LHV (kJ/kg) liq	44563.01	44197.80	44425.76
HHV (kJ/kg) gas	48443.11	47696.40	48254.69
LHV (kJ/kg) gas	44928.22	44560.68	44786.32
T_{bp} (°C)	98	121	126
h_{fg}(kJ/kg)	317.33	303.86	300.96
T_{ad}(K)	2277	2321	2278
ρ_{liq}(kg/m³)	683.8	714.9	702.8
$\overline{C}_p$(kJ/kmol-K)	224.9(l), 166(g)	178.1	254.6(l), 188.9(g)
	C_9H_{18}	C_9H_{20}	$C_{10}H_{20}$
Mol.wt(kg/kmol)	126.24	128.26	140.27
$\overline{h}_f^o$ (kJ/kmol)	−103500	−274700(l), −228200(g)	−173800
$\overline{g}_f^o$ (kJ/kmol)	112700	248800	105000
$\overline{s}^o$ (kJ/kmol-K)	501.5	505.7	425
HHV (kJ/kg) liq	47254.42	47756.88	47191.61
LHV (kJ/kg) liq	44118.7	44325.74	44053.57
HHV (kJ/kg) gas	47617.31	48119.77	47552.18
LHV (kJ/kg) gas	44479.27	44683.97	44414.13
T_{bp} (°C)	132–143	151	172
h_{fg}(kJ/kg)	287.54	287.69	275.89
T_{ad}(K)	2317	2279	2314
ρ_{liq}(kg/m³)	–	717.6	740.8
$\overline{C}_p$(kJ/kmol-K)	201.0	211.7	300.8

Table C.1 (continued)

	$C_{10}H_{22}$	$C_{11}H_{22}$	$C_{11}H_{24}$
Mol.wt(kg/kmol)	142.29	154.3	156.31
$\overline{h}_f^o$ (kJ/kmol)	−300900	−144800	−327200
$\overline{g}_f^o$ (kJ/kmol)	17500	129500	22800
$\overline{s}^o$ (kJ/kmol-K)	425.5	579.4	458.1
HHV (kJ/kg) liq	47642.9	47138.11	47552.18
LHV (kJ/kg) liq	44241.99	44000.07	44169.88
HHV (kJ/kg) gas	48003.46	47498.67	47910.41
LHV (kJ/kg) gas	44600.23	44360.63	44530.44
T_{bp} (°C)	174	–	196
h_{fg} (kJ/kg)	272.68	265.06	265.49
T_{ad} (K)	2280	2312	2278
ρ_{liq} (kg/m³)	730.1	750.3	740.2
$\overline{C}_p$ (kJ/kmol-K)	314.4	246.7	344.9
	$C_{12}H_{24}$	$C_{12}H_{26}$	
Mol.wt(kg/kmol)	168.32	170.4	
$\overline{h}_f^o$ (kJ/kmol)	−165400	−289700	
$\overline{g}_f^o$ (kJ/kmol)	137900	50000	
$\overline{s}^o$ (kJ/kmol-K)	618.3	622.5	
HHV (kJ/kg) liq	47096.24	47470.76	
LHV (kJ/kg) liq	43958.2	44111.73	
HHV (kJ/kg) gas	47454.48	47831.32	
LHV (kJ/kg) gas	44316.43	44472.29	
T_{bp} (°C)	208	216	
h_{fg} (kJ/kg)	261.40	261.15	
T_{ad} (K)	2310	2281	
ρ_{liq} (kg/m³)	758.4	749.0	
$\overline{C}_p$ (kJ/kmol-K)	269.6	280.3	

Table C.1 (continued)

	CH$_3$OH	C$_2$H$_5$OH
Mol.wt(kg/kmol)	32.04	46.07
$\overline{h}_\mathrm{f}^o$ (kJ/kmol)	−239.1(l), −201.0(g)	−277.6(l), −234.8(g)
$\overline{g}_\mathrm{f}^o$ (kJ/kmol)	−166.6(l), −162.3(g)	−174.8(l), −167.9(g)
$\overline{s}^o$ (kJ/kmol-K)	126.8(l), 239.9(g)	161.0(l), 281.6(g)
HHV (kJ/kg) liq	23778	30542
LHV (kJ/kg) liq	21030	27675
HHV (kJ/kg) gas	–	–
LHV (kJ/kg) gas	–	–
T_{bp} (°C)	64	78
h_fg(kJ/kg)	1098.62	837.85
T_ad(K)	2228	2254
ρ_{liq}(kg/m^3)	791.3	789.4
$\overline{C_\mathrm{p}}$(kJ/kmol-K)	81.2(l), 44.1(g)	112.3(l), 65.6(g)

Table C.2 Curve-fit coefficients for k_{fu} [122]—methanol and ethanol data from [19]

Coefficient	CH_4 ($100 < T < 1000$)	Coefficient	C_3H_8 ($100 < T < 1000$)
a_1	−1.3401499E-2	a_1	−1.07682209E-2
a_2	3.66307060E-4	a_2	8.38590325E-5
a_3	−1.82248608E-6	a_3	4.22059864E-8
a_4	5.93987998E-9	a_4	0.0
a_5	−9.14055050E-12	a_5	0.0
a_6	−6.78968890E-15	a_6	0.0
a_7	−1.95048736E-18	a_7	0.0
Coefficient	C_6H_{14} ($250 < T < 1000$)	Coefficient	C_7H_{16} ($250 < T < 1000$)
a_1	1.28775700E-3	a_1	−4.60614700E-2
a_2	−2.00499443E-5	a_2	5.95652224E-4
a_3	2.37858831E-7	a_3	−2.98893163E-6
a_4	−1.60944555E-10	a_4	8.44612876E-9
a_5	7.71027290E-14	a_5	−1.229270E-11
a_6	0.0	a_6	9.012700E-15
a_7	0.0	a_7	−2.62961E-18
Coefficient	C_8H_{18} ($250 < T < 500$)	Coefficient	$C_{10}H_{22}$ ($250 < T < 500$)
a_1	−4.01391940E-3	a_1	−5.8827400E-3
a_2	3.38796092E-5	a_2	3.72449646E-5
a_3	8.19291819-8	a_3	7.55109624E-8
a_4	0.0	a_4	0.0
a_5	0.0	a_5	0.0
a_6	0.0	a_6	0.0
a_7	0.0	a_7	0.0
Coefficient	CH_3OH ($273 < T < 1270$)	Coefficient	C_2H_5OH ($273 < T < 1270$)
a_1	−7.79E-3	a_1	−7.79E-3
a_2	4.167E-5	a_2	4.167E-5E-5
a_3	1.214E-7	a_3	1.214E-7
a_4	−5.174E-11	a_4	−5.174E-11
a_5	0.0	a_5	0.0
a_6	0.0	a_6	0.0
a_7	0.0	a_7	0.0

Table C.3 Curve-fit coefficients for $\mu_{fu} \times 10^6$ [122]

Coefficient	CH$_4$ (70 < T < 1000)	Coefficient	C$_3$H$_8$ (270 < T < 600)
b_1	2.968267E-1	b_1	−3.5437110E-1
b_2	3.71120100E-2	b_2	3.08009600E-2
b_3	1.21829800E-5	b_3	−6.99723000E-6
b_4	−7.0242600E-8	b_4	0.0
b_5	7.54326900E-11	b_5	0.0
b_6	−2.72371660E-14	b_6	0.0
b_7	0.0	b_7	0.0
Coefficient	C$_6$H$_{14}$ (270 < T < 900)	Coefficient	C$_7$H$_{16}$ (270 < T < 580)
b_1	1.54541200E00	b_1	1.54009700E00
b_2	1.15080900E-2	b_2	1.09515700E-2
b_3	2.72216500E-5	b_3	1.80066400E-5
b_4	−3.2690000E-8	b_4	−1.36379000E-8
b_5	1.24545900E-11	b_5	0.0
b_6	0.0	b_6	0.0
b_7	0.0	b_7	0.0
Coefficient	C$_8$H$_{18}$ (300 < T < 650)	Coefficient	C$_{10}$H$_{22}$ (300 < T < 700)
b_1	8.32435400E-1	b_1	NA
b_2	1.40045000E-2	b_2	NA
b_3	8.7936500E-6	b_3	NA
b_4	−6.8403000E-9	b_4	NA
b_5	0.0	b_5	NA
b_6	0.0	b_6	NA
b_7	0.0	b_7	NA
Coefficient	CH$_3$OH (250 < T < 650)	Coefficient	C$_2$H$_5$OH (270 < T < 600)
b_1	1.1979000E00	b_1	−6.3359500E-2
b_2	2.45028000E-2	b_2	3.20713470E-2
b_3	1.86162740E-5	b_3	−6.25079576E-6
b_4	−1.30674820E-8	b_4	0.0
b_5	0.0	b_5	0.0
b_6	0.0	b_6	0.0
b_7	0.0	b_7	0.0

Appendix D
Thermophysical and Transport Properties of Gases

The properties of gases are taken from Refs. [48, 85, 123] (Tables D.1, D.2, D.3, D.4, D.5, D.6, D.7 and D.8).

© Springer Nature Singapore Pte Ltd. 2020
A. W. Date, *Analytic Combustion*,
https://doi.org/10.1007/978-981-15-1853-9

Table D.1 Thermophysical properties of air

T (K)	ρ (kg/m^3)	c_p (kJ/kg-K)	$\mu \times 10^7$ (N-s/m^2)	$\nu \times 10^6$ (m^2/s)	$k \times 10^3$ (W/m-K)	$\alpha \times 10^6$ (m^2/s)	Pr
100	3.5562	1.032	71.1	2.00	9.34	2.54	0.786
150	2.3364	1.012	103.4	4.43	13.80	5.84	0.758
200	1.7458	1.007	132.5	7.59	18.10	10.30	0.737
250	1.3947	1.006	159.6	11.44	22.30	15.89	0.720
300	1.1614	1.007	184.6	15.89	26.30	22.49	0.707
350	0.9950	1.009	208.2	20.92	30.00	29.88	0.700
400	0.8711	1.014	230.1	26.42	33.80	38.27	0.690
450	0.7740	1.021	250.7	32.39	37.30	47.20	0.686
500	0.6964	1.030	270.1	38.79	40.70	56.74	0.684
550	0.6329	1.040	288.4	45.57	43.90	66.69	0.683
600	0.5804	1.051	305.8	52.69	46.90	76.89	0.685
650	0.5356	1.063	322.5	60.21	49.70	87.29	0.690
700	0.4975	1.075	338.8	68.10	52.40	97.98	0.695
750	0.4643	1.087	354.6	76.38	54.90	108.79	0.702
800	0.4354	1.099	369.8	84.94	57.30	119.76	0.709
850	0.4097	1.110	384.3	93.81	59.60	131.07	0.716
900	0.3868	1.121	398.1	102.91	62.00	142.97	0.720
950	0.3666	1.131	411.3	112.20	64.30	155.09	0.723
1000	0.3482	1.141	424.4	121.89	66.70	167.89	0.726
1100	0.3166	1.159	449.0	141.84	71.50	194.88	0.728
1200	0.2902	1.175	473.0	163.00	76.30	223.77	0.728
1300	0.2679	1.189	496.0	185.16	82.00	257.45	0.719
1400	0.2488	1.207	530.0	213.06	91.00	303.08	0.703
1500	0.2322	1.230	557.0	239.90	100.00	350.16	0.685
1600	0.2177	1.248	584.0	268.29	106.00	390.20	0.688
1700	0.2049	1.267	611.0	298.23	113.00	435.32	0.685
1800	0.1935	1.286	637.0	329.20	120.00	482.24	0.683
1900	0.1833	1.307	663.0	361.67	128.00	534.23	0.677
2000	0.1741	1.337	689.0	395.69	137.00	588.47	0.672
2100	0.1658	1.372	715.0	431.22	147.00	646.18	0.667
2200	0.1582	1.417	740.0	467.68	160.00	713.62	0.655
2300	0.1513	1.478	766.0	506.40	175.00	782.76	0.647
2400	0.1448	1.558	792.0	546.80	196.00	868.54	0.630
2500	0.1389	1.665	818.0	588.96	222.00	960.00	0.614
2600	0.1333	1.803	844.0	633.17	256.00	1065.18	0.594
2700	0.1280	1.978	871.0	680.43	299.00	1180.88	0.576
2800	0.1230	2.191	898.0	730.25	352.00	1306.46	0.559
2900	0.1181	2.419	926.0	783.77	414.00	1448.57	0.541
3000	0.1135	2.726	955.0	841.64	486.00	1571.21	0.536

Table D.2 Thermophysical properties of nitrogen

T (K)	ρ (kg/m^3)	c_p (kJ/kg-K)	$\mu \times 10^7$ (N-s/m^2)	$\nu \times 10^6$ (m^2/s)	$k \times 10^3$ (W/m-K)	$\alpha \times 10^6$ (m^2/s)	Pr
100	3.4388	1.070	68.8	2.00	9.34	2.54	0.788
150	2.2594	1.050	100.6	4.45	13.80	5.82	0.765
200	1.6883	1.043	129.2	7.65	18.10	10.28	0.745
250	1.3488	1.042	154.9	11.48	22.30	15.87	0.724
300	1.1233	1.041	178.2	15.86	26.30	22.49	0.705
350	0.9625	1.042	200.0	20.78	30.00	29.91	0.695
400	0.8425	1.045	220.4	26.16	33.80	38.39	0.681
450	0.7485	1.050	239.6	32.01	37.30	47.46	0.674
500	0.6739	1.056	257.7	38.24	40.70	57.20	0.669
550	0.6124	1.065	274.7	44.86	43.90	67.31	0.666
600	0.5615	1.075	290.8	51.79	46.90	77.70	0.667
650	0.5181	1.086	306.2	59.10	49.70	88.33	0.669
700	0.4812	1.098	321.0	66.70	52.40	99.17	0.673
750	0.4490	1.110	335.1	74.62	54.90	110.15	0.677
800	0.4211	1.122	349.1	82.91	57.30	121.29	0.684
850	0.3964	1.134	362.2	91.38	59.60	132.60	0.689
900	0.3743	1.146	375.3	100.28	62.00	144.56	0.694
950	0.3546	1.157	387.6	109.30	64.30	156.72	0.697
1000	0.3368	1.167	399.9	118.73	66.70	169.69	0.700
1100	0.3062	1.187	423.2	138.22	71.50	196.73	0.703
1200	0.2807	1.204	445.3	158.62	76.30	225.73	0.703
1300	0.2591	1.219	466.2	179.91	82.00	259.59	0.693
1400	0.2406	1.220	517.0	214.88	91.00	309.88	0.693
1500	0.2246	1.233	540.6	240.71	100.00	361.21	0.666
1600	0.2105	1.244	563.6	267.70	106.00	404.64	0.662
1700	0.1982	1.255	586.1	295.76	113.00	454.36	0.651
1800	0.1872	1.265	608.3	325.01	120.00	506.94	0.641
1900	0.1773	1.274	630.1	355.38	128.00	566.75	0.627
2000	0.1684	1.282	651.6	386.87	137.00	634.43	0.610
2100	0.1604	1.290	683.0	425.77	147.00	710.55	0.599
2200	0.1531	1.296	709.2	463.08	160.00	805.88	0.575
2300	0.1465	1.303	735.4	502.06	175.00	917.14	0.547
2400	0.1404	1.308	761.7	542.63	196.00	1067.31	0.508
2500	0.1348	1.313	788.0	584.76	222.00	1254.50	0.466
2600	0.1296	1.318	814.2	628.40	256.00	1499.46	0.419
2700	0.1248	1.322	840.3	673.56	299.00	1813.45	0.371
2800	0.1203	1.325	866.3	720.12	352.00	2208.41	0.326
2900	0.1162	1.328	892.0	767.89	414.00	2683.85	0.286
3000	0.1123	1.331	917.4	816.96	486.00	3252.68	0.251

Table D.3 Thermophysical properties of oxygen

T (K)	ρ (kg/m^3)	c_p (kJ/kg-K)	$\mu \times 10^7$ (N-s/m^2)	$\nu \times 10^6$ (m^2/s)	$k \times 10^3$ (W/m-K)	$\alpha \times 10^6$ (m^2/s)	Pr
100	3.9448	0.962	76.4	1.94	9.25	2.44	0.795
150	2.5846	0.921	114.8	4.44	13.80	5.8	0.766
200	1.9301	0.915	147.1	7.62	18.30	10.36	0.736
250	1.5415	0.915	178.3	11.57	22.55	15.99	0.723
300	1.2837	0.920	207.0	16.13	26.80	22.69	0.711
350	1.0999	0.929	233.5	21.23	29.60	28.97	0.733
400	0.9625	0.942	258.2	26.83	33.00	36.4	0.737
450	0.8554	0.956	281.4	32.90	36.30	44.39	0.741
500	0.7698	0.972	303.3	39.40	41.20	55.06	0.716
550	0.6998	0.988	324.0	46.30	44.10	63.78	0.726
600	0.6414	1.003	343.7	53.58	47.30	73.52	0.729
650	0.5921	1.018	362.3	61.18	50.00	82.96	0.738
700	0.5498	1.031	380.8	69.27	52.80	93.16	0.744
750	0.5131	1.043	398.0	77.57	56.20	105.02	0.739
800	0.4810	1.054	415.2	86.32	58.90	116.18	0.743
850	0.4527	1.065	431.2	98.79	62.10	128.81	0.767
900	0.4275	1.074	447.2	104.60	64.90	141.34	0.740
950	0.4050	1.082	462.1	114.09	68.10	155.4	0.734
1000	0.3848	1.090	477.0	123.97	71.00	169.29	0.732
1100	0.3498	1.103	505.5	144.52	75.80	196.48	0.736
1200	0.3206	1.115	532.5	166.09	81.90	229.1	0.725
1300	0.2960	1.125	558.4	188.63	87.10	261.53	0.721
1400	0.2749	1.127	584.3	212.56	92.30	297.85	0.714
1500	0.2565	1.138	608.6	237.24	98.15	336.15	0.706
1600	0.2405	1.148	632.7	263.09	104.00	376.58	0.699
1700	0.2263	1.158	656.7	290.15	109.50	417.77	0.695
1800	0.2138	1.167	680.8	318.49	115.00	460.91	0.691
1900	0.2025	1.176	705.1	348.19	123.00	516.52	0.674
2000	0.1923	1.184	735.0	382.13	131.00	575.06	0.664
2100	0.1677	1.192	759.0	454.35	145.00	725.09	0.627
2200	0.1487	1.200	783.0	526.57	159.00	891.23	0.591
2300	0.1540	1.207	806.5	523.24	188.50	1014.32	0.516
2400	0.1596	1.214	830.0	519.91	218.00	1125.07	0.462
2500	0.1527	1.220	854.0	560.06	276.50	1484.19	0.377
2600	0.1463	1.226	878.0	600.20	335.00	1867.25	0.321
2700	0.1398	1.232	903.5	647.27	442.50	2569.16	0.252
2800	0.1338	1.238	929.0	694.33	550.00	3320.47	0.209
2900	0.1273	1.243	957.5	752.96	727.00	4591.32	0.164
3000	0.1215	1.249	986.0	811.58	904.00	5959.55	0.136

Table D.4 Thermophysical properties of carbon dioxide

T (K)	ρ (kg/m^3)	c_p (kJ/kg-K)	$\mu \times 10^7$ (N-s/m^2)	$\nu \times 10^6$ (m^2/s)	$k \times 10^3$ (W/m-K)	$\alpha \times 10^6$ (m^2/s)	Pr
200	2.4797	0.778	101.04	4.07	9.59	4.97	0.820
250	1.0952	0.807	125.89	13.65	12.95	14.66	0.931
300	1.7730	0.851	149.48	8.43	16.55	10.97	0.769
350	1.5168	0.900	172.08	11.35	20.40	14.94	0.759
400	1.3257	0.942	193.18	14.57	24.30	19.46	0.749
450	1.1776	0.981	213.32	18.12	28.30	24.50	0.739
500	1.0594	1.020	232.51	21.95	32.50	30.08	0.730
550	0.9625	1.050	250.73	26.05	36.60	36.22	0.719
600	0.8826	1.080	268.27	30.39	40.70	42.70	0.712
650	0.7899	1.100	284.98	36.08	44.50	51.22	0.704
700	0.7564	1.130	301.28	39.83	48.10	56.27	0.708
750	0.7057	1.150	317.04	44.92	51.70	63.70	0.705
800	0.6614	1.170	332.25	50.24	55.10	71.21	0.706
850	0.6227	1.190	346.91	58.02	58.50	78.95	0.735
900	0.5945	1.200	361.30	60.77	61.80	86.62	0.702
950	0.5571	1.220	375.41	67.39	65.00	95.64	0.705
1000	0.5291	1.230	389.11	73.54	68.10	104.64	0.703
1100	0.4810	1.260	415.00	86.28	74.40	122.76	0.703
1200	0.4409	1.280	440.90	100.00	80.30	142.28	0.703
1300	0.4072	1.300	465.01	114.21	86.20	162.85	0.701
1400	0.3781	1.310	489.13	129.37	92.10	185.96	0.696
1500	0.3529	1.330	512.01	145.10	99.55	212.12	0.684
1600	0.3308	1.340	564.00	170.50	107.00	241.39	0.706
1700	0.3113	1.350	589.00	189.19	116.00	275.99	0.685
1800	0.2940	1.360	614.00	208.82	125.00	312.59	0.668
1900	0.2774	1.690	638.50	230.18	140.50	299.71	0.768
2000	0.2626	1.910	663.00	252.47	156.00	311.02	0.812
2100	0.2488	2.190	687.50	277.25	181.50	333.08	0.832
2200	0.2357	2.560	712.00	302.03	207.00	343.01	0.881
2300	0.2230	3.030	737.00	331.91	247.50	366.27	0.906
2400	0.2106	3.600	762.00	361.80	288.00	379.84	0.953
2500	0.1984	4.240	789.50	400.18	346.50	411.96	0.971
2600	0.1863	4.940	817.00	438.57	405.00	440.09	0.997
2700	0.1743	5.680	848.00	489.44	479.00	483.72	1.012
2800	0.1627	6.400	879.00	540.32	553.00	531.14	1.017
2900	0.1514	7.060	913.50	607.17	636.00	595.01	1.020
3000	0.1406	7.630	948.00	674.03	719.00	670.00	1.006

Table D.5 Thermophysical properties of carbon monoxide

T (K)	ρ (kg/m^3)	C_p (kJ/kg-K)	$\mu \times 10^7$ (N-s/m^2)	$\nu \times 10^6$ (m^2/s)	$k \times 10^3$ (W/m-K)	$\alpha \times 10^6$ (m^2/s)	Pr
100	3.445	1.077	69.15	2.01	8.68	2.34	0.857
150	2.261	1.049	100.8	4.46	13.14	5.54	0.805
200	1.6888	1.045	127.4	7.55	17.42	9.89	0.763
250	1.3490	1.043	154.1	11.42	21.43	15.2	0.751
300	1.1233	1.043	178.5	15.89	25.24	21.42	0.742
350	0.9524	1.045	201.0	21.10	28.84	28.7	0.735
400	0.8421	1.049	222.0	26.36	32.25	36.51	0.722
450	0.7483	1.055	241.9	32.33	35.53	45.00	0.718
500	0.6735	1.065	260.8	38.72	38.64	53.87	0.719
550	0.6123	1.076	278.8	45.54	41.59	63.13	0.721
600	0.5613	1.088	296.1	52.75	44.44	72.78	0.725
650	0.5181	1.101	312.8	60.38	–	–	–
700	0.4810	1.114	329.0	68.40	–	–	–
750	0.4490	1.127	344.6	76.75	–	–	–
800	0.4210	1.140	359.9	85.49	–	–	–
850	0.3962	1.152	374.8	98.24	–	–	–
900	0.3742	1.164	389.2	104.01	–	–	–
950	0.3545	1.175	403.3	113.77	–	–	–
1000	0.3368	1.185	417.3	123.93	–	–	–
1100	0.3606	1.204	444.4	123.22	–	–	–
1200	0.2806	1.221	470.9	167.78	–	–	–
1300	0.2591	1.235	496.2	191.54	–	–	–
1400	0.2406	1.247	520.6	216.41	–	–	–
1500	0.2245	1.258	544.3	242.41	–	–	–
1600	0.2105	1.267	–	–	–	–	–
1700	0.1981	1.275	–	–	–	–	–
1800	0.1871	1.283	–	–	–	–	–
1900	0.1773	1.287	–	–	–	–	–
2000	0.1684	1.295	–	–	–	–	–
2100	0.1604	1.299	–	–	–	–	–
2200	0.1531	1.304	–	–	–	–	–
2300	0.1467	1.308	–	–	–	–	–
2400	0.1403	1.312	–	–	–	–	–
2500	0.1347	1.315	–	–	–	–	–
2600	0.1295	1.319	–	–	–	–	–
2700	0.1248	1.322	–	–	–	–	–
2800	0.1203	1.324	–	–	–	–	–
2900	0.1161	1.327	–	–	–	–	–
3000	0.1123	1.329	–	–	–	–	–

Table D.6 Thermophysical properties of hydrogen

T (K)	ρ (kg/m³)	c_p (kJ/kg-K)	$\mu \times 10^7$ (N-s/m²)	$\nu \times 10^6$ (m²/s)	$k \times 10^3$ (W/m-K)	$\alpha \times 10^6$ (m²/s)	Pr
100	0.2437	8.093	42.11	17.28	67.0	0.34	0.509
150	0.1625	12.281	55.98	34.45	101.0	0.51	0.681
200	0.1219	13.632	68.13	55.90	131.0	0.79	0.709
250	0.0975	14.180	79.23	81.26	157.0	1.14	0.716
300	0.0812	14.425	89.59	110.27	183.0	1.56	0.706
350	0.0696	14.538	99.39	142.72	204.0	2.01	0.708
400	0.0609	14.591	108.67	178.33	226.0	2.54	0.702
450	0.0542	14.616	117.75	217.39	247.0	3.12	0.697
500	0.0487	14.631	126.42	259.33	266.0	3.73	0.695
550	0.0443	14.645	134.74	304.03	285.0	4.39	0.692
600	0.0406	14.662	142.90	351.76	305.0	5.12	0.687
650	0.0375	14.687	150.89	402.38	323.0	5.86	0.686
700	0.0348	14.720	158.46	455.07	342.0	6.67	0.682
750	0.0325	14.761	166.03	510.87	360.0	7.50	0.681
800	0.0305	14.812	173.35	568.95	378.0	8.38	0.679
850	0.0287	14.873	180.58	629.73	395.0	9.26	0.680
900	0.0271	14.941	187.57	692.58	412.0	10.18	0.680
950	0.0257	15.018	194.55	758.26	430.0	11.16	0.679
1000	0.0244	15.102	199.70	819.30	448.0	12.17	0.673
1100	0.0222	15.289	208.53	941.08	488.0	14.40	0.653
1200	0.0203	15.495	226.20	1113.62	528.0	16.78	0.664
1300	0.0187	15.712	238.45	1271.76	568.0	19.28	0.660
1400	0.0174	15.932	250.70	1439.94	610.0	21.99	0.655
1500	0.0162	16.150	262.20	1613.57	655.0	24.96	0.646
1600	0.0152	16.363	273.70	1796.63	697.0	27.96	0.643
1700	0.0143	16.569	284.90	1987.03	742.0	31.23	0.636
1800	0.0135	16.768	296.10	2186.62	786.0	34.62	0.632
1900	0.0128	16.958	307.15	2394.24	835.0	38.38	0.624
2000	0.0122	17.140	318.20	2610.92	878.0	42.03	0.621
2100	0.0116	17.312	330.05	2843.56	1034.0	51.46	0.553
2200	0.0111	17.476	341.90	3085.92	1190.0	61.46	0.502
2300	0.0106	17.632	354.20	3342.26	1467.5	78.54	0.426
2400	0.0102	17.780	366.50	3608.68	1745.0	96.64	0.373
2500	0.0097	17.917	379.20	3889.30	2250.0	128.80	0.302

Table D.7 Thermophysical properties of steam

T (K)	ρ (kg/m^3)	c_p (kJ/kg-K)	$\mu \times 10^7$ (N-s/m^2)	$\nu \times 10^6$ (m^2/s)	$k \times 10^3$ (W/m-K)	$\alpha \times 10^6$ (m^2/s)	Pr
400	0.5476	2.0078	134.20	24.65	27.00	24.56	1.003
450	0.4846	1.9752	152.25	31.42	31.17	32.56	0.965
500	0.4351	1.9813	170.30	39.10	35.86	41.56	0.941
550	0.3951	2.0010	188.35	47.67	40.95	51.79	0.920
600	0.3619	2.0268	206.40	57.03	46.37	63.21	0.902
650	0.3338	2.0557	224.45	67.24	52.05	75.78	0.887
700	0.3099	2.0867	242.50	76.16	57.96	89.6	0.873
750	0.2892	2.1191	260.55	90.09	64.08	104.55	0.862
800	0.2710	2.1525	278.60	102.80	70.39	120.67	0.851
850	0.2550	2.1868	296.95	116.45	76.85	137.81	0.845
900	0.2409	2.2216	317.00	131.59	83.47	155.96	0.844
950	0.2282	2.2568	338.05	148.13	90.22	175.2	0.845
1000	0.2167	2.2921	358.50	165.4	97.09	195.40	0.846
1100	0.1970	2.3621	397.20	201.6	111.15	238.86	0.844
1200	0.1806	2.4302	432.70	239.51	125.58	286.13	0.837
1300	0.1667	2.513	464.60	278.71	156.29	373.09	0.747
1400	0.1548	2.591	492.80	318.40	187.00	466.31	0.683
1500	0.1444	2.682	517.40	358.20	203.00	524.00	0.684
1600	0.1354	2.795	563.00	415.83	219.00	578.72	0.719
1700	0.1274	2.946	587.50	461.15	241.00	642.18	0.718
1800	0.1203	3.148	612.00	508.73	263.00	694.47	0.733
1900	0.1138	3.422	635.50	558.41	298.00	765.20	0.730
2000	0.1080	3.786	659.00	610.18	333.00	814.40	0.749
2100	0.1025	4.330	681.00	664.40	396.00	892.28	0.745
2200	0.0976	4.874	703.00	720.28	459.00	964.90	0.746
2300	0.0927	5.736	722.50	779.30	574.50	1080.44	0.721
2400	0.0883	6.597	742.00	840.31	690.00	1184.51	0.709
2500	0.0837	7.870	758.50	906.21	900.00	1366.30	0.663
2600	0.0796	9.142	775.00	974.62	1110.00	1525.34	0.639
2700	0.0749	10.916	786.50	1050.06	1465.00	1791.81	0.586
2800	0.0708	12.690	798.00	1127.10	1820.00	2025.7	0.556
2900	0.0660	15.060	804.00	1218.20	2380.00	2394.4	0.509
3000	0.0617	17.430	810.00	1312.80	2940.00	2733.8	0.480

Table D.8 Thermophysical properties of argon

T (K)	ρ (kg/m^3)	c_p (kJ/kg-K)	$\mu \times 10^7$ (N-s/m^2)	$\nu \times 10^6$ (m^2/s)	$k \times 10^3$ (W/m-K)	$\alpha \times 10^6$ (m^2/s)	Pr
100	4.9767	0.54373	82.98	1.67	6.6	2.44	0.684
150	3.2678	0.52804	123.1	3.77	9.60	5.56	0.677
200	2.4411	0.52359	159.9	6.55	12.60	9.86	0.664
250	1.9498	0.52221	195.1	10.01	15.20	14.93	0.670
300	1.6236	0.52157	227.1	13.99	17.70	20.9	0.669
350	1.3911	0.52121	259.2	18.63	20.00	27.58	0.675
400	1.2170	0.52098	289.1	23.76	20.50	32.33	0.735
450	1.0816	0.52084	316.2	29.23	24.35	43.22	0.676
500	0.9734	0.52075	341.6	35.09	26.60	52.48	0.669
550	0.8849	0.52067	365.5	41.31	29.50	64.03	0.645
600	0.8111	0.52063	389.1	47.97	30.70	72.70	0.660
650	0.7487	0.52055	411.5	54.96	32.40	83.13	0.661
700	0.6952	0.52055	433.1	62.30	34.10	94.23	0.661
750	0.6489	0.52053	454.0	69.97	35.80	106.00	0.660
800	0.6083	0.52050	474.4	77.99	37.40	118.12	0.660
850	0.5725	0.52048	494.3	89.74	39.10	131.21	0.684
900	0.5407	0.52048	513.8	95.02	40.60	144.26	0.659
950	0.5123	0.52048	551.4	107.64	42.20	158.27	0.680
1000	0.4867	0.52046	587.7	120.76	43.60	172.14	0.702
1100	0.4424	0.52044	622.7	140.75	46.30	201.08	0.700
1200	0.4056	0.52042	656.7	161.93	48.90	231.69	0.699
1300	0.3744	0.52042	667.9	178.41	51.40	263.83	0.676
1400	0.3476	0.52000	700.9	201.62	53.70	297.07	0.679
1500	0.3244	0.52040	733.4	226.05	56.00	331.68	0.682
1600	0.3042	0.52040	763.5	251.00	58.30	368.29	0.682
1700	0.2863	0.52040	793.6	277.20	60.50	406.08	0.683
1800	0.2704	0.52040	822.9	304.36	62.60	444.92	0.684
1900	0.2562	0.52040	850.8	332.14	64.70	485.35	0.684
2000	0.2434	0.52040	878.4	360.95	66.70	526.67	0.685
2100	0.2318	0.52040	905.5	180.47	–	–	–
2200	0.2212	0.52040	–	–	–	–	–
2300	0.2116	0.52038	–	–	–	–	–
2400	0.2028	0.52040	–	–	–	–	–
2500	0.1947	0.52040	–	–	–	–	–
2600	0.1872	0.52040	–	–	–	–	–
2700	0.1803	0.52040	–	–	–	–	–
2800	0.1738	0.52040	–	–	–	–	–
2900	0.1678	0.52040	–	–	–	–	–
3000	0.1622	0.52040	–	–	–	–	–

Appendix E
Atmospheric Data

The variations of p (bar) and T (K) with altitude H (m) can be correlated as (See Table E.1)

$$p = 1.01325 \, (1 - 2.25577 \times 10^{-5} \times H)^{5.2559} \, , \text{ and} \qquad (E.1)$$

$$T = 288.15 - 0.0065 \, H \, . \qquad (E.2)$$

Table E.1 Properties of the atmosphere [21]

Altitude (m)	T (K)	p (bar)	Pressure ratio	ρ (kg/m^3)	Density ratio	$\mu \times 10^5$ (N-s/m^2)
0	288.16	1.013	1.000	1.2250	1.0000	1.789
1000	281.66	0.899	0.887	1.1120	0.9078	1.758
2000	275.16	0.795	0.785	1.0070	0.8220	1.726
3000	268.67	0.701	0.692	0.9093	0.7423	1.694
4000	262.18	0.617	0.609	0.8194	0.6689	1.661
5000	255.69	0.541	0.534	0.7364	0.6011	1.628
6000	249.20	0.472	0.466	0.6601	0.5389	1.595
7000	242.71	0.411	0.406	0.5900	0.4816	1.561
8000	236.23	0.357	0.352	0.5258	0.4292	1.527
9000	229.74	0.308	0.304	0.4671	0.3813	1.493
10000	223.26	0.265	0.262	0.4135	0.3376	1.458
12000	216.66	0.194	0.192	0.3119	0.2546	1.422
14000	216.66	0.142	0.140	0.2279	0.1860	1.422
16000	216.66	0.103	0.102	0.1665	0.1359	1.422
18000	216.66	0.076	0.075	0.1217	0.0993	1.422
20000	216.66	0.055	0.055	0.0889	0.0726	1.422

© Springer Nature Singapore Pte Ltd. 2020
A. W. Date, *Analytic Combustion*,
https://doi.org/10.1007/978-981-15-1853-9

Appendix F
Binary Diffusion Coefficients

Assuming ideal gas behavior, the kinetic theory of gases predicts that $D_{ab} \propto (T^{1.5}/p)$, where T is in Kelvin (See Table F.1).

Table F.1 Binary diffusion coefficient D_{ab} (m^2/s) at 1 atm and T = 300 K

Pair	$D_{ab} \times 10^6$	Pair	$D_{ab} \times 10^6$
H_2O-air	24.0	CO_2-air	14.0
CO-air	19.0	CO_2-N_2	11.0
H_2-air	78.0	O_2-air	19.0
SO_2-air	13.0	NH_3-air	28.0
CH_3OH-air	14.0	C_2H_5OH-air	11.0
C_6H_6-air	8.0	CH_4-air	16.0
$C_{10}H_{22}$-air	6.0	$C_{10}H_{22}$-N_2	6.4
C_8H_{18}-air	5.0	C_8H_{18}-N_2	7.0
C_8H_{16}-N_2	7.1	C_6H_{14}-N_2	8.0
O_2-H_2	70.0	CO_2-H_2	55.0

© Springer Nature Singapore Pte Ltd. 2020
A. W. Date, *Analytic Combustion*,
https://doi.org/10.1007/978-981-15-1853-9

Bibliography

1. Abramzon, B., & Sazhin, S. (2005). Droplet vapourisation model in the presence of thermal radiation. *International Journal of Heat and Mass Transfer, 48,* 1868–1873.
2. Ahmed, S. F., & Mastorakos, E. (2006). Spark ignition of lifted turbulent jet flames. *Combustion and Flame, 146,* 215–231.
3. Ambrose, S., Kisch, G. T., Kirubi, C., Woo, J., & Gadgil, A. (2008). Development and testing of the Berkeley Darfur Stove. College of Engineering, UC Berkeley. http://www.darfurstoves.org.
4. Andrews, G. E., Bradley, D., & Lwakabamba, S. B. (1975). Turbulence and turbulent flame propagation - a critical appraisal. *Combustion and Flame, 24,* 285–304.
5. Andrews, G. F., & Bradley, D. (1972). The burning velocity of methane-air mixtures. *Combustion and Flame, 19,* 275–288.
6. Annand, W. J. D. (1963). Heat transfer in cylinders of internal combustion engines. *Proceedings of the Institution of Mechanical Engineers, 177,* 973.
7. Ballard-Tremeer, G., & Jawurk, H. H. (1996). Comparison of five rural, wood-burning cooking devices: Efficiencies and emissions. *Biomass and Bioenergy, 11*(5), 419–430.
8. Becker, H. A., & Liang, P. (1978). Visible length of vertical free turbulent diffusion flames. *Combustion and Flame, 32,* 115–137.
9. Bellester, J., & Jemenez, S. (2005). Kinetic parameters for oxidation of pulverised coal as measured from drop tube test. *Combustion and Flame, 142,* 210–222.
10. Benson, R. S. (1967). *Advanced engineering thermodynamics.* London: Pergamon Press.
11. Berger, R. J., & Marin, G. B. (1999). Investigation of gas-phase reactions and ignition delay occuring at conditions typical for partial oxidation of methane to synthesis gas. *Industrial and Engineering Chemistry Research, 38,* 2582–2592.
12. Bhandari, S., Gopi, S., & Date, A. (1988). Investigation of wood-burning stove, part-I experimental investigation. *Sadhana, 13*(4), 271–293.
13. Bhattacharya, S. C., Albina, D. O., & Salam, P. A. (2002). Emission factors of wood and charcoal fired stoves. *Biomass and Bioenergy, 23,* 453–469.
14. Boroson, M. L., Howard, J. B., Longwell, J. P., & Peters, W. A. (2006). Product yields and kinetics from the vapor phase cracking of wood pyrolysis tars. *AIChE Journal, 35*(1), 120–128.
15. Bicen, A. F., & Jones, W. P. (1986). Velocity characteristics of isothermal and combusting flows in a model combustor. *Combustion Science and Technology, 49,* 1–15.
16. Binesh, A. R., & Hossainpur, S. (2008). Three dimensional modeling of mixture formation and combustion in a direct injection heavy duty diesel engine. *World Academy of Science, Engineering and Technology, 41,* 207–212.
17. Blackshear, P. L., & Murthy, K. L. (1965). Heat and mass transfer to, from and within cellulosic solids burning in air. In *Proceedings of the 10th International Symposium on Combustion, the Combustion Institute* (pp. 911–923).

© Springer Nature Singapore Pte Ltd. 2020
A. W. Date, *Analytic Combustion,*
https://doi.org/10.1007/978-981-15-1853-9

18. Blake, T. R., & McDonald, M. (1993). An examination of flame length data from vertical turbulent diffusion flames. *Combustion and Flame, 94*, 426–432.
19. Borman, G. L., & Ragland, K. W. (1998). *Combustion engineering*. New York: McGraw-Hill.
20. Burcat, A. (2006). Ideal gas thermodynamic data in polynomial form for combustion and air pollution use. http://garfield.chem.elte.hu/Burcat/burcat.html.
21. Campen, C. F., Jr., Ripley, W. S., Cole, A. E., Sissenwine, N., Condron, T. P., Soloman, I. (1961). *Handbook of geophysics* (3rd ed.) United States Air Force, Air Research and Development Command, Air Force Research Division, Geophysics Research Directorate. New York: Macmillan.
22. Cebeci, T. (2002). *Turbulence models and their application: Efficient numerical methods with computer programs*. Berlin: Springer.
23. Channiwala, S. A. (1992). *On biomass gasification process and technology development - some analytical and experimental investigations*. Ph.D. thesis, Mechanical Engineering Department, IIT Bombay.
24. Clarke, R. (1985). *Wood stove dissemination*. London: ITDG Publication.
25. Coffee, T. P., Kotkar, A. J., & Miller, M. S. (1983). The overall reaction concept in premixed, laminar, steady-state flames - I, stoichiometries. *Combustion and Flame, 54*, 155–169.
26. Coffee, T. P., Kotkar, A. J., & Miller, M. S. (1984). The overall reaction concept in premixed, laminar, steady-state flames - II, initial temperatures and pressures. *Combustion and Flame, 58*, 59–67.
27. Craft, T. J., Launder, B. E., & Suga, K. (1996). Development and application of a cubic eddy-viscosity model of turbulence. *International Journal of Heat and Fluid Flow, 17*, 108.
28. Crawford, M. E., & Kays, W. M. (1976). STAN5 - a program for numerical computation of two-dimensional internal/external boundary layer flows. NASA CR-2742.
29. Date, A. (1989). Energy utilization pattern of Shilarwadi. *Indian Journal of Rural Technology, 1*(1), 33–63.
30. Date, A. W. (2005). *Introduction to computational fluid dynamics*. New York: Cambridge University Press.
31. De Lepeleire, G., Krishna Prasad, K., Verhaart, P., & Visser, P. (1981). A wood-stove compendium. In *UN Conference on New and Renewable Sources of Energy, Nairobi*.
32. Delichatsios, M. A. (1993). Transition from momentum to buoyancy-controlled turbulent jet diffusion flames and flame height relationships. *Combustion and Flame, 92*, 349–364.
33. DNES, Govt of India. (1988). Water boiling and evaporation test for thermal efficiency of wood stoves. *Antika*, 8–14.
34. Evans, D. D., & Emmons, H. W. (1977). *Combustion of wood charcoal. Fire Research, 1*, 57–66.
35. Faeth, G. M. (1977). The current status of droplet and liquid combustion. *Progress in Energy and Combustion Science, 3*, 191–224.
36. FAO. (1993). Improved solid biomass burning cookstoves: A developmental manual. Field document No 44, Regional Wood Energy Programme in Asia, FAO, Bangkok.
37. Fay, J. A., & Golomb, D. S. (2002). *Energy and the environment*. New York: Oxford University Press.
38. Ferguson, C. R. (1986). *Internal combustion engines*. Toronto: Wiley.
39. Filayev, S. A., Driscoll, J. F., Carter, C. D., & Donbar, J. M. (2005). Measured properties of turbulent premixed flames for model assessment, including burning velocities, stretch rates, and surface densities. *Combustion and Flame, 141*, 1–17.
40. Frenklach, M., Wang, H., & Rabinowitz, M. J. (1992). Optimization and analysis of large chemical kinetic mechanisms using the solution mapping method, combustion of methane. *Progress in Energy and Combustion Science, 18*, 47–73.
41. Gottgens, J., Mauss, F., & Peters, N. (1992). Analytic approximations of burning velocities and flame thickness of lean hydrogen, methane, ethylene, ethane, acetylene and propane flames. In *Twentyfourth Symposium (International) on Combustion, Pittsburgh* (pp. 120–135).
42. Halstead, M. P., Kirsch, L. J., & Quinn, C. P. (1977). The autoignition of hydrocarbon fuels at high temperatures and pressures - fitting of a mathematical model. *Combustion and Flame, 30*, 45–60.

43. Hautmann, D. J., Dryer, F. L., Schug, K. P., & Glassman, I. (1981). A multi-step overall kinetic mechanism for the oxidation of hydrocarbons. *Combustion Science and Technology*, *25*, 219–235.
44. Hawthorne, W. R., Weddel, D. S., & Hottel, H. C. (1949). Mixing and diffusion in turbulent gas jets. In *Third Symposium on Combustion, Williams and Wilkins, Baltimore* (p. 226).
45. Haywood, R. J., Nafziger, R., & Renksizbulut, M. (1989). A detailed examination of gas and liquid phase transient processes in convective droplet evaporation. *The ASME Journal of Heat Transfer*, *111*, 495–502.
46. Heitor, M. V., & Whitelaw, J. H. (1986). Velocity, temperature and species characteristics of the flow in a gas turbine combustor. *Combustion and Flame*, *64*, 1–32.
47. Hill, P. G., & Huang, J. (1988). Laminar burning velocities of methane with propane and ethane additives. *Combustion Science and Technology*, *60*, 7–30.
48. Hilsenrath, J., Hoge, H. J., Beckett, C. W., Masi, J. F., Fano, L., Touloukian, Y. S., et al. (1960). *Tables of thermodynamic and transport properties of air, argon, carbon dioxide, nitrogen, oxygen, and steam*. London: Pergamon Press.
49. Holman, J. P. (1980). *Thermodynamics* (3rd ed.). New Delhi: McGraw-Hill Kogakusha Ltd.
50. Honkalaskar, V., Bhandarkar, U. V., & Sohoni, M. (2013). Development of a fuel efficient cookstove through a participatory bottom-up approach. *Energy, Sustainability and Society*, *3*, 16. https://doi.org/10.1186/2192-0567-3-16.
51. Incropera, F. P., & DeWitt, D. P. (1998). *Fundamentals of heat and mass transfer* (4th ed.). New York: Wiley.
52. Jetter, J. J., & Kariher, P. (2009). Solid-fuel household cook stoves: Characterisation of performance and emissions. *Biomass and Bioenergy*, *33*, 294–305.
53. Jones, H. R. N. (1990). *Application of combustion principles to domestic gas burner design*. New York: Taylor and Francis.
54. Jones, W. P., & Launder, B. E. (1972). The prediction of laminarization with a two-equation model of turbulence. *International Journal of Heat and Mass Transfer*, *15*, 301.
55. Jones, W. P., & Toral, H. (1983). Temperature and composition measurements in a research gas turbine combustion chamber. *Combustion Science and Technology*, *31*, 249–275.
56. Jones, W. P., & Linstedt, R. P. (1988). Global reaction schemes for hydrocarbon combustion. *Combustion and Flame*, *73*, 233–249.
57. Kalghatgi, G. T. (1984). Lift-off heights and visible lengths of vertical turbulent jet diffusion flames in still air. *Combustion Science and Technology*, *51*, 75–95.
58. Kalghatgi, G. T. (1981). Blow-out stability of gaseous jet diffusion flames: Part I: In still air. *Combustion Science and Technology*, *26*, 233–239.
59. Kane, P. (2002). *Pressure-drop in a VSBK Brick-Kiln setting*, M.Tech. Dissertation, Mechanical Engineering Department, IIT Bombay.
60. Karpov, V. P., Lipatnikov, A. N., & Zimont, V. L. (1995). Influence of molecular heat and mass transfer on premixed turbulent combustion. In S. H. Chan (Ed.), *Transport phenomena in combustion* (Vol. 1, pp. 629–640). New York: Taylor and Francis.
61. Kaw, P. K. (2006). A Hymn to the goddess. *Resonance*, 22–39.
62. Kays, W. M., & Crawford, M. E. (1993). *Convective heat and mass transfer* (3rd ed.). New York: McGraw-Hill.
63. Kazantsev, E. I. (1977). *Industrial furnaces*. Translation by I. V. Savin. Moscow: Mir Publishers.
64. Keenan, J., Chao, J., & Kaye, J. (1980). *Gas tables, thermodynamic properties of air, products of combustion, compressible flow functions* (2nd ed.). New York: Wiley.
65. Kobayashi, H., Howard, J. B., & Sarofim, A. F. (1977). Coal devolatilization at high temperatures. In *16th Symposium (International) on Combustion, the Combustion Institute, Pittsburgh*.
66. Krishnaprasad, K., & Bussman, P. (1986). Parameter analysis of a simple wood burning cookstove. In *Proceedings of the International Heat Transfer Conference, San Francisco* (pp. 3085–3090).
67. Kunii, D., & Levenspiel, D. (1991). *Fluidization engineering* (2nd ed.). Burlington: Butterworth.

68. Kuo, K. K. (1986). *Principles of combustion*. New York: Wiley-Interscience.
69. Launder, B. E., & Sharma, B. I. (1974). Application of the energy-dissipation model of turbulence to the calculation of flow near a spinning disk. *Letters in Heat and Mass Transfer, 1*, 131–138.
70. Launder, B. E., & Spalding, D. B. (1972). *Mathematical models of turbulence*. New York: Academic.
71. Law, C. K., & Sirignano, W. A. (1977). Unsteady droplet combustion and droplet heating II - conduction limit. *Combustion and Flame, 28*, 175–186.
72. Lee, J. C., Yetter, R. A., & Dryer, F. L. (1995). Transient numerical modeling of carbon particle ignition and oxidation. *Combustion and Flame, 101*, 387–398.
73. Lefebvre, A. H. (1983). *Gas turbine combustion*. New York: McGraw-Hill.
74. Lefebvre, A. H. (1989). *Atomization and sprays*. New York: Hemisphere.
75. Longwell, J. F., & Weiss, M. A. (1955). Heat release rates in hydrocarbon combustion. In *International Mechanical Engineering/ASME Joint-Conference on Combustion* (pp. 334–340), quoted in ref. [113].
76. MacCarty, N., Still, D., & Ogle, D. (2010). Fuel use and emissions performance of fifty cooking stoves in the laboratory and related benchmarks of performance. *Energy for Sustainable Development, 14*, 161–171.
77. MacCarty, N. A., & Byden, K. M. (2015). Modeling household biomass cookstoves. *Energy for Sustainable Development, 26*, 1–13.
78. Magnussen, B. F., & Hjertager, B. W. (1976). On mathematical modeling of turbulent combustion with special emphasis on soot formation and combustion. In *16th International Symposium on Combustion, the Combustion Institute, Pittsburgh* (pp. 719–729).
79. Maithel, S. (2001). *Energy and environment monitoring of vertical shaft Brick Kiln*. Ph.D. thesis, Energy Systems Engineering, IIT Bombay.
80. McBride, B. J., Gordon, S., & Reno, M. A. (1993). *Coefficients for calculating thermodynamic and transport properties of individual species*. NASA technical memorandum (Vol. 4513). NASA Office of Management, Scientific and Technical Information Program.
81. Meghalachi, M., & Keck, J. C. (1982). Burning velocities of mixtures of air with methanol, isooctane, and indolene at high pressure and temperature. *Combustion and Flame, 48*, 191–210.
82. Modest, M. F. (1993). *Radiative heat transfer*. New York: McGraw-Hill.
83. Mukunda, H. S. (2017). *Understanding aerospace chemical propulsion*. New Delhi: I K International Publishing House Pvt Ltd.
84. Mullins, B. P. (1955). Studies on spontaneous ignition of fuels injected into a hot air stream: Part I and II. *Fuel, 32*, 211.
85. NIST Chemistry Webbook. http://webbook.nist.gov/chemistry/form-ser.html.
86. Nuntadusit, C., Wae-hayee, M., Bunyajitradulya, A., & Eiamsa-ard, S. (2012). Heat transfer enhancement by multiple swirling impingement jets with twisted-tape swirl generators. *International Communications in Heat and Mass Transfer, 39*, 102–107.
87. Okajima, S., & Kumagai, S. (1974). Further investigation of combustion free droplets in a freely falling chamber including moving droplets. In *15th Symposium (International) on Combustion, the Combustion Institute, Pittsburgh* (pp. 402–407).
88. Patankar, S. V. (1981). *Numerical fluid flow and heat transfer*. New York: Hemisphere.
89. Perry, R. H., & Chilton, C. H. (1973). *Chemical engineers' handbook* (5th ed.). Tokyo: McGraw-Hill-Kogakusha.
90. Peters, N. (2000). *Turbulent combustion*. New York: Cambridge University Press.
91. Peters, N., & Rogg, B. (1993). *Reduced kinetic mechanisms for applications in combustion systems*. Berlin: Springer.
92. Pitts, W. M. (1989). Importance of isothermal mixing processes to the understanding of lift-off and blow-out of turbulent jet diffusion flames. *Combustion and Flame, 76*, 197–212.
93. Pope, S. B. (2000). *Turbulent flows*. New York: Cambridge University Press.
94. Ragland, K. W., Aerts, D. J., & Baker, A. J. (1991). Properties of wood for combustion analysis. *Bioresource Technology, 37*, 161–168.

95. Roper, F. G. (1977). The prediction of laminar jet diffusion flame sizes: Part I. *Theoretical model. Combustion and Flame, 29*, 219–226.
96. Roper, F. G., Smith, C., & Cunningham, A. C. (1977). The prediction of laminar jet diffusion flame sizes: Part II. *Experimental verification. Combustion and Flame, 29*, 227–234.
97. Savarianandan, V., & Lawn, C. J. (2006). Burning velocity of turbulent premixed flame in weakly wrinkled regime. *Combustion and Flame, 146*, 1–18.
98. Schlichting, H. (1968). *Boundary - layer theory* (6th ed.) tr. Kestin J. New York: McGraw-Hill.
99. Schumacher, E. F. (1982). *Schumacher on energy*. London: Jonathan Cape.
100. Shah, R., & Date, A. W. (2011). Steady-state thermo-chemical model of a wood-burning cookstove. *Combustion Science and Technology, 183*(4), 321–346.
101. Sharma, A. K., Ravi, M. R., & Kohli, S. (2006). Modeling product composition in slow pyrolysis of wood. *SESI Journal, 16*(1), 1–11.
102. Siegel, R., & Howell, J. R. (1972). *Thermal radiation heat transfer*. Tokyo: McGraw-Hill Kogakusha.
103. Simmons, G., & Lee, W. H. (1965). *Fundamentals of thermo-chemical biomass conversion*. In R. P. Owerend, T. A. Milne, & L. K. Mudge (Eds.). Amsterdam: Elsevier.
104. Singer, F., & Singer, S. S. (1963). *Industrial ceramics*. London: Chapman and Hall.
105. Smith, K. R., Agarwal, A. L., & Dave, R. M. (1986). *Information package on cook-stove technology*. New Delhi: Tata Energy Research Institute.
106. Smooke, M. D., Millar, J. A., & Kee, J. A. (1983). Determination of adiabatic flame speeds by boundary value methods. *Combustion Science and Technology, 34*, 79–89.
107. Smooke, M. D. (1982). Solution of burner stabilised premixed laminar flames by boundary value methods. *Journal of Computational Physics, 48*, 72–106.
108. Smoot, L. D., & Smith, P. J. (1985). *Coal combustion and gasification*. New York: Plenum Press.
109. Spadaccini, L. J., & TeVelde, J. A. (1994). Ignition delay characteristics of methane fuels. *Progress in Energy and Combustion Science, 20*, 431–460.
110. Spalding, D. B. (1963). *Convective mass transfer*. New York: McGraw-Hill.
111. Spalding, D. B. (1971). Mixing and chemical reaction in steady, confined turbulent flames. In *13th International Symposium on Combustion, the Combustion Institute, Pittsburgh* (pp. 649–657).
112. Spalding, D. B. (1977). *Genmix: A general computer program for two-dimensional parabolic phenomena*. Oxford: Pergamon Press.
113. Spalding, D. B. (1979). *Combustion and mass transfer*. Oxford: Pergamon Press.
114. Spalding, D. B. (1959). *A one-dimensional theory of liquid rocket fuel combustion* (p. 445). No: Aeronautical Research Council Technical Report, C.P.
115. Spalding, D. B., & Cole, E. H. (1964). *Engineering thermodynamics*. London: Edward Arnold (Publishers) Ltd.
116. Speight, J. G. (2005). *Lange's handbook of chemistry* (16th ed.). New York: McGraw-Hill.
117. Sutar, K. B., Kohli, S., Ravi, M. R., & Ray, A. (2015). Biomass cook stoves: A review of technical aspects. *Renewable and Sustainable Energy Reviews, 41*, 1128–1166.
118. Tester, J. F., Drake, E. M., Driscoll, M. J., Golay, M. W., & Peters, W. A. (2005). *Sustainable energy - choosing among options*. Cambridge: MIT Press.
119. Tillman, D. A., Rosi, J. A., & Kitto, W. D. (1981). *Wood combustion - principles, processes and economics*. New York: Academic.
120. Tinney, E. R. (1965). Combustion of wooden dowels in heated air. In *Proceedings of the 10th International Symposium on Combustion, the Combustion Institute* (pp. 925–930).
121. Tsang, W., & Lifshitz, A. (1990). Shock-tube techniques in chemical kinetics. *Annual Review of Physical Chemistry, 41*, 883–890.
122. Turns, S. R. (1996). *Introduction to combustion - concepts and applications*. New York: McGraw-Hill.
123. Vargaftik, N. B. (1975). *Handbook of physical properties of liquids and gases, pure substances and mixtures* (2nd ed.). New York: Hemisphere.

124. Warnatz, J., Dibble, R. W., Mass, U., & Dibble, R. W. (2001). *Combustion: Physical and chemical fundamentals, modeling and simulation, experiments, pollutant formation*. Berlin: Springer.

125. Westbrook, C. K., & Dryer, F. L. (1981). Simplified reaction mechanisms for the oxidation of hydrocarbon fuels in flames. *Combustion Science and Technology, 27*, 31–47.

126. White, F. M. (1986). *Fluid mechanics* (2nd ed.). New York: McGraw-Hill.

127. Wolfstein, M. (1969). The velocity and temperature distribution in one-dimensional flow with turbulence augmentation and pressure gradient. *International Journal of Heat and Mass Transfer, 12*, 301–318.

128. Woschni, G. (1968). *A universally applicable equation for the instantaneous heat transfer coefficient in internal combustion engine*. SAE paper, 670931.

129. Zang, M., Yu, J., & Xu, X. (2005). A new flame sheet model to reflect the influence of the oxidation of CO on the combustion of a carbon particle. *Combustion and Flame, 143*, 150–158.

130. MNES, Govt of India., (2006). *Integrated energy policy*. Planning Commission, New Delhi: Report of the Expert Committee.

131. *Pocket world in figures*. London: The Economist (2007).

Index

© Springer Nature Singapore Pte Ltd. 2020
A. W. Date, *Analytic Combustion*,
https://doi.org/10.1007/978-981-15-1853-9

Printed in the United States
by Baker & Taylor Publisher Services